Epigenetic Inheritance and Evolution

Epigenetic Inheritance and Evolution

The Lamarckian Dimension

EVA JABLONKA

The Cohn Institute for the History and Philosophy of Science and Ideas
Tel-Aviv University

and

MARION J. LAMB

Biology Department
Birkbeck College, University of London

Oxford New York Tokyo
OXFORD UNIVERSITY PRESS
1995

Oxford University Press, Walton Street, Oxford OX2 6DP
Oxford New York
Athens Auckland Bangkok Bombay
Calcutta Cape Town Dar es Salaam Delhi
Florence Hong Kong Istanbul Karachi
Kuala Lumpur Madras Madrid Melbourne
Mexico City Nairobi Paris Singapore
Taipei Tokyo Toronto
and associated companies in
Berlin Ibadan
Oxford is a trade mark of Oxford University Press

Published in the United States
by Oxford University Press Inc., New York

A catalogue record for this book is available from the British Library

Library of Congress Cataloging in Publication Data
Jablonka, Eva.
Epigenetic inheritance and evolution : the Lamarckian dimension / Eva Jablonka and Marion J. Lamb.
Includes bibliographical references and index.
1. Evolution (Biology) 2. Genetics. 3. Inheritance of acquired characters. I. Lamb, Marion J. II. Title.
QH366.2.J32 1995 575.01'66–dc20 94-32108
ISBN 0 19 854062 0 (Hbk)

Typeset by Footnote Graphics, Warminster, Wiltshire
Printed in Great Britain by Biddles Ltd, Guildford & King's Lynn

To our parents, who gave us more than genes

Preface

This book was written because we believe that it is time to re-examine the role of the inheritance of acquired characters in evolution. We are aware that this statement is likely to produce an immediate negative reaction in the reader. Most evolutionary biologists believe that as a scientific issue the inheritance of acquired characters, or 'Lamarckian evolution', is dead. Its occasional bursts into life are usually assumed to be a result of fraudulent or inadequate experiments, or, more commonly, are thought to result from a sad misunderstanding of the explanatory power of neo-Darwinism. Most recent books on evolutionary biology treat Lamarckism as a theory of merely historical and didactic interest: it was once, for reasons often presented as somewhat puzzling, a rival to neo-Darwinism, and can now serve the useful role of demonstrating the strengths of neo-Darwinism by exposing its own, all too apparent, theoretical weaknesses.

What we have tried to show in this book is not that neo-Lamarckism is right and neo-Darwinism is wrong. Our view is that both neo-Darwinian and neo-Lamarckian mechanisms are important in evolution. It is a view that is in danger of pleasing no one, because it makes heredity and evolution seem very complex, messy, and confusing. We accept the Darwinian theory of evolution by natural selection of heritable variations that affect fitness, but we contend that the nature of these variations is a poorly developed part of this theory. In the present, neo-Darwinian, version, the theory is based on two assumptions: first, that all variations in the hereditary material are random, and second, that all hereditary variations are variations in DNA base sequences. We question both of these assumptions.

Molecular biology has already introduced many changes in our view of genetics, and many traditional tenets and even axioms are being re-examined. The genome is described as 'flexible', 'dynamic', or 'clever', to emphasize its role as an active response system as well as a passive information carrier. In addition to the DNA inheritance system underlying classical genetics, it is now recognized that variations can be transmitted between generations of cells in other ways. There are *epigenetic* inheritance systems, which enable cells with identical genotypes to acquire and transmit different phenotypes. Behavioural and cultural transmission, in which learned behaviour can be transmitted between generations, are therefore not the only systems of inheritance with Lamarckian features. These additional inheritance systems force us to recognize that the unit of heritable variation is not just the gene: the cellular phenotype and behavioural phenotype are also units of heritable variation.

In the first part of the book we describe the origin and nature of heritable variations, and advocate adopting a concept of biological heredity that is broader than the present DNA-centric one. The conceptual and theoretical problems raised by Lamarckian theories are discussed in Chapter 1. One serious objection to Lamarckism is that it cannot explain the evolution of adaptability, even though it may explain the evolution of specific adaptations. For example, even if Lamarckian mechanisms underlie the transmission of the blacksmith's big muscles (and we are not suggesting that they do), Lamarckism cannot explain the evolution of the ability to develop big muscles with exercise. We accept this argument, but believe that its importance has been exaggerated. It is clear that any Lamarckian theory of evolution must be fundamentally incomplete, for it cannot explain the evolution of the adaptability it requires. But the present neo-Darwinian theory is also incomplete if it ignores the Lamarckian mechanisms, such as epigenetic inheritance and behavioural transmission, which neo-Darwinian evolution has itself created. Once a mechanisms for the inheritance of acquired variations has evolved, it operates side by side with the neo-Darwinian mechanism of evolutionary change. The obvious evolutionary precedence of the neo-Darwinian mechanism is of no relevance to the arguments about the possibility or significance of inherited acquired variations in subsequent evolutionary history.

In Chapter 2 we discuss ways in which the inheritance of acquired characters has been explained within a fundamentally neo-Darwinian conceptual framework. The approaches we describe broaden rather than challenge neo-Darwinism. They are important for the arguments presented in later chapters, because they focus on development. In Chapter 3 we examine a recent challenge to neo-Darwinism—directed mutation in bacteria—and look at the way in which current molecular biology is forcing a fundamental change in our perception of the genome. It is beginning to be realized that the genome has to be thought of as a sophisticated response system, as well as a carrier of information.

We show in Chapters 4–6 that epigenetic systems play a role in inheritance. These systems are responsible for transmitting the functional and structural properties of cells. In Chapter 4 we describe three different types of epigenetic inheritance system (EIS) and the way in which epigenetic variations are transmitted in somatic cell lineages. Throughout the book we put special emphasis on the chromatin-marking inheritance system, in which chromatin marks, such as patterns of DNA methylation, are transmitted between cell generations and may determine cell phenotypes. The same DNA sequence can carry different marks; although the DNA sequence determines which marks are possible, the particular variant that is actually present and inherited depends not on DNA base sequence alone, but also on the environmental and developmental history of the lineage. The cells in a multicellular body, or the cells in lineages of unicellular organisms, show heritable variation even though they have identical DNA sequences.

In Chapter 5, we review genomic imprinting, in which the parental origin of chromosomes influences gene expression in the progeny. The reversible imprints established during oogenesis and spermatogenesis are a special case of the transmission of epigenetic variations. In Chapter 6 we present evidence showing that more permanent epigenetic variations can be transmitted through the germ line of sexually-reproducing multicellular organisms. Until recently, indisputable examples of the transmission of epigenetic variants from one generation to the next were rare, but the many clear cases that have emerged from molecular studies of development and cell memory suggest that the former scarcity may have been the result of experimental or ideological bias.

Epigenetic variations cannot be treated in isolation from DNA sequence variations. In Chapter 7 we discuss how variations in the genetic and epigenetic systems interact. DNA sequence variations affect the range and stabilities of epigenetic marks, and the epigenetic state of chromatin biases the chances of DNA sequence change. We suggest that chromosome structure has been moulded by evolutionary interactions between the genetic and epigenetic inheritance systems.

In Chapter 8 we explore some aspects of the evolution of adaptations. We argue that the evolutionary sophistication of epigenetic inheritance systems in primitive unicellular eukaryotes was an adaptation to fluctuating environments. These sophisticated EISs then played a major role in the evolutionary transition to multicellularity. In multicellular organisms they were further elaborated and had both direct and indirect effects on evolutionary changes. They had a direct effect because epigenetic variations can be selected; they had an indirect effect because the existence of cellular memory imposes constraints on various developmental strategies. We argue that the timing of the segregation of soma and germ line, the maternal control of early development, and the restructuring of chromatin during gametogenesis are all evolutionary consequences of these constraints.

The role of epigenetic inheritance in speciation is examined in Chapter 9. Here we argue that heritable epigenetic variations can have a significant effect at the initial stages of speciation in all groups of organisms, although the effects are not the same in all taxa. In stressful conditions, such as those found in small isolates, epigenetic variations and inter-related directed DNA changes are particularly common and important, and may be similar to the variations occurring under domestication. The use of a new resource, even when it occurs in sympatry, can also lead to heritable changes in gene expression, and promote speciation.

Chapter 10 summarizes the arguments presented in the previous nine chapters of the book, and points to further evolutionary implications of the epigenetic perspective. We urge a return to a wider view of heredity, which encompasses multiple inheritance systems. Although epigenetic inheritance systems evolved through selection of DNA variations, once they had evolved, they added another dimension to evolution.

Since our subject matter covers several different disciplines, we have provided a glossary of many of the terms used. We have included additional explanatory material, peripheral arguments, and references to sources of information in notes at the end of each chapter.

We started writing this book as a response to the mixture of enthusiasm and antagonism that we found for the idea that epigenetic inheritance is important in evolution. The emotional and aggressive reactions that we often encountered when presenting our views made it clear that we were touching sensitive nerves. The repeated argument that since the DNA inheritance system is the most fundamental system in living organisms, it is sufficient to explain all hereditary and evolutionary phenomena, convinced us that this common belief needed to be challenged. The reassurance of colleagues that the simple arguments we were developing were interesting and timely encouraged us to write this book. We are grateful to all of them. In particular we would like to thank Yehuda Elkana for his sustained enthusiasm. Special thanks are also due to Tamar Arbel, Fanny Doljanski, Rafi Falk, Mary Lyon, Anne McClaren, Peter McClaughlin, Aharon Razin, Tsvi Sachs, Sahotra Sarkar, and Jim Shapiro, each of whom read and criticized drafts of the chapter or chapters in which they had particular expertise. We are also grateful to Fred Meins for helping to clarify parts of Chapters 4 and 6. Michael Lachmann-Tarkhanov kindly allowed us to use his unpublished work and did the simulations for the model in Chapter 7. George Knowles gave invaluable advice and help in drawing the figures. Finally we want to thank and apologise to our long-suffering friends and colleagues Ros Brown, Lia Ettinger, and Joy Hoffman, who read and criticized drafts of the whole book, gave us a lot of constructive advice, and are still speaking to us.

The writing of this book was made easier by the generous support of Branco Weiss and the Alternative Thinking Programme of Tel-Aviv University whose financial support allowed us to overcome the geographical barriers between Israel and England.

September 1994 E.J.
M.J.L.

Contents

Acknowledgements of figure sources xiii

1 The legacy of Lamarckism 1

2 Neo-Darwinian explanations of the inheritance of acquired characters 30

3 Induced genetic variations 54

4 Cellular heredity: epigenetic inheritance systems 79

5 Genomic imprinting 111

6 The inheritance of directed epigenetic variations 133

7 Interactions between genetic and epigenetic inheritance 160

8 The role of epigenetic inheritance systems in adaptive evolution 191

9 Heredity and the origin of species 229

10 Multiple inheritance systems 272

Glossary 290

References 299

Index 331

Acknowledgements of figure sources

We are grateful to the following publishers and authors who kindly allowed us to use redrawn versions of their figures.

Fig. 2.7 was redrawn by permission of John Wiley and Sons, Ltd. from E.J. Steele *et al.* in *Evolutionary theory: paths into the future* (ed. J.W. Pollard), p. 219; © 1984 John Wiley and Sons.

Fig. 3.2 was redrawn by permission of W.H. Freeman and Co. from G.S. Stent and R. Calendar, *Molecular genetics* (2nd edn), Freeman, San Francisco, p. 161, 1978.

Fig. 4.5 was redrawn by permission of J. Beisson and Elsevier Trends Journals from M. Jerka-Dziadosz and J. Beisson, *Trends in Genetics* **6**, p. 42, 1990.

Fig. 5.5 was redrawn by permission of M.A. Surani and Cell Press from N.D. Allen *et al.*, *Cell* **61**, p. 855, 1990.

Fig. 6.4 was redrawn by permission of Company of Biologists, Ltd. from N. Fedoroff *et al.*, *BioEssays* **10**, p. 141, 1989.

Fig. 8.6 was redrawn by permission of G. Bell and Wiley-Liss Inc. from G. Bell in *The origin and evolution of sex* (ed. H. Halvorson and A. Monroy), p. 227; © 1985 Alan R. Liss, Inc.

Fig. 9.2 was redrawn by permission of Ginn and Co. Ltd. from R. Blackman, *Aphids*, p. 22, 1974.

Fig. 9.3 was redrawn by permission of L. Buss and the editor of *Palaeobiology* on behalf of the Palaeontological Society from L.W. Buss, *Palaeobiology* **14**, p. 315, 1988.

Fig. 10.3 was redrawn by permission of Sinauer Associates from S.F. Gilbert, *Developmental biology* (3rd edn), Sinauer, p. 855, 1991.

1

The legacy of Lamarckism

Every thing possible to be
believed is an image of truth.
Blake: *Proverbs of Hell*

The idea that the inheritance of acquired characters plays an important role in evolution has been the subject of controversy for over a century. Enthusiasm for the idea, which is usually associated with the name of Lamarck, has sometimes led to charlatanism and fraud, while opposition to it has led to 'Lamarckist' being used as a term of abuse. Nowadays, biologists usually regard ideas about the inheritance of acquired characters as nothing more than an interesting part of the history of biology. Lamarckian evolution is rejected on the grounds that there is no evidence for it, no mechanism that can produce it, and no need for it in evolutionary theory. Some people go even further and argue that the inheritance of acquired characters is theoretically impossible—it is incompatible with what is known about genetics and development.

The aim of this book is to show that there are now well recognized mechanisms by which some acquired characters can be transmitted to the next generation, and that such characters have probably played a significant role in evolution. We want to make it clear right at the outset that although we argue that some types of Lamarckian evolution are possible, there is nothing in what we say that should be construed as being anti-Darwinian.[1] We are firm believers in the power and importance of natural selection. What we do maintain, however, is that some new inherited variations are not quite as random as is generally assumed, but arise as a direct, and sometimes directed, response to environmental challenge, and that the effects of such induced variations deserve more recognition in evolutionary theory.

In this chapter we want to look at some of the reasons given for rejecting the idea that acquired characters can be inherited, and show why we think they are wrong. The objections to Lamarckism are based partly on the limited evidence for the inheritance of acquired characters, but also on prejudice and conservatism, and on a view of heredity that is no longer appropriate. During the past fifty years there has been a gradual narrowing of the concept of heredity. Although this was probably important and necessary for the development of genetics as a discipline, it is now a handicap to evolutionary thinking. The trend needs to be reversed, because there is more to heredity than DNA, and DNA is not just a passive information carrier, it is also a response system.

Before discussing Lamarckism and the objections to it, it is necessary to look at Lamarck's ideas and at what people have meant and mean today when they talk about the inheritance of acquired characters. One of the difficulties in discussing the subject is that Lamarck's ideas have themselves been the subject of cultural 'Lamarckian' inheritance. Terms such as 'Lamarckism', 'acquired character' and 'Lamarckian evolution' have undergone changes in meaning as they have been used and modified to fit the interests and biases of those using them. As we shall show later in this chapter, this is still the situation today: recent discussions of the inheritance of bacterial adaptations have shown that there is no general consensus about what would constitute evidence for the inheritance of acquired characters.

Lamarck's Lamarckism

Jean-Baptiste Lamarck (1744–1829) was the first consistent evolutionist.[2] His evolutionary theory was a network of different ideas, some old and some relatively new. They were discussed most fully in his book *Philosophie zoologique* published in 1809, fifty years before Charles Darwin published *On the origin of species by means of natural selection*.

Lamarck believed that the natural laws operating on living matter are the same as those operating on non-living matter. The difference between living and non-living is a consequence of the way in which the material is organized. Under certain special conditions, which allow the action of what were known in Lamarck's time as 'subtle fluids', inanimate matter can be reorganized in a way which changes it into living matter. According to Lamarck, the spontaneous generation of living organisms is a normal, and not uncommon, occurrence of both the past and the present. A certain combination of material constituents and the right environment produce a living organism which is a self-sustaining, growing, and self-complicating entity.

Once generated, simple life forms change. Lamarck thought that there are two reasons for change. One is an inevitable consequence of the organization of living matter, which differs from non-living matter because it acts as a whole. This whole is preserved, but also altered, by the movements of the fluids it contains. These movements divide, erode, etch out channels, and leave sediments in the soft parts of the body. Once the movements of fluids have established one set of structures, further movements build on and elaborate them to produce even more complex structures. Thus, according to Lamarck, the inherent tendency for a slow, gradual increase in complexity is a consequence of the basic properties of the living body. It occurs because the motions of the body fluids produce results that have a snowball effect.

If the self-complicating property of living organisms were the only factor causing change, the series from simple to complex forms would be progressive and linear. However, Lamarck believed that a second factor is involved. This is the ability of living organisms to react to their environment. Responses to the environment cause an alteration and diversion in the direction and pattern of the fluids in the body. The result is adaptive modifications of structure and function. Since environmental effects are accidental, the changes they induce cause a branching away from the main linear series. Both the inherent tendency for an increase in complexity, and the adaptive modifications, contribute to the transformation of organisms. Evolution is primarily progressive, but it is also divergent.

The linear nature of evolutionary history can be seen in Lamarck's famous diagram (reproduced in Fig. 1.1).[3] To modern eyes, which are used to phylogenetic trees, it seems upside down, since the 'lower' animals are at the top rather than at the bottom. Yet, in some ways, it is the same as modern phylogenetic trees, because the lineages that appear at the top are the most recent. According to Lamarck, since simple forms of life are transformed into more complex forms, and simple forms are created anew all the time, the simple organisms that exist today are the most recent in origin. Lamarck believed that there is no extinction in nature. The disappearance of species from the fossil record is due to their transformation into something else, not to their extinction. Lamarck's theory of transformation is illustrated in Fig. 1.2. It can be thought of as a series of escalators in which each lineage begins at the bottom with spontaneous generation, and continues up through the various forms as a result of the action of subtle fluids. Different lineages start at different times and progress upwards on parallel but independent escalators.

Lamarck's ideas about the nature of evolutionary adaptation and the mechanisms involved are summed up in the laws given in his *Philosophie zoologique*:

First law
In every animal which has not passed the limit of its development, a more frequent and continuous use of any organ gradually strengthens, develops and enlarges that organ, and gives it a power proportional to the length of time it has been so used; while the permanent disuse of any organ imperceptibly weakens and deteriorates it, and progressively diminishes its functional capacity, until it finally disappears.

Second law
All the acquisitions or losses wrought by nature on individuals, through the influence of the environment in which their race has long been placed, and hence through the influence of the predominant use or permanent disuse of any organ; all these are preserved by reproduction to the new individuals which arise, provided that the acquired modifications are common to both sexes, or at least to the individuals which produce the young. (Lamarck 1809, translated by Elliot 1914, p. 113)[4]

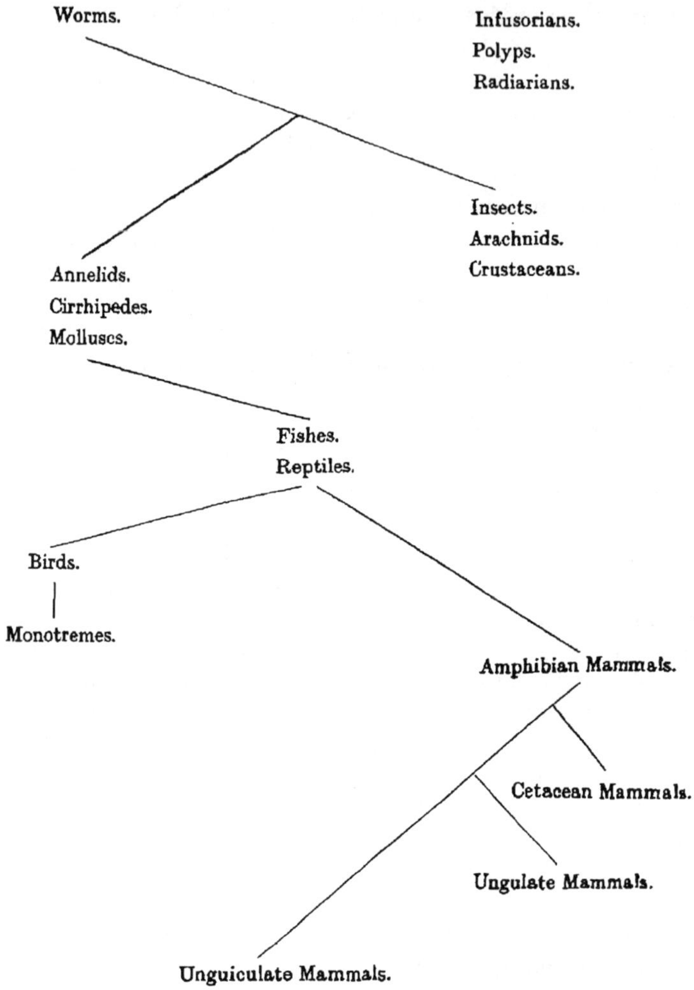

Fig. 1.1 Lamarck's figure 'showing the origin of the various animals'. (From the 1914 English translation of Lamarck's *Philosophie zoologique*, p. 179.)

The first law describes how the use and disuse of organs lead to structural modifications. Lamarck believed that the adaptive responses of animals to new environments are mediated by changes in behaviour; changes in morphology are consequences of changes in behaviour. The second law describes the evolutionary consequences of the first. It assumes that the acquired adaptive changes are inherited. The idea was not original, and Lamarck did not claim that it was. In Lamarck's day the notion that acquired characters could be inherited was almost universally accepted.[5] Lamarck believed that if environmental changes persist, acquired,

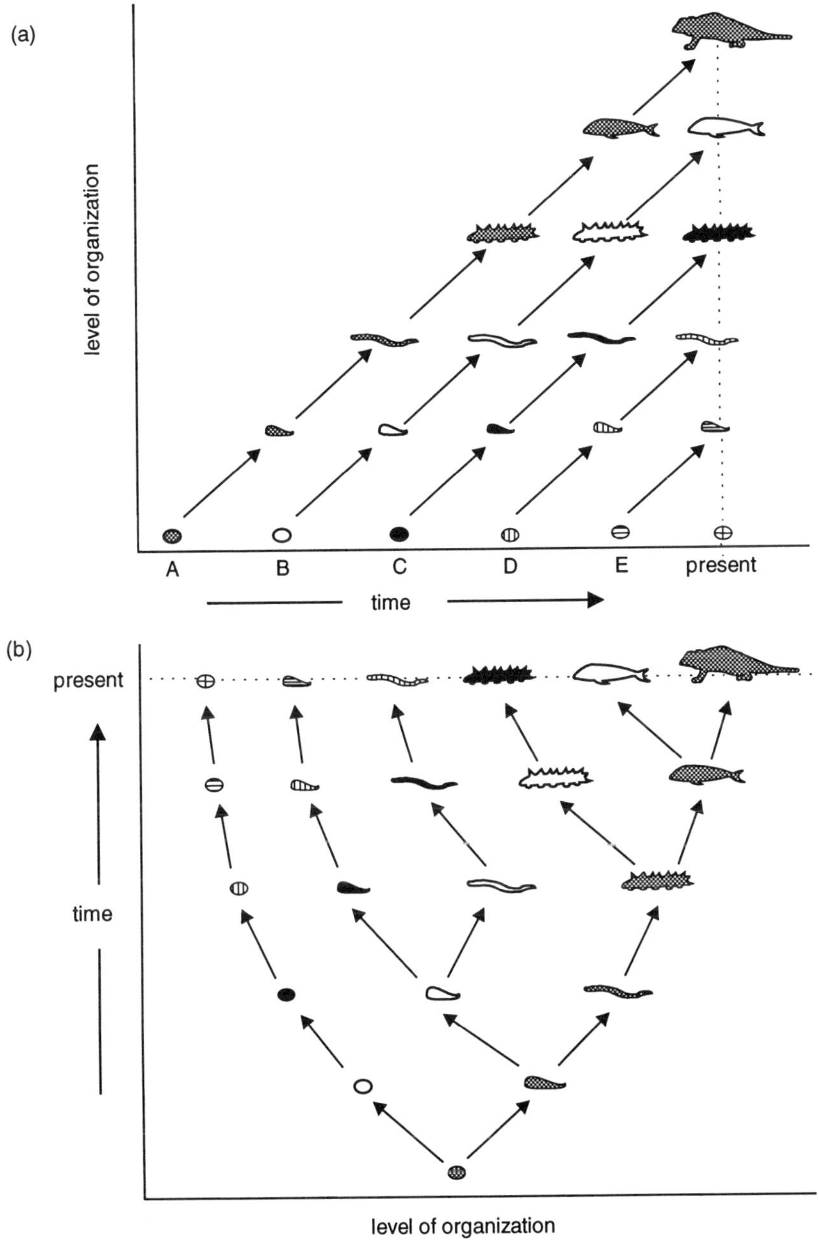

Fig. 1.2 A comparison of (a) Lamarck's theory of transformation and (b) a phylogenetic tree. (a) Lineages A–E begin with spontaneous generation, and pass through the same series of forms. The older lineages have reached more complex levels of organization. Differences in shading indicate that the forms are not identical in each lineage. (Loosely based on Bowler 1989, p. 85.) (b) A phylogenetic interpretation of the forms shown in (a), in which it is assumed there is no extinction.

adaptive, physiological changes are not only inherited, but they also accumulate over generations and become evolutionary adaptations. Evolutionary adaptations are thus simply an extension of the physiological adaptations that occur during the life of an organism.

The evolutionary origin of physiological adaptability itself did not concern Lamarck. For him it was inherent in the very essence and definition of life. Of course, by taking adaptability for granted, the explanatory power of his theory was greatly limited.[6] A complete evolutionary theory should be able to explain not only the evolution of adaptations, but also the evolution of adaptability. It is not sufficient to explain the evolutionary origin of the thick skin on the soles of the feet of a newborn child by saying that originally the skin thickened as an adaptive response to the pressure caused by walking, and this acquired character then became inherited. It is also necessary to explain how the adaptive response—the skin thickening that occurs in response to pressure and reduces the risk of injury—evolved.[7] Lamarck's theory had nothing to say about the evolution of the ability to adapt. Adaptability was taken for granted.

Lamarck's ideas about inherited changes were based on the typical, adaptive, developmental and physiological responses of the individual. He did not regard the variation between different individuals as important. According to Lamarck, all individuals have the same ability to respond adaptively, and all transmit the response to the next generation.

Post-Lamarck Lamarckism (neo-Lamarckism)

Lamarck's theories about the role of fluid movements in biology were found to be untenable and were soon abandoned. His ideas on extinction and spontaneous generation were also shown to be wrong and were forgotten. What eventually came to be termed 'Lamarckism', or 'Neo-Lamarckism', was the belief that the inheritance of acquired characters is the basis of evolutionary change. Many 'Lamarckists' also accepted Lamarck's idea that evolution was progressive, and that the use and disuse of organs was an important cause of change, but the inheritance of acquired characters became central to 'Lamarckism'. Frequently the types of acquired characters that could be inherited were very generally and loosely defined. They included not only characters that were changed by use or disuse, but also passively acquired mutilations, and characters that were directly induced by the external environment, without the behavioural mediation required by Lamarck's theory. Lamarck himself had explicitly rejected a direct effect of the environment on animal structures. He believed that new environmental conditions resulted in new activities and habits, and it was these changes in behaviour that caused changes in the body. Even in plants, which do not have 'behaviour', Lamarck stressed that the response to the environment was mediated by its effect on *internal* activities, such as those associated with nutrition or transpiration.

Rather surprisingly, Lamarck, who seems to have had a theory about everything from mineralogy to meteorology, never developed a theory of inheritance. He suggested no mechanism that would allow acquired characters to be inherited. This was in spite of the fact that Buffon, with whom he had a close association, was well aware of the need for a theory of heredity, and had himself elaborated an idiosyncratic version of a theory that had existed since the time of Hippocrates.[8] The basic idea of this theory, and of most other theories of heredity at this time, was that all parts of the body sent small representative particles to the reproductive organs where they formed 'the germ', which gave rise to the next generation. It is an irony that one of the most famous exponents of this theory, the person who gave it the name by which it is commonly known, was none other than Charles Darwin. In the version elaborated by Darwin, the particles were christened 'gemmules' and the whole hypothesis 'pangenesis'. According to Darwin's and other versions of this theory, an environmentally modified part, or a part that had become modified as a consequence of use and disuse, liberated modified gemmules into the circulation. The modified gemmules reached the germ cells and eventually participated in the formation of the corresponding modified part in the offspring. In this way, acquired characters could be passed on to the next generation.

Ideas about the nature of inheritance changed at the turn of the century when, as Zirkle put it:

> the discovery of Mendel's forgotten work put the whole matter on a new basis and pangenesis came to the end of its 2300-year career. (Zirkle 1946, p. 145)

However, it was not the rediscovery of Mendel's work in 1900 that led to doubts about the role of the inheritance of acquired characters in evolutionary change. The debate about its role began long before then, and continued long after. One of the strongest challenges to the idea of the inheritance of acquired characters came from August Weismann in the 1880s, well before the rediscovery of Mendel's work. Until he was in his mid-forties, Weismann believed wholeheartedly in the inheritance of acquired characters, but once he had changed his mind, he became the most forceful opponent of the idea. He argued that there was no evidence for this type of inheritance: all of the reputed cases could be explained in other ways. Moreover, there were many adaptations, such as those of the sterile worker castes of social insects, which, even in theory, could not be acquired through Lamarckian mechanisms. The adaptations acquired by a sterile worker ant during its lifetime cannot be transmitted to the next generation. Most important of all, Weismann maintained, there was no realistic mechanism by which acquired characters could be inherited. Weismann dismissed the two types of theory that attempted to explain how somatic changes could affect the germ line, saying that one had to assume:

> . . . either the presence of hypothetical tracks along which a modifying, though totally inconceivable, influence might be transferred to the germ-cells, or else the discharge of material particles from the modified organ, must take part in the formation of the germ-plasm. . . (Weismann 1893, p. 393)

He argued that both theories were not only unsupported by direct observation or experiment, they were also incompatible with physiological, cytological, and anatomical observations. Weismann believed:

> that *all permanent—i.e., hereditary—variations of the body proceed from primary modifications of the primary constituents of the germ*; and that neither injuries, functional hypertrophy and atrophy, structural variations due to the effect of temperature or nutrition, nor any other influence of environment on the body can be communicated to the germ-cells, and so become transmissible. (Weismann 1893, p. 395; Weismann's italics)

In other words, the germ line is unaffected by changes in the soma.

Weismann accepted Darwin's theory of evolution. According to Darwin's theory, the cause of evolution is natural selection acting on the inherited differences between individuals. Individuals with variations favourable for survival and reproduction leave most offspring, so gradually the favourable variations spread through the population. Darwin himself did not reject the idea that acquired characters could be inherited. Weismann did. Weismann believed that evolution through natural selection does not require the inheritance of acquired characters. The differences between the neo-Darwinian (Weismannian) idea of evolution through natural selection, and the Lamarckian idea of evolution through the inheritance of environmentally induced acquired characters, is shown in Fig. 1.3. Weismannian evolution operates through changes in populations; Lamarckian evolution operates through changes in individuals. It has been said that Lamarckism is an anti-Darwinian theory because it advocates directed variation: new environments elicit new, adaptive, heritable variation.[9] But Lamarckism is an anti-Darwinian theory not just because it advocates directed variations; it is anti-Darwinian also because it advocates identical directed variations in *all* the individuals of a population exposed to the new environment.

In spite of Weismann's powerfully and passionately argued case in favour of natural selection and against the inheritance of acquired characters, the debate between the neo-Lamarckians and neo-Darwinians continued until well into the twentieth century.[10] Some form of Lamarckism was generally accepted by most American and German palaeontologists who explained the evolutionary trends found in fossil series in Lamarckian terms.[11] Although their observations could not possibly provide support for a *mechanism* of evolutionary change, they claimed that they provided

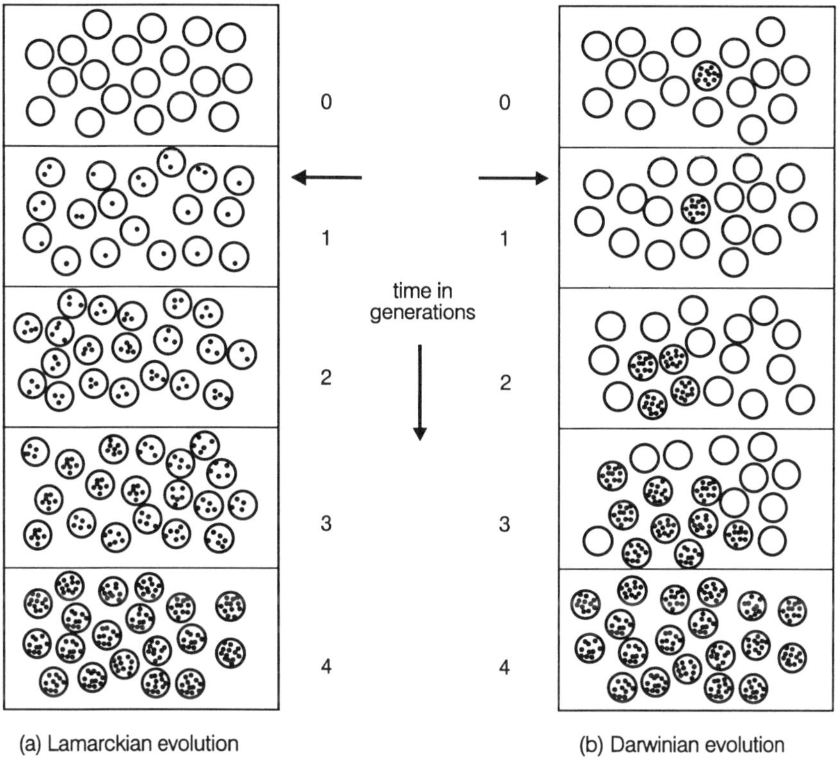

Fig. 1.3 The evolution of an adaptation to a new environment by (a) Lamarckian evolution and (b) Darwinian evolution. Individuals with the adaptation are represented by shaded circles, the intensity of shading indicating the degree of adaptation. At the time indicated by the horizontal arrows, the environment changes, and this change persists through subsequent generations. In Lamarckian evolution, all individuals acquire the adaptation, and it is gradually enhanced as individuals in subsequent generations continue to respond to the environment. In Darwinian evolution, by chance an individual has an appropriate adaptation that enables it to contribute proportionally more progeny to subsequent generations. (Loosely based on Medawar 1957, pp. 80–81.)

evidence for the inheritance of acquired characters. Hence Osborn, who was later influenced by Weismann's ideas and became more doubtful about the inheritance of acquired characters, was able to publish a paper in 1889 with the title 'The palaeontological evidence for the transmission of acquired characters'. For the palaeontologists, inherited changes brought about by the effects of use and disuse, or as a result of the direct influence of climatic factors, seemed a better explanation of the directional trends they found in their fossil series than did the selection of random variations suggested by the neo-Darwinists.

The neo-Lamarckists also could not accept that there was no direct causal relation between ontogenetic and phylogenetic characters. It seemed unreasonable to ascribe to coincidence the similarity between epidermal thickenings that develop during an animal's lifetime in response to rubbing, and the inherited epidermal thickenings that are present at birth in areas subject to pressure at a later date. Darwinian explanations of evolutionary changes that had no obvious adaptive value seemed even less reasonable and acceptable to the neo-Lamarckists. For example, Rensch (1983) described how, in 1929, he accepted a neo-Lamarckian rather than a neo-Darwinian explanation of geographic variations in size. He believed that size variation was due to direct climatic effects which, over many generations, became inherited. He followed the ideas of Cope who, at the turn of the century, had developed a theory of 'diplogenesis' to explain how environmental influences on somatic characters could be transmitted to the next generation through the germ cells.[12] This theory suggested that the change-producing influence affected the germ plasm as well as the somatic parts of the organism. Therefore, the effects could be transmitted to the next generation. This type of mechanism has also been referred to as 'parallel induction'. Parallel induction occurs when the characters of the offspring show that both the somatic and the germ line modifications in the parent were of a corresponding and equivalent type (see Fig. 1.4).

Weismann's definition of the inheritance of acquired characters was a very narrow one. He claimed that the inheritance of acquired characters can be said to occur only if first, the environmental change affects a somatic trait, and second, the modified soma itself (and *not* the environmental agent which affected it) induces a change in the germ plasm which in turn produces the same somatic modification in the following generation.[13] The phenomenon covered by this narrow definition is sometimes referred to as 'somatic induction'. Somatic induction occurs if the environment first modifies the soma, and this effect is then transmitted to the germ cells (Fig. 1.4).

A much broader interpretation of the concept of the inheritance of acquired characters was adopted by other evolutionary biologists.[14] For example, in 1909, in their book *Les théories de l'évolution*, Yves Delage and Marie Goldsmith criticized Weismann's restrictive view, and emphasized what a broad church Lamarckism is:

> Neo-Darwinism, which has found its most complete expression in Weismann's writings, constitutes a well-harmonized system of conceptions relative to the structure of living matter, ontogenesis, heredity, evolution of species, etc. Lamarckism on the other hand is not so much a system as a point of view, an attitude towards the main biological questions.
>
> Whatever theory emphasizes the influence of the environment and the direct adaptation of individuals to their environment, whatever theory gives to actual factors the precedence over predetermination can be designated as Lamarckian. (Delage and Goldsmith 1909, translated by Tridon 1912, pp. 244–245)

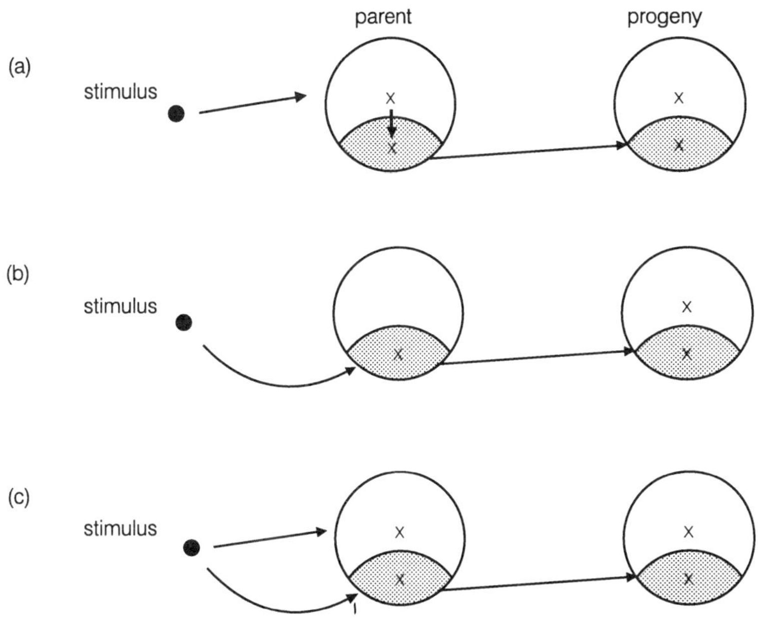

Fig. 1.4 Inherited environmental effects. Circles represent organisms with a soma (unshaded) and germ line (shaded); **X** is the effect of the environmental stimulus. (a) Somatic induction (a 'Lamarckian' mechanism): the stimulus produces an effect in the parental soma, which is transferred to the germ line and thence to the progeny. (b) Random or directed germ line variations (Weismann's mechanism): the germ line is affected directly, so the effect is passed to progeny. (c) Parallel induction: the stimulus acts on both the germ line and soma, so the effect is passed to the progeny. (Based on Fothergill 1952, p. 158.)

For people like Delage and Goldsmith, Lamarckian inheritance occurs whenever a stimulus-dependent character (i.e. a character whose appearance depends on a specific external stimulus) in one generation, becomes stimulus-independent in the following generations (i.e. appears whether or not the stimulus is present). Such a broad interpretation makes 'Lamarckism' and 'the inheritance of acquired characters' purely descriptive terms. No assumptions are made about the mechanisms that bring about the transition from stimulus-dependent to stimulus-independent adaptations. There is no restriction on the type of character involved, nor on the type of external stimuli. Many different phenomena can be described as Lamarckian. In Table 1.1 we have listed some examples of types of inheritance that have been called Lamarckian, and indicated the mechanisms suggested for the transmission of the acquired character. We shall return to many of these again and elaborate on them in later chapters.

Table 1.1 Different types and mechanisms of 'Lamarckian' inheritance

Affected cell types	Affected character	Response to the inducing agent	Mechanism leading to the character's inheritance	References
Somatic	Somatic	Phenotypic change in somatic cells	Pangenesis: transport of gemmules or other entities to germ cells	Hippocrates, Darwin, many others; see Zirkle (1946)
Somatic that can become germ line	Somatic	None—random somatic mutations	Somatic selection between cells followed by the selected mutant cells becoming gametes	Buss (1987), Klekowski (1988); see Chapter 2
Somatic	Somatic	Transcription of mRNA	Somatic selection followed by transfer of mRNA to the germ line where it is reverse transcribed into DNA and integrated into germ-cell DNA	Steele (1979); see Chapter 2
Somatic	Usually somatic, could be germ line specific	Phenotypic change (change in gene expression?)	Natural selection of the ability to respond to the stimulus	Baldwin (1896), Waddington (1942), Schmalhausen (1949); see Chapter 2
Somatic and germ line	Somatic	Parallel induction: phenotypic change in soma and corresponding mutation in germ line	Somatic change (e.g. hormonal) causes an identical genetic change in the germ line	Stempell, and others; see Rensch (1983)
Somatic and germ line	Somatic or germ line specific	Parallel induction: identical mutations in soma and germ line	Conventional genetic transmission	see Blacher (1982)
Somatic and germ line	Somatic or germ line specific	Parallel induction: identical heritable epigenetic changes in soma and germ line	Epigenetic inheritance through the germ line	Jablonka and Lamb (1989); see Chapter 6
Germ line (or unicellular organism)	Germ line	Directed mutation	Conventional genetic transmission	Weismann (1902), Cairns *et al.* (1988), Hall (1988); see Chapter 3
Germ line	Somatic or germ line	Change in chromatin structure	Epigenetic inheritance	Jablonka and Lamb (1989); see Chapters 4 and 6

Table 1.1 (*continued*)

Affected cell types	Affected character	Response to the inducing agent	Mechanism leading to the character's inheritance	References
Somatic (or unicellular organism)	Somatic	Change in architectural structures	Templating and self-propagation	Sonneborn (1964), Nanney (1968), and others; see Chapter 4
Somatic (or unicellular organism)	Somatic or germ line	Change in metabolic feedback	Self-sustaining metabolic loops	Delbrück (1949), Hinshelwood (1953); see Chapter 4

Definitions of 'Lamarckian' terms

Because of the confusion that has always surrounded the use of 'the inheritance of acquired characters', Mayr (1982*a*) has suggested that the term 'soft inheritance' should be used to cover all the different meanings that have been associated with the former concept. He defines soft inheritance as:

> Inheritance during which the genetic material is not constant from generation to generation but may be modified by the effects of the environment, by use or disuse, or other factors. (Mayr 1982*a*, p. 959)

Medawar made a useful distinction between two types of Lamarckism. He described the 'weak' form of Lamarckism in the following way:

> Modifications acquired in each member of a succession of individual lifetimes, as a result of recurrent responses to environmental stimuli, may eventually make their appearance in ontogeny even when the environmental stimuli are absent or are deliberately withheld . . . and the age of appearance of these modifications in ontogeny will eventually anticipate the age at which environmental stimuli could in any case have been responsible for them. (Medawar 1957, p. 83)

This weak form of Lamarckism has nothing to say about the mechanism underlying the inheritance of the acquired character. The 'strong' form does. The 'strong' form of Lamarckism requires that:

> The repeated induction of character-differences within the lifetimes of individuals of successive generations is accompanied by a genetic change in each individual, the change being such as eventually to reproduce the character-difference elicited by environmental stimuli even when those stimuli are withheld. (Medawar 1957, p. 91)

For the purposes of this book we define the inheritance of acquired characters in the following way. The inheritance of an acquired character has occurred if:

(1) the change in the character is induced by the environment;
(2) the induced change is specific and repeatable, although not necessarily adaptive;
(3) a specific change in hereditary information is involved;
(4) the change is transmitted to the next generation.

Essentially, this definition conforms with Mayr's definition of soft inheritance, but it avoids Mayr's phrase 'the genetic material', which is usually assumed to be synonymous with DNA. It therefore allows for the possibility that heritable information can be carried in ways other than in the sequence of bases in DNA. The definition also conforms with Medawar's definitions of both the weak and strong forms of Lamarckism, again providing that the word 'genetic' in Medawar's definition is not confined to classical DNA-based heredity. It is important to note that our definition, like those of Mayr and Medawar, is a definition of the inheritance of *acquired* characters, not of *required* characters.

Mendelian genetics and Lamarckian evolution

With the benefit of hindsight, the way in which the new science of genetics influenced ideas about evolution at the beginning of this century is strange.[15] At first many Mendelians claimed that their work showed that Darwin was wrong. They argued that mutations are much more important than natural selection in bringing about evolutionary change. On the other hand, many non-geneticists believed that the discontinuous characters studied by the geneticists are irrelevant in evolution. Natural selection works on continuous variation, which can be influenced by environmental factors. Many embryologists and physiologists thought that even if Mendelian factors in the nucleus are responsible for individual and racial characteristics, non-Mendelian hereditary factors located in the cytoplasm are responsible for the characters that determine the genus and species to which an animal belongs.[16] They believed that the pliable cytoplasm, which harbours these non-Mendelian factors, allows the inheritance of acquired characters. Initially, therefore, Mendelian genetics did not strengthen Darwinism, and did little to make non-geneticists doubt the possibility of the inheritance of acquired characters.[17]

It was not until the late 1930s that Mendelian genetics became integrated with evolutionary biology in a way that signalled the end for Lamarckian theories of evolution. In what is now known as 'The Modern Synthesis of Evolution', the Mendelian gene, a factor that is stable in heredity and immune from the influence of the environment, was accepted as the material basis of all evolutionary change. It was assumed that environmental effects on characters, which are so widespread and so striking, do not involve changes in the genotype; the phenotypic differences induced by the environment are not inherited.

The conceptual distinction between genotype and phenotype—between instructions and their implementation—was made by Johannsen in the first decade of this century.[18] It was of fundamental importance for the development of genetics, and also had important consequences for the development of evolutionary theory, because it was the concept of heredity that grew out of Johannsen's ideas that was incorporated into the Modern Synthesis of evolutionary biology in the 1930s and 1940s, and to a large extent remains with it. It is a very restricted notion of heredity, and it is this restricted view that has been the basis of many of the objections to the possibility that acquired characters can be inherited.

Johannsen's ideas were based both on the patterns of inheritance of Mendelian genes, and on his own work on pure lines of plants. Pure lines are strains produced by self-fertilization. Johannsen found that lines developed from different individuals had different characteristics. Although individuals within a line differed in appearance, the differences between them were not heritable. Selection was ineffective in pure lines.[19] Similar results were obtained by Jennings (1909) with *Paramecium*, in which pure lines derived from single individuals were found to differ in size, structure, and physiological characteristics. Although these characters were influenced by environmental conditions, the environmentally induced changes were not passed on.

In 1909, Jennings asked about the pure line idea 'Is it possibly of sufficient importance to deserve agitating a little before the American Society of Naturalists?' Clearly it was, because in the following year the Society held a symposium devoted to 'The Study of Pure Lines or Genotypes'.[20] Most papers read at this meeting supported Johannsen's ideas. Johannsen's own contribution was entitled 'The genotype conception of heredity'. In it Johannsen attempted to clarify the concept of heredity in biology. He insisted that biological heredity is not the transmission of characters, it is the transmission of what we would now say are the instructions for building characters. Johannsen distinguished between 'heredity', the passing on of 'potentialities', and 'transmission', a concept based on human practices such as the transfer of property or ideas from one person to another. He argued that in biology the physical transmission of the personal qualities of individuals to their progeny does not occur:

The *personal qualities* of any individual organism do not at all cause the qualities of its offspring; but the qualities of both ancestor and descendant are in quite the same manner determined by the nature of the 'sexual substances'—i.e., the gametes—

from which they have developed. Personal qualities are then *the reactions of the gametes* joining to form a zygote; but the nature of the gametes is not determined by the personal qualities of the parents or ancestors in question. (Johannsen 1911, p. 130; Johannsen's italics)

The appearance of an individual depends on the inherited potentialities, which Johannsen called the 'genotype', and on the environment. The character, the end-product of the interaction between environment and genotype, Johannsen christened the 'phenotype'.[21] The unit of biological heredity, the Mendelian factor which Johannsen named 'gene', was not a material model or representation of the phenotype, but a unit of information. All individuals in a pure line are genotypically the same. Their heritable genotypes are unchanged by environmental factors, although the material realization of these genotypes may be.

Johannsen stressed the implications of his conceptual distinction between genotype and phenotype for the questions concerning the inheritance of acquired characters:

The principle of pure lines or, generally, pure culture, is of importance also for elucidating the celebrated question of the inheritance of 'acquired characters'. Mendelism and pure-line researches are here in the most beautiful accordance, both emphasizing the stability of genotypical constitution; the former operating with the *constituent unities*, the latter with the behavior of the *totality* of the genotypes in question. . . . as yet no experiment with genotypically homogeneous cultures has given any evidence for the Lamarckian view, the most extreme 'transmission'-conception ever issued. (Johannsen 1911, p. 141; Johannsen's italics)

Initially, Johannsen's experiments were also seen as evidence that Darwinian natural selection could not be the basis of evolutionary change. Selection, like environmental actions, had no effect on pure lines, so Johannsen concluded that mutation was more important than selection in bringing about evolutionary change.

In the long run, Johannsen's influence on evolutionary ideas in the first quarter of this century was probably of less significance than the influence his concept of heredity had on the development of the new science of genetics. According to Johannsen:

Heredity may then be defined as *the presence of identical genes in ancestors and descendants*, or, as Morgan says in full accordance with this definition: 'The word heredity stands for those properties of the germ-cells that find their expression in the developing and developed organism'. (Johannsen 1911, p. 159; Johannsen's italics)

It was this restricted concept of heredity that was adopted by the influential American geneticists. As genetics increased in importance and influence, so did this view of heredity.[22] The mechanisms of cellular inheritance—of the inheritance of determined and differentiated states during development

—were largely excluded from the study of heredity. Only nuclear genes were important, and these were immune from cytoplasmic and environmental effects. Sapp has summed up the situation in this way:

> Ironically, Johannsen's genotype/phenotype distinction offered geneticists the conceptual space or route by which they could bypass the organization of the cell, regulation by the internal and external environment of the organism, and the temporal and orderly sequences during development. Although the genotype/phenotype distinction represented an implicit theoretical acknowledgement of the beginning and end of a production, in practice, Mendelian geneticists ignored developmental processes and the possible influence of extragenic conditions in the production of characters. (Sapp 1987, p. 49)

When in the late 1930s, after a quarter of a century of largely independent growth, disciplines such as biogeography, palaeontology, systematics, natural history, and genetics began to be integrated in the Modern Synthesis, the view of genetics brought to evolutionary studies was based on a narrow concept of heredity. It is this limited view of heredity that remains with much of evolutionary biology today.[23] There was little in the synthesis about development and differentiation.[24] It was the Mendelian hereditary unit, the gene, whose behaviour could be studied through transmission genetics, and whose frequency could be manipulated on paper by the population geneticists, that became the material basis of evolutionary change.

Molecular genetics and the inheritance of acquired characters

The 1940s and 1950s saw the growth of microbial and biochemical genetics, and the development of techniques for studying gene action. The gene was found to control the production of specific proteins. The molecular nature of the gene and the way in which it specifies proteins were quite rapidly unravelled: the genetic material is DNA, which carries the information for making proteins encoded in its base sequence; this information is first transcribed into RNA, and the RNA messages are then translated into the amino acid sequences of proteins. The *central dogma*[25] of molecular biology was established: information flow is unidirectional—it passes from DNA to proteins, but not in the reverse direction:

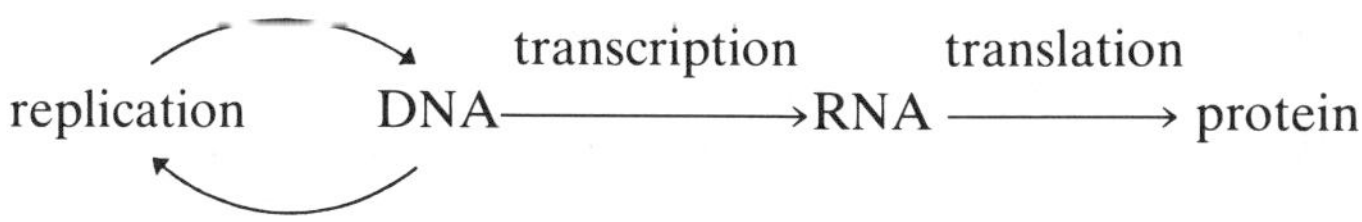

The environment can alter the instructions in DNA only accidentally. Proteins, and systems built from proteins, are highly sensitive to the environment, but since protein is not the hereditary material, modifications in protein structure and function cannot lead to inherited changes.

Molecular genetics has tended to reinforce the attitude to heredity adopted by the Mendelian geneticists. The genotype has become identified with DNA, the phenotype with proteins or the products of proteins. The genotype is regarded as a set of instructions, subject to only random changes; the phenotype is the result of the implementation of these instructions, and can be modified by the environment in an adaptive way. One of the most explicit articulations of this position has been made by Dawkins, who claimed:

> The inheritance of acquired characteristics not only *doesn't* happen: it *couldn't* happen in any life-form whose embryonic development is epigenetic rather than preformationistic. (Dawkins 1986, p. 298; Dawkins's italics)

Since embryonic development is not preformationistic—genes do not contain a description of the adult characters—Dawkins argued that if a character is modified by the environment, information about the change cannot, even in theory, be fed back into the genes. Dawkins pointed out that DNA is not, as is sometimes said, a blueprint; it is more like the recipe for a cake: a set of instructions for carrying out a process. The words of the recipe (the DNA sequences) do not correspond to crumbs of the cake (parts of the body); they represent stages in the process of making the cake. Consequently, a word difference leads to a whole-cake difference:

> 'Baking powder' does not correspond to any particular part of the cake: its influence affects the rising, and hence the final shape, of the whole cake. If 'baking powder' is deleted, or replaced by 'flour', the cake will not rise . . . There will be a reliable, identifiable difference between cakes baked according to the original version and the 'mutated' versions of the recipe, even though there is no particular 'bit' of any cake that corresponds to the words in question. This is a good analogy for what happens when a gene mutates. (Dawkins 1986, p. 297)

Using this cake analogy, Dawkins says of Lamarckian inheritance:

> We can no more imagine acquired characteristics being inherited than we can imagine the following. A cake has one slice cut out of it. A description of the alteration is now fed back into the recipe, and the recipe changes in such a way that the next cake baked according to the altered recipe comes out of the oven with one slice already neatly missing. (Dawkins 1986, p. 298)

This cake analogy is a clear and comprehensible expression of the current dogma about Lamarckian inheritance. It also exposes the weakness of that dogma. The theoretical impossibility it is meant to illustrate is an artefact of the analogy. The assumption that the instructions in DNA are isolated from their implementation, with no interactions between the products of the instructions—the protein or phenotype—and the DNA instructions themselves, is not valid. As we shall discuss in Chapter 3, there is evidence suggesting that there are mechanisms that enable the genome to sense an environmental change, respond to it, and transmit the response to descendants. The recipe is not isolated from the cake, because it is contained in and is executed by the components of each crumb of the cake. The analogy is inadequate because it is based on a rigid distinction between the genotype and phenotype—between the instructions encoded in DNA and the result of the execution of these instructions.

There is a second invalid assumption in the cake analogy. It is that there is a single inheritance system. We want to suggest a rather different analogy, which reflects the fact that there are multiple inheritance systems. This analogy shows how the distinction between genotype and phenotype becomes blurred if information is transmitted in ways additional to that involving DNA base sequence. Instead of a cake and a recipe, consider a piece of music that is transmitted from generation to generation as a written score. If the score represents hereditary information in DNA, the phenotype is a specific interpretation of this score at a certain time by certain artists. The interpretation does not affect the score. However if there is another transmission system—recordings—through which a particular interpretation can be transmitted from generation to generation along with the written score, the situation is rather different. There can then be evolution of interpretations of the score, based on the influence that one interpretation has on subsequent interpretations, and that these have on still later ones, and so on. Both the phenotype (the present interpretation) and the genotype (the written score) influence subsequent interpretations.

We believe that this music analogy is more appropriate than the cake analogy because the old notion that the DNA of nuclear genes is the sole carrier of hereditary information is incorrect. Molecular biology, which was at first centred around genetics, has now spread to encompass many other disciplines, including embryology. Studies of growth and development are no longer divorced from studies of heredity. Through molecular studies of differentiation and cellular inheritance, it has become clear that information is transmitted in ways other than through the primary base sequence of DNA. As we shall show in later chapters, these additional systems can also transmit information between generations of organisms, and permit the inheritance of acquired characters.

The arguments against the inheritance of acquired characters just discussed stem from the conventional picture of the relation between genetics and development. There is another argument, an evolutionary argument, against the idea that acquired characters can be inherited. It is that most acquired characters are detrimental: they are the consequences of injury, disease and ageing.[26] Consequently, it is argued, mechanisms allowing the inheritance of such maladaptive changes should be strongly opposed by natural selection. Although this must be true, it has to be remembered that the same argument can be applied to mutations: they, too, are frequently maladaptive, yet no one doubts that they occur and are inherited, in spite of often being counter-selected. As with other evolutionary features, including mutation frequencies, the extent to which acquired characters are inherited presumably depends on the relative costs and benefits of the presence or absence of the mechanisms involved. We shall consider this further in Chapter 8.

Why is evidence for Lamarckian inheritance so sparse?

If there are mechanisms through which acquired characters can be inherited, why has more than a century of study of heredity failed to reveal sufficient cases of 'Lamarckian inheritance' for it to have become an accepted part of evolutionary theory? We shall discuss this in some detail in Chapter 6, but for the present we want to suggest two types of reason for the apparent paucity of evidence for Lamarckian inheritance. The first is that usually people have looked in the wrong type of organism, in the wrong place, and for the wrong type of change. The second is that people have gone to great lengths to interpret all experiments and observations that might be taken as evidence for Lamarckian inheritance in non-Lamarckian terms.

Evidence for or against the inheritance of acquired characters is not to be found in the type of experiment carried out by Weismann in the last century. He cut off the tails of mice for 22 generations and showed that it had no effect on the tail length of the progeny.[27] Many generations of male and female circumcision, and the docking of sheep's tails, have shown the same thing: mutilations are not inherited. This type of experiment (which is the type illustrated by Dawkins's slice of cake analogy) would demonstrate Lamarckian inheritance only if information from the somatic parts of the adult mammalian body were transferred to germ cells. As far as we know it is not. An experiment that is more likely to demonstrate Lamarckian inheritance is one in which an induced change affects the whole organism, including the cells that produce the next generation. For example, cellular adaptations to an environmental change that affects all cell types in a species that reproduces by fragmentation are likely to be inherited. Unfortunately, the dominance of Mendelian genetics during the first half of this century meant that the study of heredity centred on organisms and characters that are unlikely to yield evidence of the inheritance of acquired characters very readily. The animals used were mainly mammals and insects, in which the germ line is segregated from the soma early in development. Most of the characters studied were stable adult features. Examples of irregular hereditary patterns were found, but as Sapp (1987) has documented, for many years the study of such oddities attracted few workers and little financial support. Studies of heredity were concentrated on nuclear genes and chromosomes, and their segregation in crosses between sexually reproducing organisms; organisms and characters that were not amenable to analysis by the techniques of transmission genetics were generally excluded.

To a large extent the concept of heredity adopted by the geneticists, and the methods and organisms they used to study it, have prevented cases of Lamarckian inheritance being found. Induced inherited changes are most likely to be detected, and are probably of more evolutionary importance, in plants, in some invertebrate groups other than insects, and in microorganisms (see Chapters 2 and 8). The inherited, environmentally induced changes that occur in organisms such as the mouse and *Drosophila* are probably small, causing only minor variations in the expression of genes, and are likely to have been explained away as the result of the action of 'modifiers', rather than being investigated seriously. As Lindegren (1949) described with reproach, even in *Neurospora*, mutations that were found to be unstable in inheritance were discarded as a matter of course.[28]

The reason for the paucity of examples of the inheritance of acquired variations may therefore be similar to that which, until relatively recently, pertained to the rarity of examples of jumping genes. For more than thirty years most people regarded jumping genes as an esoteric peculiarity of maize, a rarity of no fundamental importance for genetics. Yet now that their existence is taken for granted, and the right techniques are available, they are found almost everywhere.

In fact, and in contradiction to general belief, there are several well-recognized and quite well-understood examples of the inheritance of acquired characters. Most involve microorganisms; many involve changes in cytoplasmic DNA or cytoplasmic organelles. For example, if the chloroplasts of the protist *Euglena* are destroyed with streptomycin, the subsequent generations of their progeny lack chloroplasts; bacteria can acquire and transmit to their descendants plasmids from species that carry genes for drug resistance.[29] It is now quite generally accepted that some cell organelles such as mitochondria and chloroplasts were originally prokaryotic symbionts in eukaryotic cells. We do not intend to devote much space in this book to the inheritance of these and similar acquired characters, because they do not pose a theoretical problem for orthodox evolutionary theory. But the fact that they are not regarded as a problem is of interest, because it illustrates the general attitude to Lamarckian inheritance. Discussing examples such as those above, Fitch suggested that the reason why they are not regarded as a problem for Darwinism has nothing to do with genes or DNA *per se* because:

> Genes and DNA are the means of inheritance and both Darwinism and Lamarckism must incorporate these facts.
>
> The reason for there being little concern among Darwinists for this rather common inheritance of acquired characters is that 'the inheritance of acquired characters' is more of a slogan that captures a part of the Lamarckian spirit than a

statement of its basic distinctiveness from Darwinism. Rather, it seems to me, the crucial difference arises from the issue of cause and effect between the phenotype and the genotype. Clearly, each affects the other but we are in no danger of circularity because the question is whether an evolutionary novelty, arising at the level of the phenotype, can produce a genotype that assures the phenotype's continuance, or whether an evolutionary novelty, arising at the level of the genotype, can produce a phenotype that assures the genotype's continuance. In every one of the above examples of the inheritance of acquired characters, *it was the genetic material that changed first and was acquired*. ... Only an overly narrow definition of the source of genetic variability, more narrow than anything Darwin could have seriously proposed, could require a modification of Darwinism ... (Fitch 1982, p. 1137; Fitch's italics)

We agree with Fitch that 'the inheritance of acquired characters' is not well defined, and that the basic tenets of Darwinism are not shaken by the observations he discusses. What we find interesting is his reliance on the genotype-phenotype distinction to show why the observations have not been a problem for Darwinism. Fitch seems to have been saying that so long as the primary changes occur at the genotype level, the inheritance of acquired characters is compatible with Darwinism. Although this attitude is certainly compatible with Darwin's Darwinism, which embraced the inheritance of acquired characters, we doubt that it is readily acceptable to most contemporary Darwinians. The attempt to place observations such as acquired drug resistance within the Darwinian framework is strange for two reasons. First, as Fitch himself stressed, Darwinian evolution does not require all evolution to be Darwinian. Second, as Fitch also pointed out, molecular biology is revealing all sorts of strange genomic behaviour and genomic responses, which have to be incorporated into evolutionary theory. If the internal and external environments can direct genomic behaviour in the way it seems to, surely the genotype–phenotype distinction has become so blurred that it is no longer possible to say whether a response begins with the genotype or with the phenotype, and it is no longer useful to try to do so.

When considering the impact molecular biology would have on evolutionary theory, Fitch (1982) predicted that some organisms would be found to have mechanisms that increase the mutation rate specifically in those genes whose activity could be useful for survival. Evidence for such 'directed mutation' was soon found: some mutations in bacteria appear preferentially in the environmental conditions in which they are beneficial. This discovery came as a shock to most biologists, reared as they had been on the notion of random mutation. It led to a lively debate about the interpretation and significance of the experiments. We discuss this work in some depth in Chapter 3. What we want to highlight here is the emotional response to the possibility of directed mutation, and the way in which the debate revealed an extreme reluctance to admit the possibility of Lamarckian evolution. For example, Lenski and his colleagues (1989) claimed that even if the existence of directed mutations in bacteria were proved beyond reasonable doubt, it would not constitute evidence of Lamarckian evolution, because Lamarck thought that the inheritance of acquired characters occurs through the effects of use and disuse. This is unreasonable. First of all, some explanations of directed mutation do in fact suggest that the mutations are induced as the result of a form of 'use', namely, gene expression. More importantly, as we discussed earlier, Lamarckism has not remained unchanged since Lamarck's time, any more than Darwinism has remained unmodified since the publication of *The origin of species*. Many scientists have considered themselves, and were considered by others, to be Lamarckians because of their belief in the direct effect of the environment on heritable qualities, not necessarily through use and disuse. It is confusing if, in order to avoid the stigma of Lamarckism, Lamarckian evolution is re-defined in restrictive terms.

The debate about directed mutation also illustrates another way of avoiding the stigma of Lamarckism, and reconciling the possibility of induced inherited variations with neo-Darwinian orthodoxy. This is to shift from thinking about selection of individuals, to thinking about selection within the individual. Some interpretations of the experiments showing directed mutation were based on selection occurring between variable intracellular DNAs, RNAs, etc. In this way, instead of the individual bacterium being the unit of selection, the unit of selection became the intracellular molecule. On this basis, although directed mutation looks Lamarckian when the unit of analysis is the individual bacterium, the adaptive response is really the result of Darwinian selection between accidental molecular variants *within* the bacterium. As Keller (1992) has noted, the issue is whether directed mutation can be described in the Darwinian language of chance and selection, or whether the Lamarckian language of purpose and choice has to be used. If the individual is the unit of analysis, it seems that Lamarckian language is appropriate, whereas if considered at the level of intracellular variations, Darwinian language is suitable.

Keller has shown how, through the choice of language and use of intracellular selection, directed mutation was brought within the Darwinian framework. Although more comfortable for Darwinians, the problem with this approach is that it makes no sense to change the level of analysis if Darwinian and Lamarckian explanations are to be usefully compared and assessed. Medawar recognized and emphasized this point many years ago when discussing a possible Lamarckian mechanism for gradual adaptation in bacterial cultures:

> It may be assumed that there are alternative pathways of metabolism within each cell, i.e. alternative enzyme sequences or metabolic gearings, as there are, for

example, alternative pathways for the degradation of glucose. Such metabolic pathways may for a variety of reasons be so adjusted as to be mutually inhibitory, so that only one prevails in any one of a possible set of steady states. The inhibition of one such system therefore entails its replacement by another. In other words, as Hinshelwood (1946) has made clear, the Lamarckian transformation . . . may be Darwinian at the lower analytical level represented by the enzymic population or complex of intersecting metabolic pathways within the individual bacterial cell. Such a description would be pointless for any except explanatory purposes, but it shows that *no discussion of the rival interpretative powers of Darwinism and Lamarckism can have any useful outcome unless a certain analytical level is defined and adhered to*. Hereafter we shall be concerned with individual organisms as analytical units, for it is only in this context that the rivalry is of any moment. (Medawar 1957, p. 82; our italics)

Perhaps the foremost reason for the reluctance to accept Lamarckian interpretations is the feeling that by so doing, one is accepting purposeful evolutionary responses: that an organism has some indefinable properties that propel it towards some goal. How does the organism *know* how to change its genetic material according to new environmental specifications? Again, the debate on directed mutation yielded examples of how this problem can be circumvented. Bruce Wallace (1990) and Sydney Brenner (1992) both attempted to make the Lamarckian idea of environmentally directed mutation more palatable to Darwinians by treating it as an adaptive response, which has itself evolved through Darwinian selection of random variations. Brenner wrote of directed mutation in bacteria:

There may still be biologists who would like the phenomenon to have some trivial explanation and to go away. However, even if it stays, as seems likely, no radical alteration of our views is demanded. Bacteria spend a considerable part of their natural lives under nutritional constraints so that if there was a mechanism to promote adaptive mutation, the mechanism itself would have been selected for, and the apparent genetical intentionality of *E. coli* could have been brought about by the process of natural selection. (Brenner 1992, p. 168)

Wallace argued that, from an evolutionary point of view, it does not matter whether an adaptive response is brought about by a mechanism involving a DNA change such as the excision of a DNA segment, or by more conventional means such as the binding of a regulatory protein. Whatever the nature of the response, the ability to make the adaptive modifications is under genetic control, and the genetic programme underlying the response system must have evolved via Darwinian selection. According to Wallace, the important part of evolution is the evolution of the genetic programme, not of the specific manifestations of that programme. His argument emphasizes the point made by many previous critics of Lamarckism, that Lamarckian mechanisms of induced variation may explain the evolution of an adaptation, but they cannot explain the evolution of adaptability. Although the specific adaptive response may be the result of directed mutation or other types of genomic response to the environment, the genetic machinery that underlies it must have evolved first by Darwinian mechanisms.

The evolution of systems that can lead to Lamarckian evolution is a fascinating subject, and we shall return to it in Chapters 7 and 8. We do not doubt that the basic mechanisms underlying the inheritance of acquired variations evolved in a Darwinian fashion by the selection of accidental variations. However, why should the evolutionary origin of Lamarckian inheritance systems be of any relevance when considering the effects these systems have once they are in place? By stating that the rules of the evolutionary game must have evolved via Darwinian selection, Wallace and Brenner may have helped make the idea of directed mutation more acceptable to Darwinists, but if the rules of the game are Lamarckian, their evolutionary origin is irrelevant to the way in which they dictate the course of evolution. Consider cultural inheritance. It is reasonable to assume that the cognitive mechanisms that allow the transmission of information among human beings were the result of Darwinian evolution, but once cognitive mechanisms such as long-term memory, and the ability to imitate and to teach appeared, they dictated the course of cultural evolution. The new rules of the game help to determine the course, the direction, and the rate of evolution. The same is true for the inheritance of acquired characters. The ability to acquire and transmit changes in some characters may have evolved through Darwinian evolution, but once the response and transmission systems had evolved, they constituted a new mechanism of evolution, operating alongside Darwinian evolution by means of natural selection of random genetic variations.

Through Darwinian selection, multiple inheritance systems have evolved, generating different types of heritable variation, which now play a role in evolutionary change. The transmission system on which we shall concentrate in this book is the Lamarckian inheritance system that operates in cell lineages. We are going to present a picture of heredity which, unfortunately, is not as neat and elegant as that of classical genetics, where the unaltered gene passes from generation to generation, immune to environmental influences. The non-DNA systems that we shall discuss are sometimes referred to as *epigenetic inheritance systems*, although the distinction between genetics and epigenetics, like the distinction between genotype and phenotype, has become rather blurred. Epigenetic inheritance systems are responsible for transmitting determined and differentiated states during ontogeny. The attitude of evolutionary biologists to this type of inheritance was summed up with characteristic honesty by John Maynard Smith in 1966:

> The view generally taken by geneticists of differentiation, when it is not simply forgotten, is that the changes involved are too unstable to be dignified by the name 'genetic', or to be regarded as important in evolution. I tend to share this view, although I find it difficult to justify. (Maynard Smith 1966, p. 71)

The mechanisms underlying differentiation and the role of epigenetic inheritance in development are now beginning to be unravelled. We hope to show that the epigenetic inheritance systems, which are so important in development, are also important in evolution, and that evolution and development are far more directly intertwined than is usually supposed. When epigenetic systems are considered, the environment is more than a mere selective agent, it is also an inducer of specific heritable variations.

Summary

The ideas developed by Lamarck are very different from what is now commonly referred to as Lamarck's theory of evolution. 'Lamarckism' is generally considered to be more or less synonymous with 'the inheritance of acquired characters', but there is no agreement about exactly what this phrase means. For some people it is appropriate only if the characters that are acquired are adaptive, while for others any repeatable change induced in one generation and transmitted to the next can be regarded as an inherited acquired character. Since the 1940s, most people have doubted that Lamarckian evolution can occur at all, believing that changes in the phenotype cannot be transmitted to the genotype, and therefore that Lamarckian inheritance is impossible in principle. Belief in the one-way flow of hereditary information between genotype and phenotype was reinforced when the central dogma of molecular biology was formulated in the late 1950s, proclaiming as it did that information passes from DNA to proteins, but never in the reverse direction.

In recent years, molecular biology has shown that the genome is far more fluid and responsive to the environment than previously supposed. It has also shown that information can be transmitted to descendants in ways other than through the base sequence of DNA. Even so, most people still deny that Lamarckian evolution occurs. On the one hand, there are those who accept that the genome is a response system and not just a passive information carrier, but argue that because it is the genotype or DNA that is modified, the induced changes cannot be regarded as Lamarckian. On the other hand, there are those who recognize the existence of non-DNA inheritance systems (such as cultural inheritance), but argue that since only the genotype is inherited in a biological sense, the transmission of non-DNA variations does not qualify as inheritance and is therefore irrelevant in biological evolution. Critical evaluation of the role of the inheritance of acquired variations in evolutionary change is thus avoided by using arbitrary definitions of heredity and acquired variations.

Adhering to the idea that evolutionary change cannot be the result of the inheritance of environmentally induced changes is misguided. Not all adaptive changes are the result of Darwinian selection of random variations created by the shuffling of genes and rare chance mutations. The nature of different types of heritable variation is now beginning to receive closer attention, and there is a growing realization not only that some DNA variations can be environmentally induced, but also that there are non-DNA heritable variations that play a crucial part in development. The importance of the inheritance systems underlying the variations seen in development has only recently been recognized. Nevertheless, as we show in the next chapter, there was interest in epigenetics and development, and an awareness that the mechanisms underlying development needed to be integrated into evolutionary theory, long before molecular biology began to uncover the types of mechanism that may be involved.

Notes

1. We feel it necessary to stress our belief in Darwinian evolution because recent history has shown that any argument suggesting that Darwinian evolutionary theory should be modified or amended is liable to be used by Creationists as evidence that the theory of evolution is wrong. Like most Darwinians, we believe that Darwinian evolutionary theory is a flexible theory, quite capable of accommodating modifications and amendments.
2. See Mayr (1982*a*, p. 352). Other accounts of Lamarck's ideas and his place in the history of biology can be found in Blacher (1982), Bowler (1983), Jordanova (1984), and Oldroyd (1983).
3. Dr Peter McLaughlin has directed our attention to the fact that in the original French edition, Lamarck's figure, and the addition of which it is a part, appear at the very end of the book, and not immediately after Chapter 8 as in the English translation published in 1914. He suggests that this placement reflects the interpretation given to Lamarck's ideas in the post-Darwinian climate at the beginning of this century, when one of the main concerns was with phylogenetic trees.
4. In his *Histoire naturelle des animaux sans vertèbres* published in 1815, Lamarck gives four, rather than two, laws: the first law describes the inevitability of increase in size; the second and third describe how the need to cope with the environment leads to changes in organs through use and disuse; the fourth deals with the inheritance of the acquired character.
5. An account of the history of the idea that acquired characters can be inherited is given in Zirkle (1946).
6. The criticism that Lamarckism does not explain adaptability has been made many times, e.g. see Blacher (1982, pp. 154–155), and Dawkins (1986, p. 299).
7. The example of skin thickening and calluses has been a favourite with evolutionary biologists seeking to explain the relation between adaptation and adaptability, and we shall refer to it in other parts of this book. Darwin used the prenatal thickening of the sole of the human foot as an example of the inheritance of an acquired character (Darwin 1871, Vol. 1, Chapter 4).
8. Accounts of early theories of inheritance are given in Zirkle (1946), Blacher (1982), and Mayr (1982*a*).
9. For example, see Gould (1982, p. 381).

10. Detlefsen (1925) gives a contemporary account of Lamarckism in the 1920s.
11. Pfeifer (1965) and Rensch (1980) give accounts of American and German neo-Lamarckism in the early part of this century.
12. A detailed exposition of this theory is given in Cope (1904).
13. In later life Weismann accepted that environmentally induced characters could be inherited if they affected the germ line itself directly. He also accepted the idea of parallel induction—the parallel effect of the environment on both germ line and soma. Weismann did not consider these to be cases of the inheritance of acquired characters because there was no *transfer* of information from soma to germ line. These aspects of Weismann's theory are discussed more fully in Chapter 2.
14. See Hull (1984, p. xliii) for a discussion of how 'Nearly every type of hereditary phenomenon has been termed at one time or another Lamarckian'.
15. See Bowler (1988) Chapter 5 for a discussion of the impact of Mendel's findings on evolutionary biology. Falk (1994) gives an interesting discussion of the way in which de Vries, one of the 're-discoverers' of Mendel's laws, gave a secondary role to Mendelian characters in his theory of evolution.
16. See Sapp (1987, p. 17). Sapp's book gives a fascinating account of the history of the debate about the relative importance of the nucleus in inheritance and development.
17. One of the most important consequences of Mendelian genetics for theories about the inheritance of acquired characters was that it introduced new and strict methodological requirements for experiments designed to test them. It became clear that it was necessary to use pure lines in order to exclude the possibility that hidden recessives and altered epistatic interactions are responsible for newly acquired inherited characters.
18. Although Johannsen was at pains to disassociate himself from Weismann's ideas (e.g. see Johannsen 1923), in many ways his genotype–phenotype distinction follows naturally from Weismann's insistence on the difference between effects on the soma and those on the germ line (see Chapter 2).
19. In fact, as Provine (1971) has documented, almost immediately after publication in 1903, Johannsen's data were claimed by Pearson and Wheldon to be at variance with his conclusion that selection is ineffective in pure lines.
20. Jennings paper suggesting that the pure line idea was important was published in the *American Naturalist* 1910; papers presented at the symposium held in December 1910 are to be found in the *American Naturalist* of the following year.
21. Originally, Johannsen applied the terms phenotype and genotype to populations, rather than individuals. 'Genotype' was almost synonymous with 'pure line', and 'phenotype' referred to the range of variation in a pure line over several generations. However the terms were rapidly adopted for the genetic constitution (genotype) and appearance (phenotype) of individuals. For a discussion of the history of the genotype–phenotype distinction and its role in the history of biology see Churchill (1974) and Allen (1979).
22. For an interesting discussion of the route by which development became detached from genetics and the reasons for it, see Falk (1994). Harwood (1993) has stressed that in Germany genetics was not divorced from studies of development and evolution as it was in the USA, and the view of heredity adopted by most German geneticists was much broader.
23. Hull (1988) gives a modern version of biological inheritance which almost echoes

Johannsen's: 'In biological evolution, inheritance counts as "Lamarckian" if adaptive changes in the phenotype of an organism were transmitted to the genetic material and thereafter inherited by the organism's progeny. Acquired characteristics must be *inherited*, not just transmitted'. (p. 37, Hull's italics). Hull makes the distinction between transmission and heredity in the same way, and indeed in the same context, as that made by Johannsen in his 1911 paper. He argues that the analogy between the inheritance of cultural products and the inheritance of genes is only metaphorical, and can be misleading, because genes transmit instructions, whereas usually in the cultural context one thinks of transmission of overt qualities.

24. For a discussion of the role of embryology in the evolutionary synthesis, see Hamburger (1980).
25. Crick (1958). The central dogma of 1958 did not explicitly include mRNA, and of course did not include the possibility of reverse transcription from mRNA to DNA. When reverse transcription was discovered in 1970, no conceptual change in the central dogma was necessary, because information still could not flow from proteins to DNA, from phenotype to genotype.
26. Lamarck was aware of this problem and rejected direct influences of the environment on hereditary characters. He stressed that only active responses to the environment would lead to heritable changes.
27. Weismann (1904, Vol. 2, p. 66). Weismann's experiment was not, of course, designed to refute Lamarck's version of the inheritance of acquired characters, which was based on use and disuse. Like Dawkins' cake analogy, the results of the experiment are relevant only to the primitive type of idea about the inheritance of acquired characters that originated long before Lamarck's time.
28. Lindegren wrote: 'The genetical data on which the modern conception of the gene is based are intensively selected data . . . The search for precisely segregating genes compels the selection of genetical material. In our own work on Neurospora we were unable to classify the progeny of over two-thirds of our matings'. (Lindegren 1949, Chapter 20, pp. 6–7)
29. For a more detailed discussion of these and similar examples, see Landman (1991).

2

Neo-Darwinian explanations of the inheritance of acquired characters

We have, in my opinion, to return to Darwin's concern with the nature of phenotypic variation—a subject about which he was always complaining that there was, in his day, a total absence of any understanding. Nowadays not only has the development of genetics given us some insight into genetic variation, but the development of epigenetics is giving us at least a few hints as to the nature of phenotypic variation.

Waddington 1968, p.20

By the end of the 1940s, neo-Darwinian ideas were almost universally accepted. The Mendelian basis of continuous variation was recognized, the mathematical treatment of natural selection had begun, and there was an increasing awareness of the wealth of genetic variation present in natural populations. The 'Modern Synthesis' of evolution had arrived: Darwin's theory had been reconciled with Mendelian genetics. Lamarckism, which had become almost synonymous with the inheritance of acquired characters, was generally rejected. In the first edition of his book *Evolution: the modern synthesis*, published in 1942, Julian Huxley was scathingly dismissive of Lamarckists, Lamarckian reasoning, and most experiments claiming to show the inheritance of acquired characters. He wielded Occam's razor at those who invoked Lamarckian explanations 'when known agencies would suffice'. He wrote:

> . . . the majority of biologists, including the very great majority of those who have experience of actual genetic work, repudiate lamarckism, or, at best, assign to it a subsidiary and unimportant role in evolution. (Huxley 1942, pp. 464–465)

During the subsequent two decades, more and more evidence supporting neo-Darwinian interpretations of evolutionary phenomena accumulated. There seemed little doubt that chance mutations and recombination are the source of variation, and that evolutionary changes occur gradually as a result of changes in gene frequency. Natural selection directs evolutionary change and brings about adaptation. Inheritance is hard; it seemed to most people that not only was there no evidence for soft inheritance, evolutionary theory did not require it. The inheritance of acquired characters was an unnecessary hypothesis because natural selection of random genetic variation could explain almost all evolutionary changes. Of course, there were exceptions to the general rejection of Lamarckism, Graham Cannon being one of the most prominent in the English speaking world during the 1950s.[1] In the Soviet Union the situation was very different, with Lamarckian inheritance being the required and generally accepted dogma from 1948 until the 1960s. However, even there, as Blacher has documented so well, many of the leading geneticists were openly or covertly anti-Lamarckian.[2]

Neo-Darwinian inheritance of acquired characters

For some people, any lingering doubts about the adequacy of the neo-Darwinian interpretation of certain types of evolutionary change were dispelled by the kind of ideas developed by Schmalhausen and Waddington. These ideas were not completely new, having been foreshadowed by those of Baldwin, Lloyd Morgan and Osborn at the end of the nineteenth century. In 1896 each of these biologists had independently suggested an evolutionary mechanism that, they believed, helped to reconcile the Lamarckian and Darwinian positions. The mechanism became known as the 'Baldwin effect' or 'the principle of organic selection'.[3] According to this 'principle', environmentally-induced somatic modifications can result in hereditary changes, not because they affect the hereditary material itself, but because they enable the population to survive long enough to allow the accumulation and selection of similar hereditary changes. This process simulates Lamarckian evolution, since acquired characters eventually become hereditary characters, but of course the induced somatic modification has no direct effect on the germ line. Natural selection is the cause of the hereditary change.

Until the 1940s, little interest seems to have been shown in these early ideas about the relation between environmentally-induced characters and similar inherited characters. However, in the 1940s, several people, notably Gause and Schmalhausen in the USSR, Hovasse in France, and Waddington in Great Britain, began to re-examine the evolutionary significance of the interrelations between genotype, phenotype and environment. They were able to show not only that natural selection could lead to a character that was originally induced by the environment becoming an inherited character, but also that there is a causal connection between the environmentally-induced changes and subsequent genetic changes. They argued that since adaptability—the ability to acquire an adaptive variation during an individual's lifetime—has a genetic basis, the genes underlying flexible adaptive variations may ultimately be responsible for the evolution of fixed adaptations to new environmental conditions. Their ideas made it possible to explain how an inherited character such as the thickened epidermis on the soles of the feet is causally related to the environmentally-induced thick skin produced in response to rubbing. Waddington and Schmalhausen stressed that the environment has a far more active role in evolution than that of a mere sieve selecting chance mutations.

Waddington's ideas can best be illustrated by considering one of his experiments on what he termed 'genetic assimilation'.[4] He defined this term as:

> . . . a process by which characters which were originally 'acquired characters', in the conventional sense, may become converted, by a process of selection acting for several or many generations on the population concerned, into 'inherited characters'. (Waddington 1961, p. 257)

In his experiments Waddington used environmental stimuli to produce phenocopies. Phenocopies are developmental abnormalities which simulate the effects of known mutant genes. In one experiment Waddington gave a 2–4 h 40°C heat shock to pupae of a wild-type strain of the fruit fly *Drosophila melanogaster*. Approximately 40% of the adults that emerged after this treatment were phenocopies of the crossveinless mutation, i.e. the whole or part of one of the cross veins on the wing was missing (Fig. 2.1). Waddington set up two selection lines: in one he bred only from flies that showed the phenocopy (upward selection) and in the other he bred only from flies that did not show the phenocopy (downward selection). Both lines responded to selection: the proportion of flies showing the phenocopy in the upward selected line increased, and in the downward selected lines it decreased (Fig. 2.1). What was remarkable about the results was that from the twelfth generation onwards, flies from the upward selected line began to show the crossveinless phenotype even in the absence of a temperature shock (Fig. 2.2). Matings between such flies resulted in strains in which some crossveinless flies were produced without a temperature shock. Thus, the environmentally induced character had become an inherited character. In Waddington's jargon, genetic assimilation of the heat-shock induced crossveinless phenotype had occurred.

How did Waddington interpret his results? What had been selected in these experiments was the capacity to respond to the environmental stimulus. Waddington suggested that responsiveness is variable and is controlled by many genes. The original strains used in the experiments contained allelic variants of the genes producing the response. Selection acted on the variability, and the upward selected line accumulated alleles causing an increased response to the stimulus. The possibility of direct 'Lamarckian' inheritance of the acquired character was ruled out when it was shown that inbred lines, which lack genetic variability, did not respond to selection.[5] Ultimately the acquired character became genetically assimilated, not because the phenotypic changes had caused the genetic material to change, but because selection had resulted in the alleles producing the response accumulating to a point beyond the developmental threshold for normal

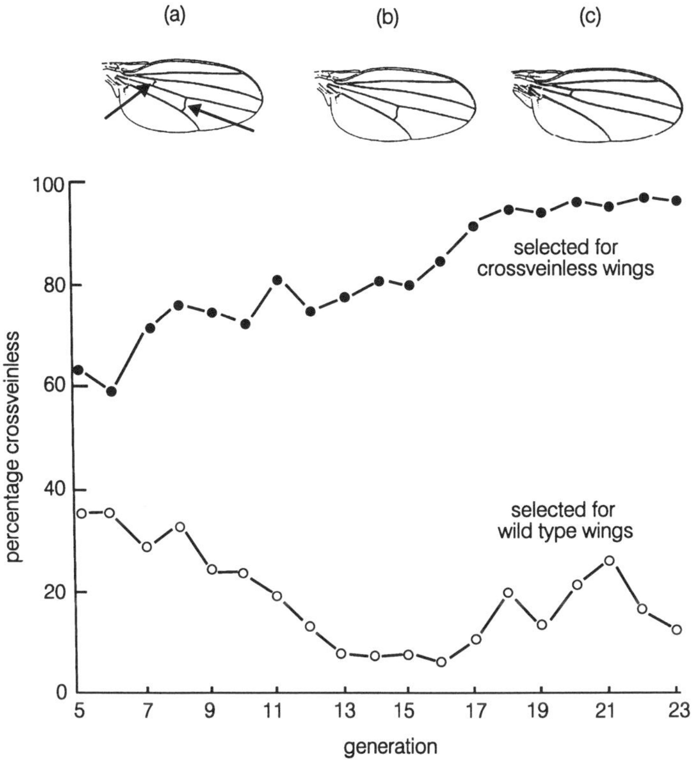

Fig. 2.1 Selection for and against the ability to produce the crossveinless phenocopy. Examples of wild-type wings (a), and crossveinless wings (b and c) are shown at the top of the figure. Arrows point to the crossveins. (Based on Waddington 1953, pp. 119 and 120.)

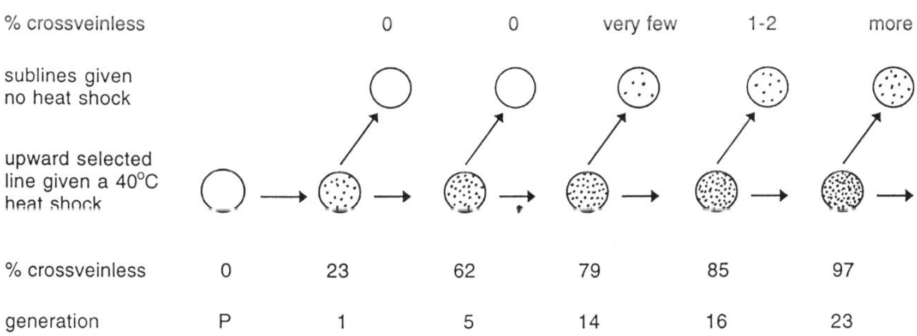

Fig. 2.2 The design and results of part of the genetic assimilation experiment shown in Fig. 2.1. Only the upward selected line is shown. Pupae were given a 40 °C heat shock and adults were selected for the absence of crossveins. The intensity of shading reflects the proportion of individuals showing the crossveinless phenotype.

wings set by the rest of the genome. In more modern terms we would say that the variations being selected were the result of different combinations of the alleles of the many genes that are involved in the regulation of development; as a result of selection, the frequency of the initially rare combinations that contribute to an enhanced response to the stimulus increased. Eventually, selection resulted in the production of those originally extremely rare combinations that produce the crossveinless phenotype even in the absence of the temperature stimulus.

A simple illustration of the type of changes in allele frequencies which could be involved in genetic assimilation may be helpful. Imagine that for the character of interest (wing venation) there are three unlinked interacting loci (A, B and C) each with two alleles. There are $3^3 = 27$ possible combinations of alleles (see Table 2.1, column X). Assume that when there is no external stimulus such as a heat shock, 26 of the 27 combinations produce the normal, wild type, wing venation (N), and only one produces the stimulus-independent or 'constitutive' (CON) crossveinless phenotype (column Y in Table 2.1). Now assume that in a new environment (i.e. when the heat shock is applied), in addition to the one in 27 that is constitutive (CON), an additional 10 of the 27 combinations produce the crossveinless phenotype. These inducible (IND) combinations are shown in column Z of Table 2.1 The remaining 16/27 combinations produce normal (N) wild type wings. The proportions chosen in this example are quite arbitrary, but the rules used to determine whether a particular allelic combination produces the normal, inducible, or constitutive phenotype are given at the bottom of the table. The probability that a strain will contain individuals that show the crossveinless phenotype in the absence of a heat shock will depend on the proportion of individuals of the genotype $A_1A_1B_2B_2C_2C_2$. This in turn will depend on the frequency in the population of the two alleles of each gene. Suppose that the allele frequencies (f) before any selection takes place are:

$$f\{A_1\} = 1/2,\quad f\{A_2\} = 1/2,$$
$$f\{B_1\} = 9/10,\quad f\{B_2\} = 1/10,$$
$$f\{C_1\} = 4/5,\quad f\{C_2\} = 1/5.$$

With these allele frequencies, only one in 10000 flies will have the genotype $A_1A_1B_2B_2C_2C_2$ and produce crossveinless wings without an inducing heat shock. By applying the heat shock and selecting the induced phenotype for several generations, the frequencies of the initially rare alleles B_2 and C_2 are increased in the upward-selected lines. Hence the frequency of the $A_1A_1B_2B_2C_2C_2$ genotype, which produces the crossveinless phenotype without the environmental stimulus, is increased. More and more flies will show the crossveinless phenotype without the heat shock. By selecting and breeding from these flies, a line that produces crossveinless constitutively (i.e. are all $A_1A_1B_2B_2C_2C_2$) could be produced.

Table 2.1 Genotypic combinations and their phenotypes in the original (column Y) and a new environment (column Z)

X Genotypes	Y Phenotypes in the original environment	Z Phenotypes in the new environment
$A_1A_1B_1B_1C_1C_1$	N	N
$A_1A_1B_1B_1C_1C_2$	N	N
$A_1A_1B_1B_1C_2C_2$	N	N
$A_1A_1B_1B_2C_1C_1$	N	N
$A_1A_1B_1B_2C_1C_2$	N	IND
$A_1A_1B_1B_2C_2C_2$	N	IND
$A_1A_1B_2B_2C_1C_1$	N	N
$A_1A_1B_2B_2C_1C_2$	N	IND
$A_1A_1B_2B_2C_2C_2$	CON	CON
$A_1A_2B_1B_1C_1C_1$	N	N
$A_1A_2B_1B_1C_1C_2$	N	N
$A_1A_2B_1B_1C_2C_2$	N	IND
$A_1A_2B_1B_2C_1C_1$	N	N
$A_1A_2B_1B_2C_1C_2$	N	IND
$A_1A_2B_1B_2C_2C_2$	N	IND
$A_1A_2B_2B_2C_1C_1$	N	N
$A_1A_2B_2B_2C_1C_2$	N	IND
$A_1A_2B_2B_2C_2C_2$	N	IND
$A_2A_2B_1B_1C_1C_1$	N	N
$A_2A_2B_1B_1C_1C_2$	N	N
$A_2A_2B_1B_1C_2C_2$	N	IND
$A_2A_2B_1B_2C_1C_1$	N	N
$A_2A_2B_1B_2C_1C_2$	N	N
$A_2A_2B_1B_2C_2C_2$	N	IND
$A_2A_2B_2B_2C_1C_1$	N	N
$A_2A_2B_2B_2C_1C_2$	N	N
$A_2A_2B_2B_2C_2C_2$	N	N

The rules governing inducibility or non-inducibility are:

(1) $A_1A_1B_2B_2C_2C_2$ is CON, i.e. it is constitutive and produces the modified phenotype in both environmental conditions.

(2) All genotypes A_1–B_2–C_2– (except $A_1A_1B_2B_2C_2C_2$) are IND, i.e. are inducible and produce the modified phenotype only in the new environment. Dashes represent either of the two alleles.

(3) Genotypes A_2–B_1–C_2C_2 are also IND.

(4) Homozygosity for C_1, and all other combinations not included in (1)–(3), always result in the normal (N) phenotype in both environments.

Many people welcomed and accepted Waddington's hypothesis that the genetic assimilation of inducible characters has contributed significantly to adaptive evolution. It provided a neo-Darwinian explanation of facts that previously had been difficult to explain in non-Lamarckian terms. However, others criticized his ideas. For example, Williams (1966) claimed that genetic assimilation cannot play a creative role in adaptive evolution, because most changes resulting from extreme environmental challenges are not adaptive. Moreover, he argued, the fixation of one response where two or more were previously possible corresponds to a decrease in genetic potential, because less information is needed to specify the fixed response.[6]

How valid are these arguments? It is certainly true that many responses to environmental stimuli are, like phenocopies, detrimental. Of course, as Williams notes, this argument is also true for classical mutations. Nevertheless, if even a small proportion is adaptive, it is enough for these responses to be of evolutionary significance. Clearly, only those responses that confer an advantage can be subject to assimilating selection. Williams's second argument, that selection resulting in a stimulus-independent response causes a reduction in phenotypic and genetic flexibility because some genetic information is lost, is also correct. It is, after all, a type of specialization, and almost by definition specialization narrows the range of possible adaptations. However, even if some alleles are selectively eliminated during genetic assimilation, the selection involved could ultimately result in the number of *loci* taking part in a response increasing. For example, suppose, as Waddington (1942) suggested, that the external stimulus needed to bring about a phenotypic modification was superseded by some internally produced substance, such as an intermediate in another developmental pathway. In such a case, selection for a stimulus-independent phenotype could affect the frequency of alleles at additional loci. It could lead to functional interdependence between previously independent developmental pathways; as a consequence, the production of the modified character would involve a larger number of interacting loci than previously. Genetic flexibility might actually increase, not decrease, as a result of genetic assimilation.[7]

At about the same time as Waddington was developing his ideas on genetic assimilation, Schmalhausen in the Soviet Union was thinking along similar lines. However, since his work was published in Ukrainian and Russian, his ideas did not become known to most Western scientists until the English translation of his book *Factors of evolution* was published in 1949. Schmalhausen, like Waddington, emphasized the fact that characters are not rigidly determined by the genotype. The genotype determines the range within which a character can develop. Environmentally induced 'acquired characters' occur within this range, but natural selection acting on the genetic system can increase or decrease the range of phenotypic responses. Schmalhausen suggested that at first the relationship between an environmental stimulus and the adaptive response may be proportional, but commonly, as a result of selection, the response becomes 'autoregulated'. By this he meant that, providing the external stimulus is above a certain threshold, an adaptive response of fixed magnitude is produced whatever the size of the stimulus. Autoregulation is essentially the same as

Waddington's 'canalization': the adjustment by natural selection of developmental reactions so as to bring about a clear-cut end result in spite of small environmental and genetic differences between individuals. Schmalhausen, like Waddington, argued that the adaptive response can become even more rigid if internal factors take over from the external stimulus: development then becomes 'autonomous', and the response is obtained quite independently of any environmental stimulus. A response that was originally environmentally induced has become an inherited character.

Although the evolutionary processes postulated by Waddington and Schmalhausen can be regarded as having a Lamarckian outcome, they are examples of what Medawar (1957) called the 'weak' form of Lamarckism. Waddington and Schmalhausen did not suggest, as is required in the 'strong' form of Lamarckism, that the acquired characters were accompanied by the origin of adaptive genetic changes in the individuals in which they were induced. The environment affects the expression of genes, not the genes themselves. The acquired characters result from changes in gene frequencies in populations.

Weismann's doctrine and somatic selection

Waddington's and Schmalhausen's explanations of the inheritance of acquired characters were very much within the traditional neo-Darwinian framework. In fact, from the 1940s until the 1980s, there were few real challenges to neo-Darwinism, or to the assumptions on which it was based. However, during the past decade, several people have pointed out that one of the most basic assumptions of classical neo-Darwinism—the early and rigid segregation of the germ line—is of only limited validity. Consequently, in some organisms, a rather different and important type of selection, somatic selection, can take place.

The assumption that has been questioned is embodied in what has become known as 'Weismann's doctrine', although Weismann's ideas, like those of Lamarck, are commonly misrepresented.[8] Weismann's 'doctrine of the continuity of the germ plasm' was part of his wide-ranging theory of inheritance, ontogeny and differentiation.[9] This theory, which involves a complex hierarchy of hypothetical structures, was modified and added to by Weismann as his ideas developed. Although it is a theory of heredity, it provided the framework for Weismann's ideas about evolution. In fact, he dedicated his book *The germ plasm: a theory of heredity*, published in 1892, to 'the memory of Charles Darwin'.

Weismann based his theory on the new information which he and others had obtained from studies of cells and cell division. He believed that the nucleus contained the hereditary substance—the germ plasm. He knew that gamete formation involves a reduction division in which the germ cells lose half their chromosomes, and that fertilization restores the chromosome number. According to Weismann's theory, the nucleus carries hereditary information in the form of *ids*. Chromosomes are aggregates of ids. Each id contains all the information needed for the development of the entire organism. It is what we would now call a haploid genome, although Weismann believed that each fertilized egg contains many equivalent, although not identical, ids. Within each id there are thousands of smaller units, the *determinants*, which are arranged in a specific way. The id therefore has a definite architecture. Determinants are aggregates of *biophores*, the fundamental vital units of all living matter. Each type of cell is controlled by one particular type of determinant, which contains several kinds of biophore.

How did Weismann suppose that somatic differentiation occurred? Here he made a mistake: he assumed that the chromatin of different cell types was not all the same, and that ontogeny depended on a series of qualitative changes in the contents of the nuclei. He believed that although some cell divisions involve an exact duplication and equal division of the chromatin, in others there is an unequal, but regulated, distribution of the determinants between the daughter cells. Gradually, as ontogeny proceeds, differential division leads to each nucleus containing fewer and fewer types of determinant. Weismann argued that the 'germ track', the lineage leading from the zygote to the germ cells, is the only lineage in which the nucleus retains and passes to daughter cells a full complement of inactivated and unalterable determinants. It is these latent sets of determinants which constitute Weismann's 'germ plasm'. Weismann believed that the germ track is separate and distinct from the somatic track right from the beginning of development. Contrary to common belief, Weismann did not believe in the complete segregation and continuity of the germ *cells*; he knew from his own work on hydroids that germ cells can originate from somatic tissues quite late in development.[10] He distinguished continuity of the germ cells from continuity of the germ plasm, arguing that continuity of the germ plasm occurs because all cells capable of giving rise to gametes retain complete sets of inactivated determinants arranged as ids. It is this inactive set of determinants which is passed on to the next generation in the germ line.

Weismann assumed that some somatic cells that normally do not contribute to the germ line also retain a partial or complete set of inactive determinants, in addition to their active complements. This reserve germ plasm contains all the information necessary to rebuild the structure or organism, even though it is in a somatic cell. Germ plasm is not limited to germ cells. This assumption meant that his theory could accommodate, albeit somewhat uncomfortably, phenomena such as regeneration in animals, or the growth of whole plants from leaves.

Weismann used very few explanatory diagrams in his books and essays. He did give a diagram showing how the many ids of an individual are derived from their ancestors, and others showing the basic behaviour of determinants and of the germ line during development (reproduced in Fig. 2.3a and b). However, he provided no comprehensive diagram showing the composition and behaviour of ids and determinants during ontogeny. We have tried to partially remedy this omission and show the main elements of the germ plasm theory in Fig. 2.4, although we are well aware of the difficulty of avoiding interpreting Weismann's theory in the light of modern knowledge.[11]

There is no doubt that early in this century Weismann's ideas and advocacy were influential in exposing the weaknesses of Lamarckism.[12] Although there was little enthusiasm for Weismann's theory of heredity and ontogenesis, and it was soon shown to be largely incorrect, the idea of a segregated germ line remained an important foundation for much evolutionary thinking. This is surprising for several reasons. First, once it became clear that ontogenesis probably does not involve a differential distribution of the hereditary material between cells,[13] the main theoretical reason for 'Weismann's doctrine' disappeared: it is unnecessary to postulate a segregated germ line containing a complete set of hereditary information if *all* cells retain a complete complement. Second, in his discussion of the reasons why acquired characters cannot be inherited, Weismann himself put very little emphasis on the role of the segregation of the germ line.[14] Third, Weismann did not claim that the germ line was segregated in *all* organisms.

Over the years, 'Weismann's doctrine of the continuity of the germ plasm' has been modified and simplified. The familiar version now is that illustrated in Fig. 2.5b, which shows that during development there is an early and irreversible separation of the germ line from the soma.[15] Somatic cell lineages exist only for the length of the organism's life; they are developmentally distinct from the potentially immortal germ line through which genetic information is passed from generation to generation. A somatic cell cannot be converted into a germ-line cell and differentiate into a gamete in the way illustrated in Fig. 2.5a. Consequently, heritable information cannot pass from somatic cells to germ cells by direct inheritance. Furthermore, other ways of transmitting heritable information from the soma to the germ line, such as via the gemmules or similar hypothetical entities that are postulated in various versions of pangenesis, are also excluded by Weismann's doctrine. The inheritance of somatically acquired characters is therefore impossible.

In spite of the widespread belief in the importance of Weismann's doctrine, for many years it has been clear that for *most* organisms it simply is not true.[16] For example, in 1932 Haldane stated quite clearly that since plants did not have a segregated germ line, if Lamarckian evolution occurred,

(a)

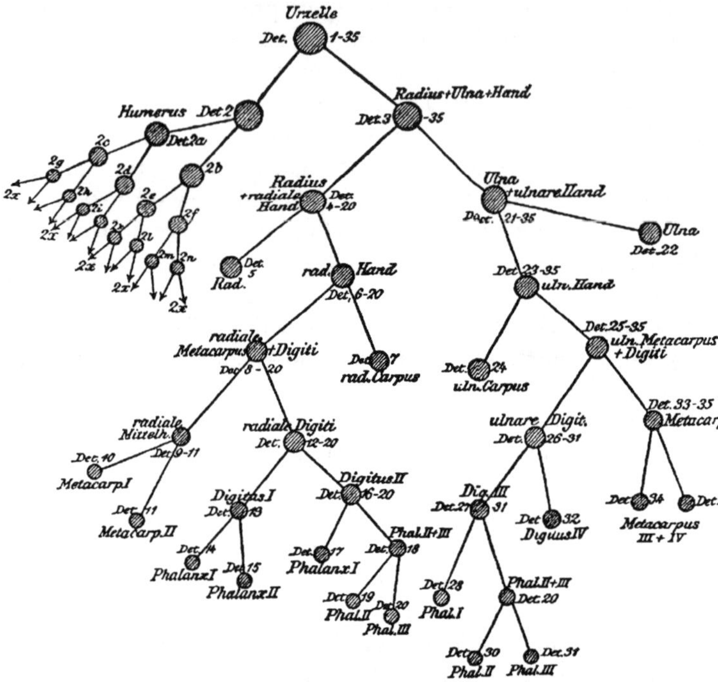
Urzelle
Det. 1-35
Humerus Det 2
Det 2a
Radius+Ulna+Hand
Det 3 -35
Radius +radiale Hand
Det. 4-20
Ulna +ulnare Hand
Det. 21-35
Ulna
Det. 22
Rad.
rad. Hand
Det. 6-20
Det. 23-35
uln. Hand
radiale Metacarpus +Digiti
Det. 8-20
rad. Carpus
Det. 24
uln. Carpus
Det. 25-35
uln. Metacarpus +Digiti
radiale Mittelh.
Det. 9-11
radiale Digiti
Det. 12-20
ulnare Digit.
Det. 26-31
Det. 33-35
Metacarpus
Det. 10
Metacarp. I
Det. 11
Metacarp. II
Digitus I
Det. 13
Digitus II
Det. 16-20
Dig. III
Det. 27-31
Det. 32
Digitus IV
Det. 34
Det. 35
Metacarpus III + IV
Det. 14
Phalanx I
Det. 15
Phalanx II
Det. 17
Phalanx I
Det. 18
Phal. II+III
Det. 19
Phal. II
Det. 20
Phal. III
Det. 28
Phal. I
Phal. II+III
Det. 29
Det. 30
Phal. II
Det. 31
Phal. III

(b)

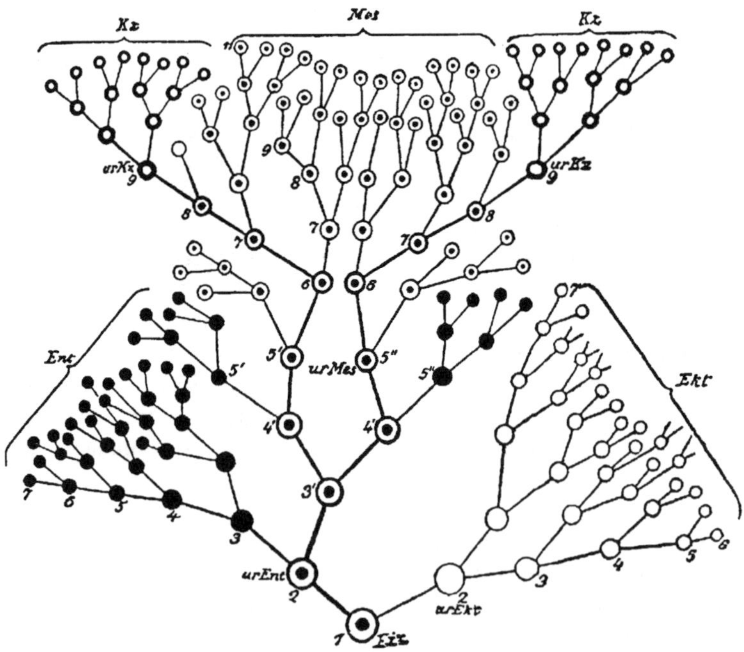
Kz
Mes
Kz
urKz
urKz
urMes
Ent
Ekt
urEnt
urEkt
Eiz

Fig. 2.3 Weismann's diagrams. (a) The way in which determinants (*Det.*) segregate at cell division during the development of the forelimb of the amphibian, Triton. The initial cell in the lineage has 35 determinants, but is controlled only by determinant 1. After the first division, one cell has *Det.2*, the other *Det.3–35*, and so on. (b) The germ line of the parasitic nematode *Rhabditis nigrovenosa*. The translation of Weismann's own legend reads 'The various generations of cells are indicated by Arabic numbers, the cells of the germ-track are connected by thick lines, and the chief kinds of cells are distinguished by various markings: the cells of the germ track by black nuclei, those of the mesoblast (*Mes*) by a dot in each, those of the ectoderm (*Ekt*) are white, those of the endoderm (*Ent*) black; in the primitive germ-cells (*ur Kz*) the nuclei are white. The cells are only indicated up to the twelfth generation.' (Taken from Weismann 1893, p. 102 and p. 196.)

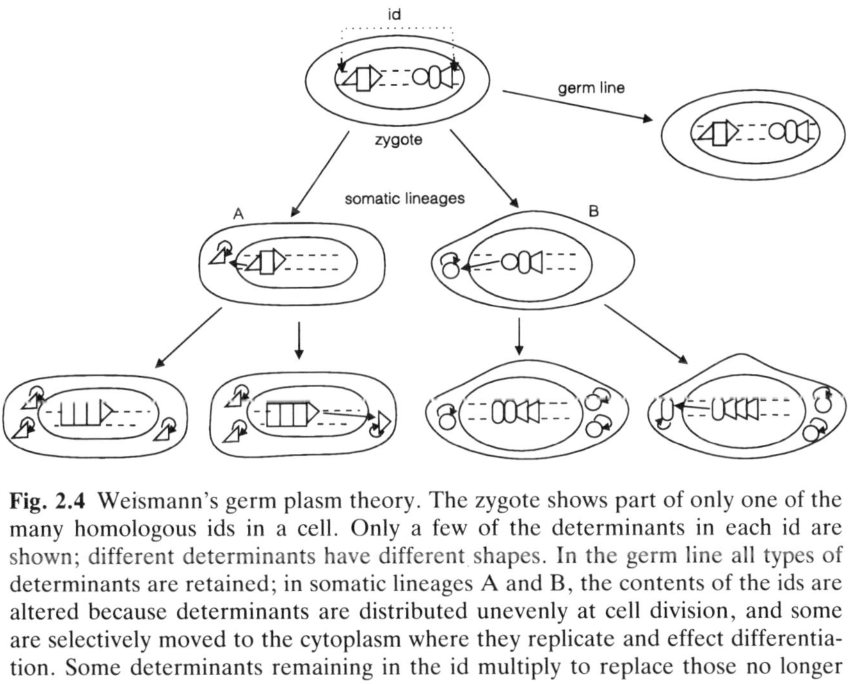

Fig. 2.4 Weismann's germ plasm theory. The zygote shows part of only one of the many homologous ids in a cell. Only a few of the determinants in each id are shown; different determinants have different shapes. In the germ line all types of determinants are retained; in somatic lineages A and B, the contents of the ids are altered because determinants are distributed unevenly at cell division, and some are selectively moved to the cytoplasm where they replicate and effect differentiation. Some determinants remaining in the id multiply to replace those no longer present.

it should be found in plants. In 1948, Berrill and Liu wrote about the theory of the continuity of the germ plasm:

To many geneticists it still seems to have an odor of sanctity, to most embryologists it has an old-fashioned association with what are now regarded as problems or phenomena of development pure and simple, while many botanists are but vaguely aware that Weismann ever existed. (Berrill and Liu 1948, p. 124)

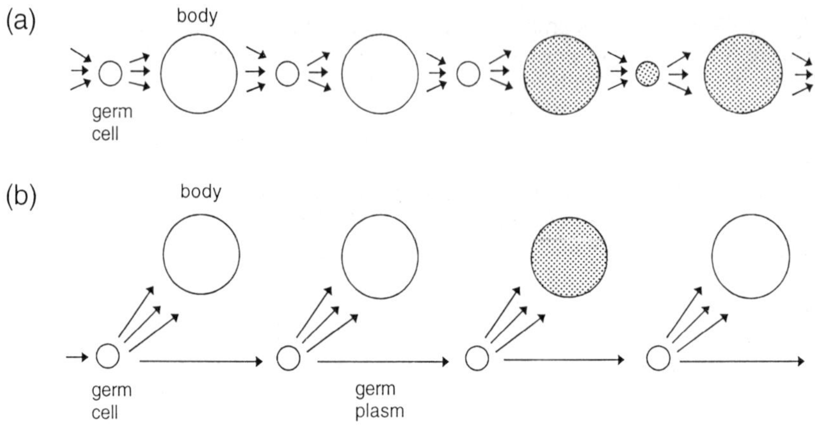

Fig. 2.5 Weismann's doctrine and the inheritance of acquired characters. (a) The inheritance of acquired characters: somatic cell lineages contribute to germ cells, so the somatic change in generation 3 is inherited by generation 4 onwards. (b) Inheritance according to Weismann's doctrine: the segregation of the germ line prevents the somatic variant in generation 3 from being transmitted to the next generation. (After Darlington 1953, p. 89.)

Berrill and Liu recognized that the validity and importance of Weismann's doctrine depend on the organism being considered, and they continued:

> Thus according to the nature of the living material with which one is most familiar, the theory of germplasm continuity may appear to be obvious, plausible, doubtful, or even absurd. Perhaps some of the passion with which it is often upheld by geneticsts comes from the suspicion that those who leave the safety of the germplasm fall into the heresy of Lamarckianism. (Berrill and Liu 1948, p.124)

For an anti-Lamarckian, the difficulty of accepting that Weismann's doctrine is not true is this: if some somatic cells can be converted into germ cells, there is no reason to believe that heritable variations acquired by somatic cells are evolutionary dead ends. In groups in which there is no clear distinction between the soma and germ line, the first obstacle to the inheritance of a somatically acquired variation—the transfer of the somatic variant to the gametes—is not present. Genetic events that happen in the soma may have a direct effect on the next generation. Whether or not they do, and how important they are, depends on the type of development and timing of germ-cell formation.

Differences in the timing and stability of the segregation between the soma and germ line have been used by Nieuwkoop and Sutasurya (1981) and by Buss (1983, 1987) to distinguish three modes of germ-cell formation. Although this division into categories is convenient, as Nieuwkoop and Sutasurya emphasized, there is probably a more or less continuous series from organisms such as the sponges and coelenterates, in which there is no distinct germ line, through organisms like echinoderms, in which there is a germ line but it segregates late in embryonic development, to organisms such as nematodes, which have a germ line that segregates very early in embryogenesis. We shall refer to the three types of germ-cell formation as *early and rigid germ-line determination, late germ-line determination, and somatically derived germ cells.*[17] They are illustrated schematically in Fig. 2.6. Table 2.2 shows their distribution in the various phyla of the animal kingdom.[18] Note that for more than a third of the phyla, no data on the mode of germ-cell development are available.

Early and rigid determination of the germ line (Fig. 2.6a) is found in 14 phyla in the animal kingdom and one phylum of protists. It is characterized by an irreversible segregation of the soma and germ line early in embryogenesis. Often this is under maternal control, the germ-line cells being determined by material deposited by the mother in the egg cytoplasm. During embryogenesis, germ-line cells tend to have a low rate of division.

In organisms that have *late germ-line determination* (Fig. 2.6b), the germ line is distinct, but differentiates at an advanced developmental stage, after the formation of several other major tissues. Often it seems to be formed as the result of the inductive influences of the surrounding tissues. This mode of development is found in some groups of platyhelminths, arthropods, hemichordates and chordates, and in all groups of echinoderms and molluscs.

When germ cells are *somatically derived* (Fig. 2.6c), there is no clear distinction between the soma and germ line. Throughout development there are cells that are multipotent, capable of giving rise to both somatic and germ-line cells. Sometimes somatic cells can 'dedifferentiate' and produce both stem cells and germ cells. This mode of development is commonly associated with asexual propagation. It is characteristic of all plants, all fungi, and all multicellular protists except the Volvacales. At least six animal phyla also show this type of germ-cell formation.

It is clear from the information collated by Nieuwkoop and Sutasurya and by Buss that only a minority of groups have early determination of the germ line. For the majority, Weismann's doctrine is not true. Consequently, variations that arise in germ-line stem cells, or in somatic cells that can be converted into germ cells, may have rather greater evolutionary significance than Weismann's doctrine allows. Selection at the level of cells and cell lineages, as well as at that of whole organisms, has to be taken into account. If a new variant cell arising in a cell lineage is more successful than other cells, either because its replication rate is higher, or because it exploits resources more effectively, its descendants may come to dominate the lineage. If that lineage then contributes to the germ line, the variant will be transmitted with a high frequency to the next generation.

The possible significance of such within-individual selection in organisms

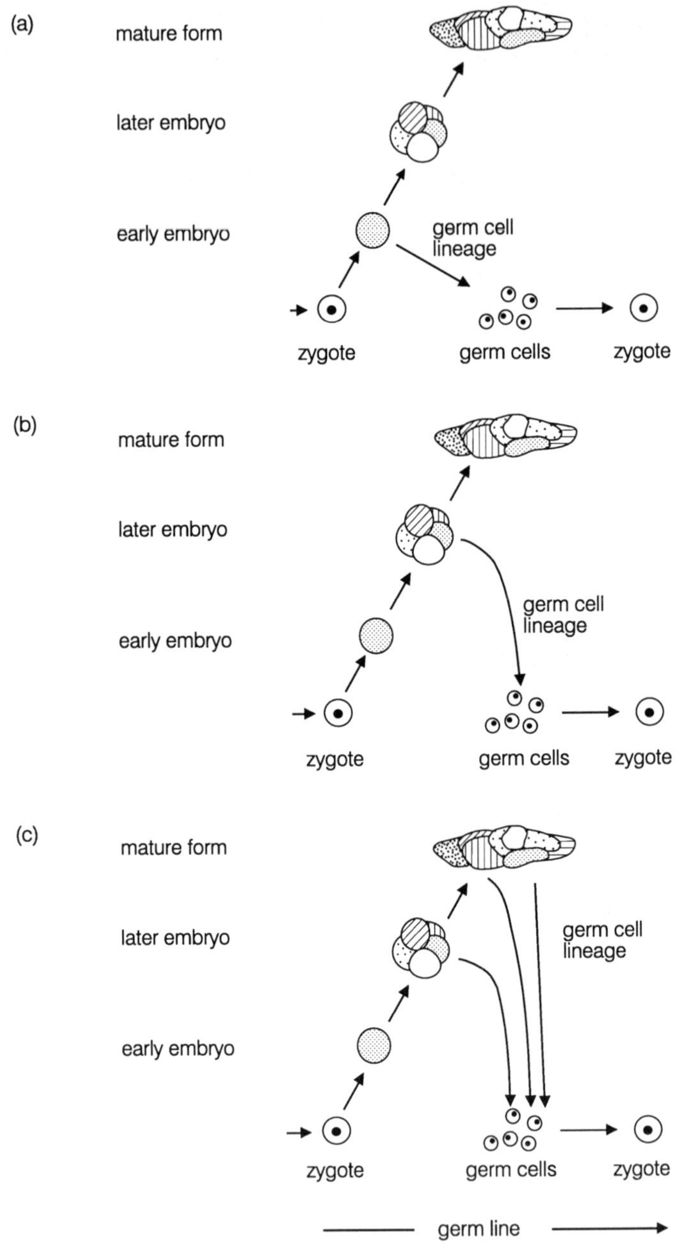

Fig. 2.6 Three modes of germ cell formation. (a) The germ line is determined early in embryogenesis, often as a result of cytoplasmic factors present in the egg. (b) The germ line is determined late in development, after the differentiation of some of the major tissues. (c) Germ cells can be produced from late embryonic or adult somatic cell lineages. Different shading represents different tissues.

Table 2.2 Mode of germ cell development and number of living species for animal phyla (based on Buss 1988)

Early germ line determination		Late germ line determination		Somatically derived germ cells		More than one type		Unknown	
Mesozoa		Echinodermata	6,000	Bryozoa	4,000	*Early or late*		Placozoa	2
Orthonectida	18	Mollusca	100,000	Cnidaria	9,000	Arthropoda		Priapulida	10
Dicyemida	65			Porifera	10,000	Crustacea	75,000	Phoronida	13
Onychophora	70					Chelicerata	100,000	Pentastomida	90
Chaetognatha	70					Uniramia	800,000	Gnathostomulida	100
Ctenophora	80							Pogonophora	100
Kinorhyncha	125					*All three modes*		Hemichordata	100
Gastrotricha	500					Annelida	8,700	Entoprocta	130
Tardigrada	550					Platyhelminthes	12,700	Echiura	130
Acanthocephala	1,150					Chordata	39,000	Nematomorpha	230
Rotifera	1,800							Sipuncula	320
Nematoda	10,000							Brachiopoda	330
								Nemertinea	800

without a segregated germ line has been discussed and modelled by several workers. Klekowski (1988) claimed that in plants most of the mutations that are transmitted to the next generation originate as somatic mutations. However, although plants lack a segregated germ line, not all somatic cells can contribute to germinal tissues, so most somatic mutations do not affect the next generation. On the other hand, mutations in meristematic tissues may result in chimeric plants, with some parts of the old genotype and some of the new. Competition can then occur not only between cells within tissues, but also between equivalent structures such as the buds or flowers of the same plant. Variants that are somatically successful may contribute more to the floral structures, and hence to the gametes, than those that are less successful. In this way, mutations are tested somatically before they are passed to the next generation. This may enable plants to make rapid adjustments to changes in the environment at very little cost: harmful meristematic mutations can be eliminated, since the death of a bud does not affect the whole plant, and beneficial mutations come to form an increasing proportion of the whole plant, and eventually of the gametes for the next generation (Whitham and Slobodchikoff, 1981). In this way, a new adaptive somatic character acquired within one generation can be inherited by the next.[19] This is, of course, a form of Lamarckian evolution, but it is only the 'weak' form. It is not the 'strong' form of Lamarckism because there is no suggestion that the new mutation is a direct consequence of environmental change.

Klekowski (1988) has emphasized the negative side of somatic mutations in plants. He argued that since most somatic mutations are deleterious, selection has favoured various developmental and genetic mechanisms that reduce the frequency with which they occur, and minimize their detrimental effects. Often they are removed before they can be passed to the next generation.

An even more important outcome of somatic selection has been proposed by Buss (1987). He suggested that competition between the variants arising in ontogeny has had a profound effect on the moulding of metazoan development and evolution. The starting point of his argument is that in primitive multicellular organisms, which have no distinct segregated germ line, there is competition between cell lineages carrying different variants. Like Klekowski, Buss believes that most new variants are likely to be disadvantageous to the individual, even if advantageous to the cell lineage. Consequently, selection has favoured mechanisms that prevent cells with mutations that enhance their own replication, but are detrimental to the organism as a whole, from forming part of the germ-cell lineage. According to Buss, early segregation and sequestration of the germ line, and maternal control of early ontogeny, are such mechanisms. These aspects of Buss's thesis are fairly conventional and readily accepted. The more important and interesting part of his argument is the suggestion that competition between cell lineages has been responsible for the evolution of developmental programmes. Buss argued that new variant cells that favour their own survival and replication by inhibiting or directing the activity of other cell lineages can, providing they are passed to the germ line, lead to new epigenetic interactions between tissues. According to Buss:

Variants may arise in the course of ontogeny which simultaneously favor their own replication and that of the individual harboring them. If these variants find their way to the germ line, they may effectively establish an epigenetic program. In the myriad details of metazoan development is written a partial record of the interactions between variant cell lineages, particular sequences of which gained access to the germ line to be revealed today as the epigenetic programs by which metazoan embryos developed. (Buss 1987, p. 81)

Several people have pointed out that there are difficulties with this argument.[20] Unless additional assumptions are made about the nature of the new variant and its cellular environment, it is difficult to see how the relationship between cell lineages can be changed. After all, if a successful somatic variant is transmitted via the germ line to the next generation, then *all* somatic cells of that generation will carry the new mutation, and they should not compete with each other. If Buss's mechanism is to work, the mutation must be in a gene that controls tissue-specific or stage-specific functions, and it must, first, cause the cell lineage harbouring it to outcompete other cell lineages in the same tissue; second, cause the phenotype of the tissue to change in a way that affects its interactions with other tissues; third, have a net beneficial effect at the level of the whole organism. Only if a cell carrying a mutation that fulfils all three independent requirements reaches the germ line could the mutation become fixed in the population.

Although they have met with criticism, the ideas of people such as Buss and Klekowski have led to a renewed interest in the role of somatic selection in evolution. The suggestion that selection occurs between units smaller than individuals is not new. It was recognized by Weismann nearly a century ago. He emphasized that the struggle for existence occurs at all levels, from biophores to colonies of organisms, and distinguished three types of selection:

The three principal stages of selection, that of *personal* selection as it was enunciated by Darwin and Wallace, that of *histonal* selection as it was established by Wilhelm Roux in the form of a 'struggle of the parts', and finally that of *germinal selection* whose existence and efficacy I have endeavoured to substantiate in this article—these are the factors that have co-operated to maintain the forms of life in a constant state of viability and to adapt them to their conditions of life, now modifying them *pari passu* with their environment, and now maintaining them on the stage attained, when that environment is not altered. (Weismann 1902, pp. 65–66; italics Weismann's)

Weismann clearly realized that natural selection can act at many different levels, and built some of these levels into his theory of evolution. 'Germinal' selection was his way of solving many of the evolutionary problems, such as the continued atrophy of vestigial organs, for which personal (individual) selection seemed inadequate. He argued that although primary variations are random and accidental, once they have occurred there is competition between variant determinants in the germ plasm, and this compels variations to go on changing in a definite direction. 'Histonal' selection is Weismann's version of somatic selection. It is selection between cells, tissues, and organs. Since he believed that selection can occur between units within an organism, Weismann realized that if the germ line is not segregated, a successful variant can be passed to the next generation. However, he believed that histonal (somatic) selection was likely to have been important only in the very early stages of metazoan evolution.

Steele's somatic selection hypothesis

The essence of Weismann's doctrine, both in the original 'continuity of the germ plasm' form and in the derived 'continuity of the germ line' form, is that there is a barrier between the soma and the germ line, which prevents newly acquired information in somatic cells from reaching germ cells. Weismann argued that there are strong theoretical reasons for suspecting that adaptive events occurring in the soma cannot affect the germ plasm in a way that would lead to the inheritance of the adaptation by the next generation:

> If we were now to try to think out a theoretical justification we should require to assume that the conditions of all the parts of the body at every moment, or at least at every period of life, were reflected in the corresponding primary constituents of the germ-plasm and thus in the germ-cells. But, as these primary constituents are quite different from the parts themselves, they would require to vary in quite a different way from that in which the finished parts have varied; which is very like supposing that an English telegram to China is there received in the Chinese language. (Weismann, 1904, Vol. 2, p. 63)

In other words, the language in which adaptive changes are expressed in the soma is different from the language used to transmit information through the germ line. This argument is also a strong argument against the likelihood of the form of parallel induction in which an external stimulus modifies the somatic phenotype and the germ-line genotype in a concordant manner. Weismann believed that the flow of hereditary information is unidirectional: from the nucleus to the cytoplasm, but never back from the cytoplasm to the nucleus.

Weismann's ideas were in some ways remarkably similar to those that emerged when the molecular nature of the genetic material and the way in which it directs the activities of cells began to be unfurled in the 1950s, when the *central dogma* was established. Information passes from DNA to proteins via RNA, and cannot flow from proteins to nucleic acids. The later discovery of reverse transcriptase and reverse transcription, i.e. that RNA can sometimes act as a template for DNA synthesis, did not alter the basic belief that the flow of information between nucleic acids and proteins occurs in one direction only. The DNA of the germ line became the equivalent of Weismann's germ plasm; even if somatic cells enter the germ line, any information they have acquired cannot be passed to the next generation unless it is in the DNA language.

In 1979, Steele proposed a new and provocative hypothesis suggesting a way in which acquired characters can be inherited. The hypothesis involves somatic selection, but is very different from the hypotheses that we have considered so far. According to Steele, the barrier between the soma and germ line is traversed not by the conversion of a variant somatic cell into a germ-line cell, but by the transmission of the *information* present in the variant cells to the germ line. This takes place as a result of the following sequence of events: first, mutations occur in somatic cells; second, cells with a favourable mutation proliferate and eventually come to dominate in the tissue or organ, i.e. clonal selection occurs; third, an endogenous viral vector picks up copies of the abundant new 'mutant' mRNA produced by the successful cells, and transports it to the germ line; finally, the new genetic information carried by the retrovirus is integrated into the DNA of the germ-line cells by a process involving reverse transcriptase. This hypothesis is illustrated diagrammatically in Fig. 2.7.

Steele's hypothesis is, of course, a curious and interesting version of pangenesis. If true, it is important, because the mechanism suggested could be significant in the evolution of all organisms, regardless of their mode of germ-cell formation. The mechanism is compatible with current concepts in molecular biology: somatic selection of genetic variants is known to occur in the immune system, and there is increasing evidence that RNA can be copied back to DNA. In fact it has been estimated that as much as 20% of some mammalian genomes has arisen as a result of reverse transcription (Lewin 1983). Many mammalian pseudogenes certainly look as if they have been produced from an RNA template.[21]

Although Steele's hypothesis is attractive, there is no convincing experimental evidence showing that the mechanisms he proposed actually occur. Experiments designed to see whether induced immune tolerance in mice can be transmitted from fathers to offspring seemed to show that induced tolerance could be inherited, but other workers failed to replicate Steele's results.[22] The theory demands a sequence of seemingly independent events (somatic mutation, somatic selection, transmission of RNA to the germ

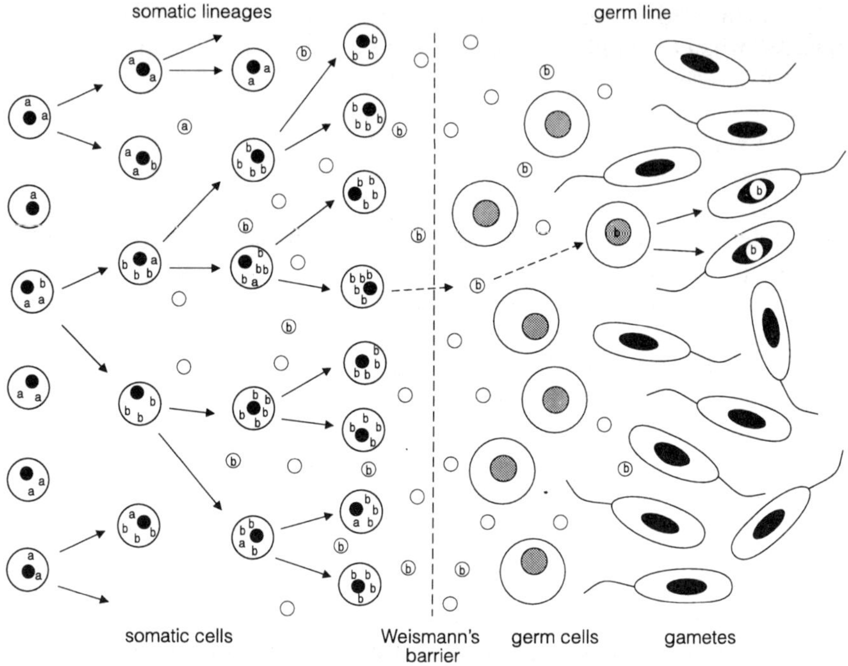

Fig. 2.7 Steele's somatic selection hypothesis. A new variant somatic cell proliferates and comes to dominate a cell lineage. The predominant mRNA, **b**, of the successful cell type is picked up by a retrovirus (small circle) and carried to the germ-line cells where it is integrated into germ-line DNA through the action of reverse transcriptase. (Modified from Steele *et al.* 1984.)

cells, reverse transcription, appropriate integration), each of which has a low probability of occurring. The chances that they will all occur must be extremely low. Steele's theory also suffers from the drawback that it is limited to genes that are transcribed. This restricts the kind of information that could be transmitted to the next generation. It is a serious limitation, because often adaptations are associated with genetic changes in regulation that do not involve altered polypeptide coding sequences at all. Furthermore, many regulatory changes are known to involve protein products that are present in only low concentrations, and Steele's version of pangenesis can hardly apply to genes producing low amounts of mRNA.

In plants, grafting has been a common practice for hundreds of years, so if the type of mechanism that Steele proposed exists in plants, evidence of it should have been found in graft hybrids. Romanes, a contemporary of Darwin, attempted to test the theory of pangenesis by making graft hybrids, but although he continued his experiments for at least eight years, he found no evidence that the graft affected the stock or vice versa.[23] However, there have been reports of work suggesting that hereditary traits *are* changed by grafting. The claims by Lysenko and other workers in the USSR that grafts affect the stock are generally doubted and usually discounted, but work done in Japan by Sinoto, and in Belgrade by Glavinic, has been accepted by some biologists as evidence that information can be transferred between stock and graft.[24] How this occurs is not clear, but it is possible that the type of mechanism proposed by Steele operates in plants and can bring about the observed changes.

Summary

In this chapter we have considered some of the ways in which somatically acquired characters may be converted into inherited characters. In each case, although the outcome is 'Lamarckian' evolution, the mechanism involved is based on Darwinian natural selection. The first type of apparently Lamarckian evolution described in this chapter, genetic assimilation, occurs as a result of Darwinian selection for the capacity to respond to environmental stimuli. Conventional selection between individuals is all that is involved. In the other types of Lamarckian evolution discussed here, the selection that is important is that which occurs between cell lineages within individuals. In many animals, the germ line and soma are not as rigidly separated as is commonly supposed, and successful somatic variants may come to form part of the germ line. This is even more true in plants. The genetic events that occur in somatic cells can have a direct effect on the inherited characters of the next generation (see Chapters 6 and 8). Even in organisms with an early and rigid segregation of the germ line, there may be ways in which information from somatic cells can be transferred to germ cells. Steele's hypothesis suggests that a viral vector could carry hereditary information from a successful variant somatic cell lineage to the germ line. How important for adaptive evolution this and the other mechanisms considered in this chapter are is unknown. What is clear is that 'Weismann's barrier' between the soma and the germ line is not absolute. In some organisms it does not exist, and in others, at least in theory, it can be breached or circumvented, so that information from somatic cells can be transmitted to the next generation. Although if this occurs it is brought about through Darwinian selection operating at the intra-individual level, this does not detract from the fact that at the level of individual organisms, what would be taking place is Lamarckian evolution through the inheritance of acquired characters.

Notes

1. See Cannon 1959.
2. Blacher's fascinating history of the concept of the inheritance of acquired

characters was written in the post-Lysenko period and published in Moscow in 1971; the English translation edited by F. B. Churchill was published in 1982.

3. See Simpson (1953) for a discussion of the history and significance of the Baldwin effect.
4. Waddington's ideas and experiments relating to genetic assimilation are described in Waddington (1957, 1961). Scharloo (1991) gives an interesting up-to-date assessment of experiments on genetic assimilation and their interpretation. Polikoff (1981) and Hall (1992*a*) discuss Waddington's contribution to evolutionary theory.
5. The experiments purporting to show the absence of a response to selection in inbred lines were carried out in Waddington's laboratory by Bateman (1959). Ho *et al.* (1983) have criticized the design of these experiments, and, by using alternative procedures, found that induced characters in inbred lines of *Drosophila* can be assimilated, even in the absence of selection (see Chapters 6 and 7).
6. Williams (1966, pp. 75–83) gives an extensive and critical discussion of the significance of Waddington's approach to the evolution of adaptations.
7. For a fuller discussion, see Jablonka Tavory (1982).
8. For a discussion of the misrepresentation of Weismann's views, see Van Valen (1987).
9. The first detailed account of this theory was in an essay published in 1885 with the title *The continuity of the germ-plasm as the foundation of a theory of heredity*, (English translation published in 1889). Weismann's book *The germ-plasm: a theory of heredity* was published in German in 1892, and in English in 1893.
10. In early accounts of his theory, Weismann did suggest that there was continuity of the germ-cell lineage from generation to generation, but he soon realized his error, and in his 1885 essay (*The continuity of the germ-plasm as the foundation of a theory of heredity*, p. 205 in the 1889 English translation) apologetically corrects himself and makes clear that it is not the germ cell, but the germ plasm, that shows continuity: '. . . only the nuclear substance passes uninterruptedly from one generation to another'.
11. Weismann revised and modified his theory in response to criticism and new knowledge, so there are differences in the way he presents his theory. The figure is based on the account given in *The evolutionary theory* (1904).
12. See Mayr (1985, 1988) for a detailed discussion of Weismann's role in the development of evolutionary thought.
13. There are a few organisms in which there is a regular unequal division of the genetic material during early ontogeny. One example, which was known to Weismann, is the nematode *Ascaris*, where chromatin diminution occurs during the formation of somatic cell lineages. However, according to Nieuwkoop and Sutasurya (1981, p. 173), genome reduction of this type is very rare, and when it occurs, *all* somatic cells have the same reduced chromosome complement. More recently Spradling *et al.* (1993) have suggested that DNA elimination may occur during differentiation in *Drosophila* and other organisms.
14. Maynard Smith (1989) discusses the lack of emphasis on a segregated germ line in Weismann's arguments against the inheritance of acquired characters.
15. Griesemer and Wimsatt (1989) have given an interesting account of the way in which Weismann's diagrams have been copied, modified, simplified and distorted by subsequent authors. In particular, they show how 'A diagram origin-

ally representing the continuity of the germ-plasm comes to convey as its primary message the mistaken idea of the discontinuity and mortality of the phenotype, and provide iconic support for the views of Williams and Dawkins' (Griesemer and Wimsatt 1989, p. 127).

16. De Vries and other botanists who were contemporaries of Weismann always disputed his basic assumption that there was a sharp distinction between soma and germ track. For an account of this argument, see Robinson (1979, Chapter 8).
17. These categories correspond to those described by Buss (1987) as preformistic (early and rigid determination), epigenetic (late determination) and somatic embryogenesis (somatically derived). We have avoided using Buss's terms because they have been used by other authors in different ways. For example, Nieuwkoop and Sutasurya (1981) use 'epigenetic' for groups in which asexual reproduction and sexual reproduction alternate, and germ cells are formed under the inductive influences of other cells or under the influence of environmental factors; they therefore include groups that Buss would regard as having somatic embryogenesis.
18. Table 2.2 is based on Buss (1988). It differs slightly from the tables in Buss's earlier publications, but is in general agreement with the data of Nieuwkoop and Sutasurya (1981). Differences in the various tables are not surprising, since for many groups the information available is very limited.
19. Slatkin (1984) has attempted to model this type of evolutionary process in plants. He concluded that somatic mutations are likely to have had a significant effect on evolutionary change only if they occur at very high frequency and if selection is very strong.
20. For example, see Raff (1988), Van Valen (1988), Wolpert (1990).
21. Pseudogenes are homologous to functional genes, but are not exact duplicates. They seem to have no function, and usually are not transcribed. Many resemble messenger RNA in that they lack introns and have poly-A regions at one end. Such 'processed' pseudogenes are thought to be the result of the transfer of information from cellular mRNA to genomic DNA. Some gene families, such as the globins, which as far as is known are not expressed in the germ line, nevertheless have processed pseudogenes. This may mean that their mRNA was transported by a viral vector from the somatic cells where it was produced to the germ cells.
22. Results of the attempts to replicate Steele's results are presented in Brent *et al.* (1981, 1982).
23. Robinson (1979, p. 22) discusses correspondence between Romanes and Charles Darwin in which Romanes refers to his extensive unpublished work with grafts.
24. Sinoto's experiments have been described and evaluated by Lindegren (1966) who believed that they probably did demonstrate that grafting can produce changes which are transmitted through seeds to the next generation. Similarly, Michie (1958) believed that Glavinic's work showed that the stock can affect the germ cells of the graft. Blacher (1982, Chapter 19) has reviewed much of the work on grafting from Eastern Europe and elsewhere, and could find no clear evidence of effects of the graft on stock or *vice versa*.

3

Induced genetic variations

. . . it follows from this theory that *somatogenic or acquired characters cannot be transmitted*. This, however, does not imply that external influences are incapable of producing hereditary variations; on the contrary, they always give rise to such variations when they are capable of modifying the determinants of the germ plasm.

Weismann 1893, pp. 462–463

In the previous chapter we considered how new genetic information present in somatic cells could cross 'Weismann's barrier' and become part of the germ line. In each of the cases we looked at, the new information was assumed to have arisen through random mutation. Cells with new mutations were subject to Darwinian selection before they, or their DNA, entered the germ line. Although the outcome was Lamarckian, the mechanism was not. The inheritance of acquired characters requires not only that information from the soma is transferred to the germ cells, but also that the environment is instrumental in producing the new variations. Mutations should arise as specific responses to environmental change. They should be directed, not random.

Even Weismann, one of the most influential critics of Lamarckism, was forced to accept the Lamarckian idea that some new variation is directed.[1] In his experiments with butterflies he had found that a temperature change induced a consistent, repeatable heritable alteration in coloration. He wrote:

It is thus certain that there are external influences which cause particular determinants to vary in a particular manner. (Weismann 1904, Vol. 2, p. 137)

Weismann explained such directed variation in terms of his germinal selection theory. According to this theory, in the germ line the numerous replicas of particular determinants compete for a limited supply of nutrients. Local fluctuations in nutrition mean that some determinants become stronger, whereas others become weaker. Once such chance differences between determinants are established, competition for nourishment makes the weak weaker, and the strong stronger. The system is self-reinforcing.[2] Weismann accepted that this germinal selection would occasionally lead to the non-adaptive trends known as orthogenesis.[3] He also conceded that an environmental stimulus could determine the direction of new variation by affecting nutritive conditions in the germ plasm, thereby influencing *all* copies of particular determinants in the same way.[4]

Weismann's germinal selection theory had few adherents, and as Mendelian genetics grew in influence, it was rejected along with most other aspects of his theory of inheritance. However, Mendelian genetics had no solution to the problem of the randomness or otherwise of new variation. Many of the early Mendelian geneticists, some of whom did not accept the role of selection in evolution, favoured the idea that new variation is directed. They believed that mutations tend to produce variation in a particular direction, and that 'mutation pressure' could drive evolution in directions that are not necessarily adaptive.[5] In the 1920s and 1930s, the directing role of environmental factors such as temperature was one of the Lamarckians' favourite explanations for the apparently non-adaptive trends found in the palaeontological record. They also believed that the parallelism in size and colour seen in different geographical races of birds and mammals is not the result of selection, but of climatic factors that induce traits that are of no selective importance (Rensch 1983).

As studies were made of agents that increased mutation frequency, the randomness or non-randomness of the induced mutations became a subject of debate and experimentation. In the 1930s Jollos produced experimental evidence suggesting that temperature-induced mutations in *Drosophila* were not entirely random.[6] He heat-treated larvae and found that as well as producing non-inherited somatic modifications of the adult, the treatment also increased mutation frequency. Moreover, he observed parallel induction: the somatic variations were similar to the induced mutations. Although there was no reason to think that the induced mutations were adaptive, they did appear to be directed. Not only did particular mutations at the same locus recur, he found that when flies that already carried a heat-induced mutation were treated again, a more extreme manifestation of the same mutant gene was produced. There seemed to be a step-by-step orthogenetic trend.

Plough and Ives (1935) repeated Jollos's experiments but could only partially confirm his results. They agreed that heat stress increases mutation frequency, and that it induces mutations in some genes more than in others. Like Jollos, they also found some evidence of parallel induction, but, unlike him, they did not see orthogenetic trends. Consequently, Plough and Ives argued that their experiments supported the view that induced mutations have no role in evolution beyond that of increasing variation. In spite of finding recurrent mutations, they denied that this was evidence of directed mutation, because the most mutable genes in their experiments were not the same as those found by Jollos. They cited the growing body of theoretical work by Fisher, Haldane and Sewall Wright as support for their view that mutation pressure is unlikely to determine the direction of evolutionary change.

Random or directed mutation?

With the coming of the Modern Synthesis, the belief that all new heritable variations are produced randomly and continuously became, and remained, one of the cornerstones of neo-Darwinian theory. The idea that environmentally-induced mutation is important in evolution was largely abandoned as more and more evidence of selection was found in both laboratory experiments and natural populations. It is now firmly believed by almost all biologists that the heritable, directed variations needed for Lamarckian inheritance do not occur. The concept of random mutation has become so pervasive that it has influenced other, related, areas of knowledge. For example, in evolutionary epistemology, where it is assumed that selection plays a major role in moulding and maintaining hypotheses, fashions and ideologies, it has been argued that the formation of hypotheses and the growth of knowledge is governed by the selection of random variations.[7] But how sound is the basis of the belief that all mutations are random? Is there any reason why an environmental stimulus should not, as Weismann and others believed, induce specific hereditary changes? Are there stimuli that, rather than acting as simple mutagens that increase the probability of mutations in many parts of the genome, preferentially and reproducibly induce mutations in only some specific genes, or even, in extreme cases, in one particular gene? Are some mutations directed?

The answer to these questions may depend on what is meant by the terms 'directed' and 'random'. What exactly does the term 'random' mean when it refers to mutation? It is well known that in many ways mutations are *not* random: genes are not all equally likely to mutate; backward and forward mutations of a gene are not equally likely; mutagens influence mutation rates, and affect some genes more than others. Dawkins has suggested:

> It is only if you define 'random' as meaning 'no general bias towards bodily improvement' that mutation is truly random. (Dawkins 1986, p. 307)

> Mutation is random with respect to adaptive advantage, although it is non-random in all sorts of other respects. It is selection, and only selection, that directs evolution in directions that are non-random with respect to advantage. (Dawkins 1986, p. 312)

Even if this is true, we doubt that it is really what most Darwinians have in mind when they talk of random mutation. It is a very narrow definition, and poses some terminological problems. For example, it is possible to imagine a situation in which a particular mutagen induces a high rate of mutation in just one or two genes. Even if the induced mutations are not adaptive, can we really talk about such mutation as 'random'? According to Dawkins, we should, yet most biologists probably would not. A broader definition such as 'mutation is random because it is not possible to predict which new variation will be produced at any moment' is probably closer to what most people mean when they refer to random mutation, although this definition also has its limitations.[8]

The terminological problems become even more complicated when we consider non-random, or 'directed' mutation. The term 'directed mutation' seems to have been used in two rather different ways. Sometimes it has been used to mean simply that a particular environmental challenge produces specific and repeatable changes in the hereditary material. The changes can be advantageous, detrimental, or neutral. In this case mutation is said to be 'directed' because the environment determines the nature and direction of the variation. In other cases, however, 'directed mutation' is used to mean rather more than this. The new variation is 'directed' towards an adaptive end: the mutations produced in response to an environmental challenge are both specific and selectively advantageous. The difference between the two usages of 'directed mutation' is important because showing that new mutations are not directed in the second sense, i.e. showing that they are not specifically advantageous, does not mean that they are not directed in the first sense. Mutations that confer no advantage can still be the result of non-random mutational processes.

We have attempted to illustrate some of the problems associated with the terms 'random' and 'directed' mutation in Fig. 3.1. In the figure we show the two extreme situations for which 'random' (Fig. 3.1a) and 'directed' (Fig. 3.1b) are clearly appropriate, and a third, intermediate, situation (Fig. 3.1c). It is difficult to think of this intermediate type as being 'random' mutation, yet is not 'directed' in the sense of producing only adaptive mutations. Because it seems to be closer to common usage of the terms 'random' and 'directed', and also because of the historical precedents, we shall refer to this type of mutation as 'directed'. We have suggested previously that the term 'guided' variation might be appropriate for any change that is specifically induced by a particular environment, regardless of whether or not it enhances fitness (Jablonka *et al.* 1992). Hall (1991) has suggested that 'adaptive mutations' may be preferable to 'directed mutations' when referring to mutations that arise in, and are adaptive to, the selective conditions. Although these terms might be preferable, in this book we will retain the established convention and call such mutations directed.

Whatever definitions are used, the general belief is that inherited variations in the information carried in DNA arise only through non-directed processes. This belief is a little puzzling, because there is ample evidence from both eukaryotes and prokaryotes that some transmitted variations in DNA are not random. Much of this evidence has been available for years, but the debate over the randomness or otherwise of mutations was rekindled in 1988 by some studies of mutation in bacteria. Because of the

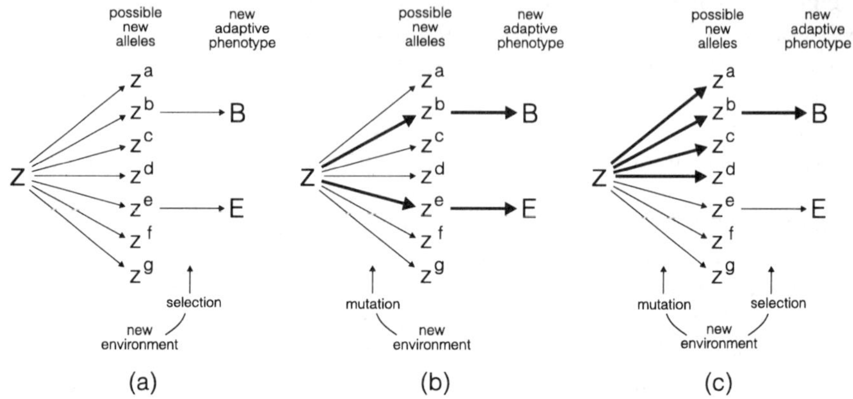

Fig. 3.1 Random and directed mutation. Gene **Z** is able to mutate to seven different alleles at rates indicated by the thickness of the arrows. Alleles Z^b and Z^e produce a phenotype that is adaptive in the new environment. (a) Random mutation in which the new environment has no effect on the mutations produced, only on those that are retained. (b) Directed mutation in which the new environment increases mutation to those alleles that are beneficial in that environment. (c) Directed mutation in which the new environment increases mutation to some alleles more than others. Notice that following selection this leads to one adaptation being more likely than the other.

controversy they generated, the background to these studies, as well as their results and conclusions, are important.

Random and directed mutation in bacteria

One of the reasons for the almost universal acceptance of the idea that variations are random was a series of experiments with bacteria reported by Luria and Delbrück in 1943. Bacteria were one of the last bastions of Lamarckism. In bacteria the whole organism divides, so the distinction between the soma and germ line has no relevance, and environmental influences on any part of the organism can be passed to the next generation. Some adaptation to antibacterial agents was known to occur rapidly and repeatedly, but in a step-wise fashion, so it had every appearance of Lamarckian inheritance. The importance of Luria and Delbrück's experiments was that they showed conclusively that adaptive mutations in bacteria can arise spontaneously in the absence of a selective agent. Their experiments provided no support for the Lamarckian belief that mutations are a direct and specific response to an environmental challenge. However, as has recently become clear, the experiments were really of rather limited scope, and some of the generalizations that have been based on them cannot be justified on either theoretical or experimental grounds.

Luria and Delbrück studied the occurrence of phage-resistant mutants in cultures of the bacterium *Escherichia coli*. They considered two possibilities: first, that the phage-resistant mutants are induced by the presence of the phage, i.e. that the mutation process is directed; second, that the mutations arise spontaneously all the time, and are merely selected by the presence of the phage, i.e. the mutation process is random. These two possibilities give different predictions about the distribution of mutant colonies in parallel cultures of the bacteria. In the type of experiment Luria and Delbrück carried out (now known as the fluctuation test), a series of identical cultures, each founded by a few genetically similar bacteria, are grown initially in normal culture conditions. When each culture reaches a specified density, it is plated on agar medium containing the phage. Sensitive cells die; resistant cells survive and form colonies. The number of resistant colonies on each plate is counted. The way the surviving colonies are distributed between the culture plates depends on whether resistance is due to induced mutations, or to continuously occurring spontaneous mutations. If mutants arise in response to the presence of the phage, then after plating, the number of resistant colonies on each plate will not show much variation. Since the probability that the phage induces a heritable change to resistance is likely to be low, the number of mutant colonies on the different plates should follow a Poisson distribution, with the variance equal to the mean. This outcome of induced mutation is illustrated in Fig. 3.2a. However, if mutations are not induced, but occur spontaneously before the bacteria are exposed to the phage, then the different cultures will have a highly variable number of mutants, depending on the stage in the culture's history at which the mutant arose. Plates from cultures in which a mutation occurred early will have many resistant colonies, whereas those from cultures in which mutations arose late will have few colonies (Fig. 3.2b). The variance of the distribution of the number of mutants will be much greater than the mean.

Luria and Delbrück found that the number of mutant colonies varied greatly from culture to culture. The distribution was that predicted by the hypothesis that mutations are not induced by the environment, but arise spontaneously. The phage-resistant mutations existed prior to exposure to the phage.

Lederberg and Lederberg (1952) confirmed Luria and Delbrück's conclusion by using their replica plating technique to show directly that relevant mutations are present at low frequency before the application of selective conditions. Replica plating involves growing colonies of bacteria on an agar master plate containing normal, non-selective medium, and then transferring exact copies of the colonies to identical positions on a series of plates of selective medium. When transferred to the selective

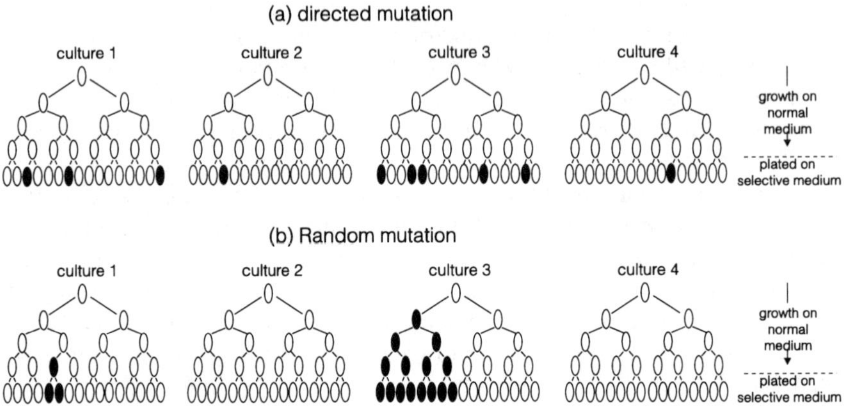

Fig. 3.2 Luria and Delbrück's experimental design. (a) The distribution of mutant colonies if all mutation is directed; (b) the distribution if all mutations arise randomly. (Modified from Stent and Calendar 1978.)

plates, only mutant colonies grow. The Lederbergs found that the pattern of mutant colonies was the same on every replica of the master plate (Fig. 3.3a), showing conclusively that the mutations had been present before the colonies were transferred to the selective medium. Phage-resistant mutants were present before exposure to phage, and mutants resistant to streptomycin were present before the bacteria ever came into contact with the drug. The mutations did not arise as an adaptive response to the selection pressure.

For over 30 years these experiments with bacteria provided one of the main sources of evidence for the argument that mutation is a random process.[9] Yet although the experiments provided clear experimental proof of the spontaneous, non-induced, origin of the mutations in the cultures studied, they did not prove that mutations cannot be induced by the selection agents. In fact, with the experimental protocols used, it is impossible to detect newly-induced directed mutations, because the selection applied is so intense, it kills sensitive bacteria almost immediately. Consequently, even if directed mutation does occur, there is not enough time before cell death for the mutated genes to express themselves and counter the effects of the selection agents. A much less intense form of selection is necessary if cells are to live long enough to have a chance to undergo directed mutation, express their new mutant phenotype, and produce the types of distribution shown in Fig. 3.2a and 3.3b.[10] In fact, experiments using gentler selection were carried out by Ryan in the 1950s and 1960s.[11] He showed that bacteria can accumulate mutations in the absence of growth. Although it seems not to have been widely recognized at the time, these experiments made the generalization that all mutations arise before bacteria are exposed to selection invalid.

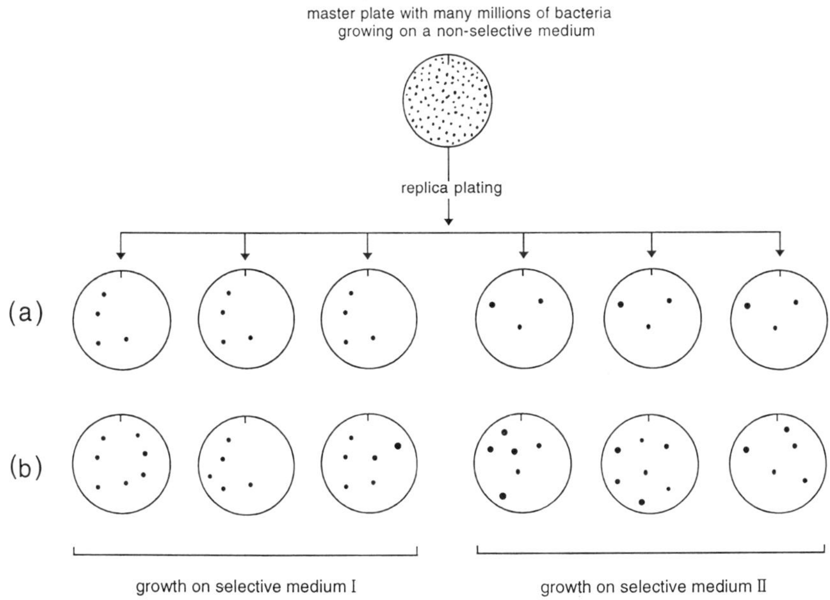

Fig. 3.3 Lederberg and Lederberg's replica plating experiments. (a) The expected result of random mutation: when colonies are replicated onto plates containing selective medium I, surviving mutant colonies are in identical positions on all plates, showing that the mutations were already present on the master plate; other colonies on the master plate contain mutants enabling survival on medium II. (b) The expected result if directed mutation also occurs: additional mutant colonies appear after plating on the selective medium. (This result was not obtained in Lederberg and Lederberg's experiments.) (After Stent and Calendar 1978.)

The renewed interest in directed mutation in the 1980s was the result of experiments similar to those of Ryan in which a gentle form of selection was used.[12] The selective media stopped the growth of the bacteria, but did not kill them. The results showed that, contrary to the prevailing belief, some new adaptive mutations in bacteria do seem to appear more frequently if the bacteria are exposed to the selection agent. They seem to be induced by the imposed conditions.

In many of the experiments, *E. coli* carrying various *lac* (*lactose*) mutations were used. Mutations in the *lac* operon make the bacteria unable to use the sugar lactose. However, when plated on a medium with lactose as the only source of sugar (the selective medium), lac^- bacteria do not die immediately; the cells are starved and do not divide, but they do metabolize. The strain Shapiro (1984) used lacks a functional copy of the *lac* gene, and also has a non-functional copy of *ara*, the gene that enables the bacterium to use the sugar arabinose; the regulatory part of the arabinose operon and the structural part of a lactose operon are present, but they are separated by a 38 kb stretch of DNA from the bacteriophage Mu, which contains transcriptional and translational stop signals. If the Mu fragment is deleted, the genes fuse, thereby allowing expression of the *lac* gene and growth on lactose, providing arabinose is also present to induce transcription. Shapiro found that the frequency of viable cells with fused genes was at least two orders of magnitude higher after incubation on the selective (lactose and arabinose) medium than it was after incubation on non-selective medium. Furthermore, the excision of Mu did not begin until several days after the cells were plated on the selective medium. Quite clearly, in these experiments pre-existing mutations were not being selected. The excision necessary to allow growth occurred *after* exposure to the selection medium.

Cairns and his associates (1988) repeated, confirmed, and extended Shapiro's experiments. They also followed Luria and Delbrück's experimental protocol to study reversion in a strain of *E. coli* that had a nonsense mutation in the *lac* operon. When they plated the bacteria on lactose-containing medium, they found that the distribution of viable, dividing, revertant cells fell between the distributions predicted by the two models tested by Luria and Delbrück. This composite distribution indicated that whereas some revertants appeared independently of exposure to the selective environment, most appeared only in the presence of the selection agent. In other words, although some were random, most mutations seemed to be directed. Cairns and his colleagues also monitored the accumulation of the *lac* revertants in cultures unable to grow through lack of other nutrients. They found that more mutations accumulated when lactose was present. Moreover, lactose did not increase mutation rates in a gene that had no selective importance on the medium. The mutations produced on the lactose-containing medium were apparently very specific—they were mutations that were beneficial in that particular environment.

Hall's (1988) experiments were rather different from those just described, and in some ways the results were more remarkable. In Hall's system, two mutations, one a point mutation and the other the excision of an insertion element, were required before the bacteria could use the sugar salicin as an energy source. In normally growing cultures, the spontaneous mutation rates for the two mutations were very low, 10^{-8} and 10^{-12}, yet in the presence of salicin, the combined mutation rate was 12 orders of magnitude higher than that expected from the combined spontaneous rates. The possibility that the selective medium produced a general increase in mutation frequency was ruled out by showing that the mutation rate of resistance to valine, a mutation that had no selective advantage on salicin, was unaffected by growth on salicin-containing medium. Hall found that the two mutations which together enabled the cells to grow on salicin were produced sequentially. Deletion of the insertion element occurred first, and at the remarkably high frequency of 1–10%. Hall claimed that this deletion by itself conferred no selective advantage on the cells; what it did was to create the potential for the second, selectively advantageous, mutation to occur.

In each of the three sets of experiments we have summarized, the selective agent apparently triggered mutations in certain specific genes, and the mutations induced were appropriate for the environment that induced them. The results seem to point towards the idea that the mutation of specific genes may be regulated by the environment. Not all mutations are random; some are directed. However, the interpretation of some of these experiments has been challenged.[13] It has been argued that what was being seen was non-specific responses to stress, or the results of selection between cells following limited growth on the selective medium, or various other experimental artefacts. In spite of these doubts, subsequent work using different strains has largely confirmed the original conclusions.[14] In some way, cells that have stopped dividing are able to generate those mutations that will allow them to resume growth.

Mechanisms for directed mutation in bacteria

The publication in 1988 of the paper in which Cairns and his colleagues presented their results suggesting that some mutations are directed caused quite a stir in the scientific community. There was a marked reluctance to accept the possibility that the respectable dogma that all mutations are non-directed could be wrong. This reluctance led to considerable speculation about the mechanisms that could underlie the production of what appeared to be directed mutations. In their paper, Cairns and his co-workers proposed a mechanism based on reverse transcription of the mRNA molecules produced in conditions of starvation. According to their hypothesis, when cells are starved, a variety of different mRNA molecules are transcribed from each active gene. Intracellular selection occurs for those variant mRNAs having protein products that promote growth of the cell, and these mRNAs are then reverse transcribed to DNA and integrated into the genome. This scheme requires some kind of coupling between the mRNAs and their protein products, since the cell has to reverse transcribe and integrate only the 'best' mRNA. It therefore needs a mechanism for identifying the mRNA that encodes the most useful protein product. A slightly different alternative suggestion offered by Cairns and his associates was that a non-specific process of reverse transcription is temporarily switched on as soon as a cell resumes growth. Growth in the presence of the selective agent occurs only if the cell accidentally produces an appropriate product. Hence, a cell that by chance produces an adaptive product enabling it to grow will automatically acquire a reverse transcript of the corresponding mRNA.

The difficulty with these hypotheses based on selection of randomly generated mRNA variants is that they make a number of assumptions for which there is little evidence, and they do not explain the results of experiments in which two mutations are needed for an advantageous phenotype. Not surprisingly, following the publication of the paper by Cairns, Overbaugh, and Miller, several alternative explanations of their results were proposed. Some of these suggested other molecular mechanisms for directed mutation, but many were attempts to explain the data within the orthodox framework of random mutations. The alternative explanations fell into three broad categories. In the first category were interpretations based on random mutation and Darwinian selection between mutant and non-mutant cells.[15] The evidence for the induction of specific adaptive mutations rests on the assumption that mutant and non-mutant cells have the same fitness prior to plating on the selective medium. If this is incorrect and there are differences in the growth rates of mutant and non-mutant cells, the distribution of colonies will resemble that which was interpreted as evidence of directed mutation.[16] Deviations from the distribution expected for random spontaneous mutations would also be found if the adaptive genotypes were the result of two sequential mutations rather than one, with the intermediate genotype having a slight selective advantage. Not only would the distribution of mutant colonies be less variable than expected from random mutation, but sequential mutations would also account for the delay in the appearance of mutations when cells were plated on the selective medium. There is evidence that sequential mutations probably did occur in some of these experiments.[17]

The second category of explanations for apparently directed mutations includes those that suggest molecular mechanisms leading directly to a preferential increase in the mutation frequency of specific genes.[18] With these mechanisms, although not all mutations are adaptive, there is a strong bias in the loci at which they occur. Davis (1989) suggested that this bias occurs because the selective agent induces transcription, and transcription is mutagenic. When a segment of DNA is transcribed, the two strands separate and the exposed single-stranded DNA is more susceptible both to damage and to error-prone repair. Therefore, when the selective agent induces transcription, it enhances the chances of mutations in that particular region. Unfortunately, there is little evidence supporting this hypothesis, because it has been found that even when the *lac* gene is constitutive, i.e. always transcriptionally active, revertants still accumulate only in the presence of lactose (Cairns and Foster 1991).

The third category of explanations for the controversial results includes those that proposed molecular mechanisms causing a general, rather than specific, increase in mutation rate.[19] Barry Hall (1990, 1991) suggested one mechanism by which a general and random increase in mutation rate could lead to a disproportionately high frequency of advantageous mutations.

This mechanism requires that a cellular stress such as starvation induces some cells to enter a hypermutable state in which extensive damage and error-prone repair occurs in many genes. If this results in a mutation appropriate for the environmental conditions, the cell leaves the hypermutable state and survives. An advantageous 'directed' mutation is recovered. If, however, none of the induced mutations is appropriate for survival, the cell is likely to die. Hence, non-adaptive mutations will not be recovered. In this way the frequency of specific mutations increases, whereas that of non-specific mutations does not. Although the mutations that are recovered are advantageous, they are the result of a fundamentally random process. One of the interesting and potentially important consequence of this mechanism is that simultaneous mutations should be quite common: because they have been in a hypermutable state, cells carrying an induced advantageous mutation are likely to carry other, non-specific, mutations. There is some evidence that non-specific mutations are indeed more common in cells with newly induced directed mutations, but unfortunately Hall's later experiments did not support his hypermutable state hypothesis.

A similar type of explanation for the apparent specificity of mutation in non-growing cultures is based on the slow repair of DNA in non-dividing cells.[20] Cells are constantly replacing stretches of their DNA, and repairing regions where there are mismatched bases. During prolonged starvation, this repair probably occurs rather slowly, so mismatches persist for longer than usual. It is suggested that, by chance, before there is time for it to be repaired, one of the mismatches may solve the cell's growth problem because it enables a functional gene product to be produced. If so, instead of being repaired, the mismatch will become fixed because growth will allow the DNA to replicate; the mismatch will be recovered as a mutation. In some ways this explanation resembles the hypermutable state hypothesis, but it differs in that it does not predict a genome-wide increase in mutation frequency in starving cells. Mismatches that do not solve a cell's problem do not become fixed because there is no growth and no replication; they will eventually be repaired. The evidence for this hypothesis is limited, but mismatch repair is one of the most favoured explanations of directed mutation.[21] A not dissimilar hypothesis is Foster and Cairns's (1992) suggestion that cells continuously produce transient gene amplifications, and mutations in these amplifications may become fixed when they are useful. The reason for favouring this type of hypothesis is that it helps to explain why repair- and replication-deficient strains are less efficient at producing adaptive mutations (Foster 1992).

Quite clearly, there is no dearth of explanations for the experiments that seem to show that mutations can occur as specific responses to environmental challenges.[22] No doubt more experimental evidence supporting or disproving the existing hypotheses will be obtained, and more interpretations of the experiments will be proposed. At present no firm conclusions about the precise nature of the mechanism underlying the directed mutations can be made. However, some of the experiments show that the mechanism is one that can produce excisions and rearrangements of the genetic material; the mutations were not simple DNA base changes. Such rearrangements of DNA in response to environmental challenge have great evolutionary implications.

In some ways, one of the most interesting features of the controversy that has surrounded the recent work on directed mutation is not the disagreement over the results and interpretation of the experiments, which is a commonplace in science, but the reluctance to discard the prevailing dogma that all mutations are random. It is generally agreed that the original belief that mutations in bacteria are random was based on an over-interpretation of the experiments carried out by Luria and Delbrück and others in the 1940s and 1950s, yet people are still unwilling to accept the alternative possibility that there may be processes that produce mutations that are needed, rather than just generating random mutations that confer no advantage. Selection is still regarded as the only role of the environment in evolution. The environment's role in inducing variation has been largely ignored. In 1990, Cairns summed up the situation in this way:

> The development of living things has depended on variation plus natural selection. The second of these has received a huge amount of attention since the days of Darwin and Wallace, but the first has hardly been investigated at all. I can think of scarcely a dozen experiments that bear upon the circumstances of what one might call normal spontaneous mutation. (Cairns 1990, p. 213)

Directed changes in the DNA of eukaryotes

Inevitably, the excitement caused by the discovery that in bacteria some mutations appear to be directed led to a search for comparable events in eukaryotes. Evidence for adaptive mutations was found in the budding yeast *Saccharomyces cerevisiae* (Steele and Jinks-Robertson 1992). Reversion of a frameshift mutation that made cells unable to grow without lysine was more frequent when the medium lacked lysine than when it lacked tryptophan, leucine, or both tryptophan and lysine. In other words, as in the experiments with bacteria, revertants accumulated only under conditions in which they were at a selective advantage. Mutation seems to be directed.

In some ways, resistance to the idea that mutation can be directed is surprising because there are many well-established cases of non-random changes in the amount, sequence, and arrangement of DNA in both unicellular and multicellular eukaryotes.[23] Some of the ways in which eukaryotic DNA is changed are illustrated in Fig. 3.4. Although the evi-

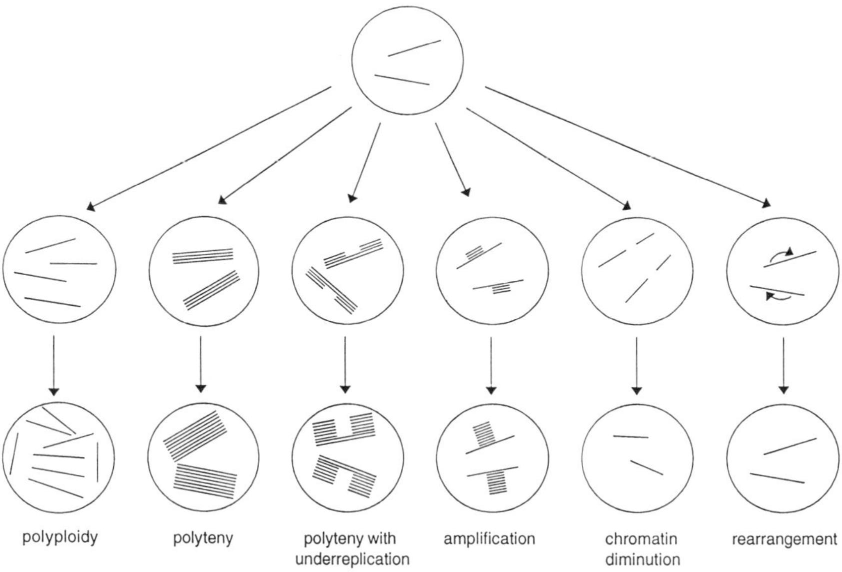

Fig. 3.4 Non-random DNA variations in eukaryotes. Each long bar represents a haploid genome, each short bar a gene or DNA sequence. (After Nagl 1978.)

dence that these changes occur is beyond doubt, the interpretation of them is just as complicated and challenging as that of the non-random DNA changes in bacteria. Some changes are directly induced responses to developmental or environmental stimuli, whereas others are probably the result of genomic challenges that, for example, affect the activity of mobile genetic elements. In some multicellular eukaryotes somatic selection between accidental DNA variants may lead to what appear to be directed changes in the next generation. This is analogous to the way in which, according to the hypermutable state or delayed repair models, advantageous changes occurring in a few bacterial cells are fixed by selection so that the mutations recovered seem to be directed. In many cases, the mechanisms responsible for the non-random changes in the DNA of eukaryotes can only be guessed at.

Some of the most extraordinary changes occur in the ciliated Protozoa. Ciliates are large and complex cells with two nuclei: a diploid micronucleus which is transcriptionally silent except during sexual reproduction, and a polyploid macronucleus which is responsible for all transcriptional activity during growth and asexual reproduction. After sexual reproduction, the old macronucleus degenerates, and the new zygote nucleus divides, eventually giving rise to a new micronucleus and an immature macronucleus. The DNA in the latter replicates many times, and in species such as *Oxytricha* and *Stylonychia*, chromosomes become polytene. The chromosomes then fragment, and most of the DNA is degraded. Fragments that remain then replicate, producing even more copies of the genetic information necessary for growth and asexual reproduction. During binary fission, the contents of the macronucleus probably divide rather unevenly, because although the fragments of chromosomes acquire telomeres, they lack centromeres and cannot participate in a normal mitotic segregation. These specific directed changes in DNA are a normal and regular feature of the ciliate life cycle: development of the macronucleus demands amplification, chromatin diminution, and the addition of telomeric sequences.

No less remarkable are the regular and specific genetic changes that have been found in the flagellated protozoan *Trypanosoma brucei*. These parasites are able to vary their surface antigens by switching transcriptional activity between a large number of variant antigen-producing genes. Most of these genes are present in tandem arrays within the chromosomes and are transcriptionally silent. The actively transcribed gene occupies one of the 'expression sites' in the telomeric region of some of the chromosomes. However, the transcriptionally silent copies of the gene can be moved to the expression sites by a process of directed gene conversion, thereby enabling the trypanosomes to switch antigen at a frequency of 10^{-6} to 10^{-7} per cell generation. Clearly, although they are not environmentally directed, in trypanosomes specific non-random changes are occurring at a very high rate. The changes are adaptive, since through switching their surface antigens, trypanosomes are able to evade the host's immune system.

In the yeast *Saccharomyces cerevisiae* there are even more frequent non-random changes in DNA. The allele at the *MAT* (mating type) locus determines whether a cell is *a* or α mating type. Almost every generation, lineages switch between the two mating types by replacing the genetic information in the *MAT* locus with that copied from the silent neighbouring genes termed *HMLα* and *HMLa*. The replacement process involves double-strand DNA breaks made by a site-specific endonuclease. It is a highly specific and directed change, since it occurs only in cells inheriting a particular regulatory protein. This protein is unevenly distributed between daughter cells, so a single cell is able to produce both α and *a* progeny. These progeny can mate and the diploid cell produced can undergo meiosis.

Specific directed changes in DNA are also a feature of the development of many multicellular organisms. One of the first cases to be described was that of chromatin diminution in the parasitic nematode *Ascaris*. Weismann used the way in which large parts of this animal's chromosomes are lost in somatic cell lineages as evidence of his theory of differentiation through unequal distribution of determinants during cell division (see Chapter 2, p. 38). Chromosome or chromatin diminution has also been found in a number of dipteran species. One well-known example is found in *Sciara coprophila*, where in both sexes the large L chromosomes are eliminated from somatic cells during the fifth or sixth cleavage division of the embryo; of the three X chromosomes present in the embryo, one (in females) or two (in males) are eliminated during the seventh to ninth division. In addition, during spermatogenesis in the male, the whole paternally derived chromosome set is eliminated (Crouse 1960). We discuss this strange chromosome behaviour further in Chapter 5 (p. 111), where we also describe the elimination of paternal chromosomes that occurs during male development in some coccid species.

Rather more subtle directed variations in nuclear DNA are found in other species. Sometimes localized gene amplification occurs and provides extra gene copies in cell types where there is a high demand for particular gene products. For example, in amphibian oocytes the build up of the vast number of ribosomes needed for development is associated with the presence of extrachromosomal nucleoli. These are produced through the excision and selective amplification of tandemly repeated rRNA genes. A different mechanism of selective gene amplification is found in the follicle cells surrounding the developing egg of *Drosophila*. Follicle cells are responsible for producing the proteins of the chorion—the complex shell formed around the egg. Two gene clusters containing coding sequences for the chorion proteins are selectively replicated several times, each round of replication being less extensive than the previous one. The result is an increase in the number of copies of the genes through a localized increase in polyteny.

The most remarkable example of non-random DNA changes accompanying differentiation is that seen in the mammalian immune system. Complex developmentally-regulated DNA rearrangements occur to produce the genes for the immunoglobulins and antigen-specific cell-surface proteins that mediate immune responses. In the germ line the genes for the light and heavy chains of antibodies are found in segments, and each segment is present in multiple copies. Antibody diversity is generated in somatic cells through the rearrangement and joining of these germ-line segments, by the addition of nucleotides to them, and by hypermutation in some regions of the assembled genes. As a result of these processes, the immune system has the ability to produce an enormous number of different antibodies. Several different types of non-random DNA changes, followed by somatic selection, bring about the apparently directed response to a particular antigenic challenge.

It is obvious that some eukaryotes have very elaborate mechanisms for producing non-random variations in their genomes during development. But are there comparable mechanisms that bring about DNA variations in response to environmental challenges? Several cases are known where cells in culture selectively amplify specific genes in response to treatment with drugs or other toxic substances. For example, if mammalian cell cultures are treated with increasing levels of methotrexate, a drug that inhibits the essential enzyme dihydrofolate reductase (DHFR), they become increasingly resistant. Eventually, resistant cells are able to produce hundreds of times the normal level of DHFR. The DHFR gene, which is usually present as only a single copy, is amplified many hundreds of times. The induced resistance is either stable, being retained even in the absence of methotrexate, or unstable, being lost if methotrexate is not present. In stably resistant cells the amplified copies of the DHFR gene, and a variable amount of DNA around it, are present as tandem repeats, integrated into the genome. In unstably resistant cells, the additional DHFR genes are located on extra-chromosomal elements which, because they lack centromeres, are gradually lost during cell division. How is the gene amplification found in methotrexate-treated cultures brought about? It seems like a directed mutational change. However, a low level of amplification of the DHFR gene occurs even in the absence of treatment with the drug: spontaneous tandem duplications occur at the relatively high frequency of 10^{-3}. Consequently, the step-wise increase in the number of copies of the DHFR gene could be the result of selection between cells carrying spontaneous amplifications. How common gene duplication is in somatic cells is not known, but selective amplifications of genes for aspartate transcarbamylase and for metallothionin-I have also been found in mammalian cells *in vivo*.

In plants, gene amplification in response to herbicide selection has been found in alfalfa and petunia cells in culture. Other plant cells in culture show chromatin diminution. One interesting example of this has been found in *Scilla siberica* where cells that show chromatin diminution are able to regenerate and form whole plants (Deumling and Clermont 1989). The DNA of the chromatin lost in culture is largely repetitive sequences, and during growth of the regenerated plants in natural conditions, these sequences are selectively amplified. The significance of these changes is unknown, but they show yet again that non-random changes in DNA sequences are not uncommon.

It might be thought that since most of the variations in DNA considered in this section have been detected in somatic cells, non-random changes do not occur in the germ line. If this were true, only random mutations would be significant for phylogenetic changes in genome structure. However, a few cases are known that suggest that this conclusion is incorrect. For example, a 100-fold amplification of a defective cholinesterase gene has been found in a human family repeatedly exposed to organophosphorus insecticides (Prody *et al.* 1989). These insecticides inhibit the action of cholinesterase, so amplification of the gene probably improves resistance to organophosphate poisoning. The amplification was found in a man and his son, and was believed to have originated during spermatogenesis in the man's father. The gene is known to be active during spermatogenesis, so the inheritance of the amplified gene could have resulted from selection in the germ line between cells carrying spontaneous amplifications.

The most well-known case of inherited induced genomic changes is found in flax.[24] Many years ago it was found that heritable and reproducible changes in the phenotype of some inbred flax varieties could be induced by growing plants for a single generation in particular nutrient conditions. The induced changes were initially detected as heritable differences in the height and weight of the plants, but the large and small 'genotrophs', as they were called, were also found to differ in morphological and biochemical features, and in their nuclear DNA. The amount of DNA, the number of copies of genes coding for rRNA, and the copy number of most highly repetitive DNA fractions, all differed. Although there is no obvious adaptive significance in any of these genomic changes, they are environment-specific and repeatable. The amount of rDNA in different parts of the plant varies during growth in the inducing environment, and the changes are transmitted to the progeny. It is not clear whether somatic selection occurs prior to gamete production. Since some of the DNA variations characteristic of the genotrophs are found in other flax varieties and other species of *Linum*, it may be that the DNA alterations that are possible are limited and rather specific.

A comparable restricted repertoire of genetic variations has been found in the soybean *Glycine max*, an obligate inbreeder (Roth *et al.* 1989). Although inbred and generally homozygous, many different phenotypic and genotypic variants are present in cultivars. It has been found that for some genetic markers there are usually only two alleles, and that the difference between the alleles is the result of a rearrangement of the DNA. What is particularly interesting is that in tissue cultures that have been maintained for many generations, the frequent newly-generated variations of the markers are always the same as those found in intact plants. This suggests that the same specific, reproducible changes occur in both somatic and germ-line cells. The workers who made this fascinating discovery suggest that the repeated generation of particular alleles may mean that obligate inbreeders such as soybean have evolved mechanisms that are internal generators of genetic variation, possibly as a response to stress.

One of the most remarkable and intriguing cases of non-random changes in DNA is that found in the germ line of some fungi. The events that take place are known as RIP[25] or ripping. Ripping is a process that detects and alters newly duplicated sequences in the haploid genome. It occurs after fertilization but before meiosis. Some duplicated sequences are removed through a pre-meiotic recombination process, but if both copies of duplicated sequences remain, frequently both are inactivated through DNA modification or mutation. Selker comments:

> The existence of processes such as somatic hypermutation of antibody genes and RIP suggests that we should reexamine the meaning of 'spontaneous' (in the sense of accidental) mutation. Cells can *cause* mutations to occur. Thus, some mutations

thought to result from environmental insults to DNA, chemical instability of the nucleotides, or mistakes in DNA replication or repair, may actually result from enzyme-catalyzed conversions. (Selker 1990*a*, p. 607; Selker's italics)

The flexible genome

During the debate on directed mutation in bacteria, Grafen (1988) argued that environmentally directed mutation in bacteria should be regarded as equivalent to physiological adaptation. He suggested that some loci probably routinely and specifically alter their DNA sequences in response to environmental changes. The loci involved in this type of directed mutation should be regarded as heritable soma rather than germ plasm, since their information is short-lived, not potentially immortal. This argument leads to the intriguing conclusion that some DNA is germ plasm, but some is not.

Grafen is not alone in believing that it is incorrect to regard all genes as similar and passive entities, whose structure is changed by chance events and whose frequency is determined by selection and drift. Campbell in particular has repeatedly stressed that many genes are dynamic structures, sometimes differing from one another in fundamental ways, and that cells can actively manipulate their genomes for both physiological and evolutionary ends.[26] Cells have a 'genetic engineering kit'.[27] They contain a whole battery of enzymes that are capable of changing gene structure in precise ways. There are enzymes that correct one copy of a gene by matching it with another, enzymes that excise DNA segments, enzymes that join segments together, enzymes that copy segments, and so on. Furthermore, many genes contain what Campbell calls 'sensing devices', which inform the gene about its environment. For example, a simple bacterial operon has an operator region and a region coding for a repressor protein, as well as the region specifying the actual structure of the enzyme. Through the interaction of repressor protein, substrate and operator, the gene is turned on only if the substrate for the enzyme is present. The repressor protein informs the gene via the operator about the concentration of substrate in the cell. In eukaryotes, genes sometimes receive information about their environment from hormone molecules. According to Campbell, through using the various sensing devices, organisms can obtain information about their environment, and respond to it for evolutionary, as well as for physiological, ends. He points out that many bacteria and higher organisms have elaborate sensory systems through which they respond to stress conditions by triggering heritable changes in their genomes. For example, an operon may be integrated within a transposable element —a jumping gene—in such a way that the operon's regulatory sequences also regulate replicative transposition. When an environmental shift induces transcriptional activity in the operon, not only is the operon's gene product produced, but the transposable element is induced to jump. The original copy remains in its existing site, while a new copy is inserted elsewhere in the genome. In this way an environmental change results not only in adaptive gene expression, but also in an increase in the number of copies of the gene producing the adaptive response.

Campbell has gone so far as to put forward the somewhat heretical notion:

> Some genetic structures do not adapt the organism to its environment. Instead, they have evolved to promote and direct the process of evolution. They function to enhance the capacity of the species to evolve. (Campbell 1985, p. 137)

Introns, satellite DNA, pseudogenes, the mechanism for DNA methylation, and many other features of the genome may, according to Campbell, function for evolutionary ends.

Few people have taken Campbell's extreme position that mechanisms for genome restructuring have evolved for evolutionary ends, but there is a growing realization that understanding genome organization and genome re-organization systems is essential for understanding evolution. The genetic engineering kit that is used during development is also used during evolution. Shapiro has highlighted three properties of genetic systems that have been revealed by recent molecular studies. First, there are mechanisms that enable natural genetic engineering processes within cells to occur in response to developmental and external stimuli. Second, genomes and genetic loci have a mosaic structure. They are built around a number of basic motifs: DNA sequences that are repeated and are scattered around the genome. These motifs combine together to form functional protein-coding or regulatory units. Third, some proteins and individual sequence motifs are highly conserved among different taxonomic groups, but the way that they are combined varies between species. Taken together, these observations suggest that evolutionary change is the consequence of genome reorganization that brings motifs together in different ways. Random point mutations are usually irrelevant to adaptive evolution. The evolutionarily important unit of variation is the motif, not single DNA bases. Random, accidental mutations are normally very rare, because cells have sophisticated error detection and correction systems. Evolution is based mainly on variation resulting from new combinations of motifs brought together by the cell's enzymatic machinery, which may be modulated by environmental and developmental signals. The same kind of genetic engineering that operates during development also operates during evolution. According to Shapiro:

> Although it may be objected that development and evolution are quite different processes, the point is that evolutionary thinking will have to [be] based on what we know cells can do. Thus, if we see examples of highly sophisticated regulation of DNA changes during development, it cannot be realistic to base evolutionary theories on the concept that such regulation is not possible. (Shapiro 1992, p. 105)

Cells not only engineer their genome, they often do it in a coordinated progressive way. The immune response is based on a series of interdependent genomic changes; the reorganization of the ciliate macronucleus is a genome-wide coordinated change involving several interdependent steps. It is possible, Shapiro argues, that some evolutionary changes are also based on extensive, coordinated, genome-wide reorganizations which produce radically new genotypes that may lead to the emergence of new taxa.

Arguments such as Campbell's and Shapiro's are a response to the new information that is emerging about the molecular nature of the genome. This information certainly seems to demand a re-thinking of the conventional neo-Darwinian 'blind mutation' approach to evolution. It looks as if some changes may involve internal processes, rather than being simply the result of selection by the external environment of favourable variations resulting from chance mutation and recombination. There may be a randomness in the outcome of the cell's genetic engineering, for example in the way particular genetic motifs are combined, even though this is probably constrained by as yet unknown 'rules' of combination. But since the use of the genetic engineering kit is influenced by environmental signals, the environment will affect the type and origin of genetic variations, not just their effect on fitness. In other words, genome reorganization may well be directed.

If the probability of the occurrence of some mutations is controlled by the environment, it has another consequence for interpreting evolutionary divergence: the number of differences between two DNA sequences may not be related to the time since these sequences diverged (Hall 1988). This may make interpretations of phylogenies based on DNA sequence data much more difficult. The existence of directed mutation also complicates even further the already complex distinction between evolutionary homology and analogy.

Shapiro (1991) has written of 'genomes as smart systems', and Wills (1991) of 'the wisdom of the genes'.[28] In the conclusion to a review of her studies with plants, presented at the time she received her Nobel prize in 1983, Barbara McClintock wrote:

> In the future, attention undoubtedly will be centered on the genome, with greater appreciation of its significance as a highly sensitive organ of the cell that monitors genomic activities and corrects common errors, senses unusual and unexpected events, and responds to them, often by restructuring the genome. (McClintock 1984, p. 800)

Summary

Recent interest in whether mutations are random or directed has brought to light not only the inadequacy of the experimental basis for the traditional view that mutation is a random process, but also the difficulty of deciding whether or not changes in DNA are directly induced adaptive responses. At one extreme are the incidental and accidental random DNA alterations that occur at low frequency, affect any and every site in the genome, and are usually non-adaptive. At the other extreme are programmed directed changes that occur at specific loci in response to specific stimuli and are an essential part of normal development. Between the two are changes such as those underlying trypanosome antigen switching, which affect particular loci but occur at low frequency and at random times; there are changes such as those in flax, which affect many loci in apparently non-adaptive ways but are induced by specific environmental factors; there are changes such as those in the mammalian immune system, which are responses to developmental signals but are not rigidly specified, and so on. There is a whole spectrum of DNA changes, some of which are clearly selectively advantageous, but others for which there is no obvious immediate benefit to the organism.

The new information about the flexibility of the genome has so changed ideas about the nature of the gene and its stability, that several biologists have argued that it is time to look again at some of our basic assumptions about the origin of genetic variation and the role of new variation in evolutionary processes. If some new variation is directed, then even if it is not adaptive, it could affect the direction of evolutionary change. Consequently, it may be necessary to modify orthodox interpretations of evolutionary trends and rates of evolution.

Various types of regulated variations in DNA sequence occur as part of the adaptive response to environmental challenges, but it is not known how common these regulated variations are, and the extent to which they occur in the germ line. Their importance in adaptive evolution depends on whether they are transmitted to descendants and how stable they are. How malleable by the environment is the genotype? How persistent are the modifications? Do they occur in the germ line? At present we have only fragmentary answers to these questions, but they are clearly important for understanding adaptive evolution.

Notes

1. According to Weismann, '. . . if we are forced by the facts on all hands to the assumption that the useful variations which render selection possible are always present, then *some profound connection must exist between the utility of a variation and its actual appearance*, or, in other words, *the direction of the variation of a part must be determined by utility*, and we shall have to see whether facts exist that confirm our conjecture'. (Weismann 1902, p. 33. The italics are Weismann's.) Weismann concluded that 'definitely directed variation exists' and proposed his theory of germinal selection to explain its origin.

2. Some of Weismann's ideas about selection between determinants within the germ line are remarkably similar to modern ideas about selfish DNA.
3. Orthogenesis was a late nineteenth century rival to Darwin's theory of evolution. The theory took several forms, but the different versions had in common the idea that the parallel, linear, often apparently non-adaptive trends in evolution had an internal cause. Bowler (1983, Chapter 7) gives a full discussion.
4. Mayr (1985) has pointed out that, although Weismann called this aspect of his theory 'induced germinal selection', it is no longer a selection theory. It is similar to Geoffroyism, which admits a direct effect of the environment on the germ plasm.
5. See Bowler (1989, p. 280).
6. An English summary of Jollos's work can be found in Jollos (1934).
7. For example, Popper wrote of Darwinian natural selection: 'More than forty years ago I proposed the conjecture that this is also the method by which we acquire our knowledge of the external world: we produce conjectures, or hypotheses, try them out, and reject those that do not fit. This is a method of critical selection, if we look at it closely. From a distance it looks like instruction or, as it is usually called, induction.

   'What a painter does is often strikingly similar. He puts on his canvas a spot of colour and steps back to judge the effect, in order either to accept it, or to reject it and go over the spot again. It does not matter for my discussion whether he compares the effect with an object painted, or with an inward image, or whether he merely approves or disapproves of the effect. What is important here has been described by Ernst Gombrich by the excellent phrase "making comes before matching". This phrase can be applied with profit to every case of selection, in particular to the method of producing and testing hypotheses, which includes perception, and especially *Gestalt* perception. Of course, the phrase "making comes before matching" can be applied also to Darwinian selection. The making of many new genetic variants precedes their selection by the environment, and thus their matching with the environment. The action of the environment is roundabout because it must be preceded by a partly random process that produces, or makes, the material on which selection, or matching, can operate.' (Popper 1978, pp. 347–348)
8. Sarkar (1991) has given a useful extended discussion of the meaning of 'random' and 'directed' mutation, and stresses 'The notion of randomness that is appropriate for evolutionary biology, however, is quite murky' (p. 237). When considering the controversy over the randomness or otherwise of mutations in bacterial cultures, he suggests 'The type of variation that arises through a mutation of a gene is random if and only if the probability of its occurrence in an environment has no correlation with the fitness of the phenotype induced by it in that environment' (p. 237). For the reasons given in the text, we believe this definition is inappropriate, because it includes mutations that historically have been called 'directed'—mutations that are not adaptive, but are produced as a consistent and repeatable response to a particular environmental challenge. The terms 'random' and 'directed' reflect the specificity of the response to the environment, not the adaptiveness.
9. The two sets of experiments we describe are those most often cited as evidence for the randomness of mutations. In fact, more sophisticated experiments involving selection and other methods were carried out around 1950, and were

influential in hardening the belief in the pre-existence of adaptive mutations. A description of these is given by Hayes (1968, Chapter 9).

10. This point was realized perfectly well by some microbiologists many years ago. Lindegren commented on the fluctuation experiments: 'However, it would be difficult to detect if phage were capable of affecting or inducing mutations by this method. While the occurrence of spontaneous mutations prior to exposure to the phage can account for the adaptation in the above instance, it is important to note that in this particular case, because of the lethal effect of the phage, it is not possible to determine the mutation rate in the continued presence of the phage, but only prior to its application. Lysis takes place very quickly. Only bacteria which are in the process of division at the time of plating may sometimes complete the division, and the division time is 19 minutes (in broth). Therefore, if the phage were able to induce resistance in bacteria, it would have to be by an almost instantaneous induction'. (*The yeast cell: its genetics and cytology*, 1949, Chapter 22, pp. 2–3).
11. Ryan (1955, 1959) and Ryan *et al.* (1961). A brief discussion of this work is given in Symonds (1991).
12. See Shapiro (1984), Cairns *et al.* (1988), and Hall (1988). Opadia-Kadima (1987) mentions earlier experiments with bacteria which, although they did not directly address the question of environmentally-induced mutation, nevertheless indicated quite strongly that it did occur.
13. See Symonds (1989), Lenski (1989), Lenski *et al.* (1989), Mittler and Lenski (1990, 1992), and MacPhee (1993) for criticisms of some of the early experiments.
14. Details of the experiments and the various interpretations can be found in Drake (1991), Foster (1992), Stahl (1992), and Lenski and Mittler (1993).
15. See Charlesworth *et al.* (1988), Partridge and Morgan (1988), Lenski *et al.* (1989), and Lenski (1989).
16. Sarkar (1990 and 1991) discusses the ways in which deviations from the Luria-Delbrück distributions can be caused by factors other than directed mutation.
17. See Lenski *et al.* (1989) and Lenski and Mittler (1993) for a fuller discussion of this idea and its limitations.
18. The most detailed description of such a mechanism is that given by Davis (1989). We independently suggested a rather similar mechanism which was based on the observation that in all of the cases studied, the selective agent (lactose, arabinose, salicin) was an inducer of the gene in which the mutations occurred, and could therefore alter the binding affinity of regulatory proteins. We suggested that this could alter the conformation of the DNA sequence in a way that makes the region a preferential target for the activity of error-prone repair enzymes (Jablonka and Lamb 1989).
19. See Holliday and Rosenberger (1988), Symonds (1989), and B. G. Hall (1990). MacPhee (1993) has argued that the rate of mutation increases in the absence of glucose, and this can explain most cases of 'directed' mutation in bacteria.
20. See Stahl (1988) and Boe (1990).
21. See, for example, Cairns and Foster (1991) and Symonds (1991).
22. Full discussions and evaluations of the hypotheses associated with directed mutation are given by Drake (1991), Foster (1992), and Lenski and Mittler (1993).
23. Examples of non-random changes in eukaryote DNA are so well known that they are included in most standard textbooks of genetics and molecular biology,

e.g. Watson *et al.* (1987), *The molecular biology of the gene*. Additional information and discussion of some of the examples can be found in Borst and Greaves (1987). *Trends in Genetics* **8**(12) pp. 403–462 (1992) is devoted entirely to programmed DNA rearrangements.

24. Reviewed in Cullis (1984) and Schneeberger and Cullis (1991).
25. Two rather different definitions of RIP can be found in the literature. In his review, Selker (1990*a*) uses RIP for both 'repeat-induced point mutation' and 'rearrangements induced premeiotically' whereas others define RIP in terms of rearrangements (Fincham *et al.* 1989) or mutations (Kricker *et al.* 1992).
26. See Campbell (1982, 1985, 1987).
27. Various names have been given to the sophisticated systems by which cells manipulate their DNA. Wills (1991) refers to the 'molecular toolbox' or 'evolutionary toolbox', Shapiro (1992) to 'cellular genetic engineering'.
28. Wills's (1991) book *The wisdom of the genes* gives an interesting and readable account of the way the new molecular biology demands changes in the way we think about evolution.

4

Cellular heredity: epigenetic inheritance systems

... not everything that is inherited is genetic.
Boris Ephrussi 1958, p.49

It is commonly assumed that all heritable information resides in DNA base sequences, and that the only mechanism by which information is transferred from one generation to the next is through the semi-conservative replication of DNA. However, these assumptions are not correct. There are other self-sustaining, information-containing systems that function as inheritance systems. The most obvious of these are the systems through which culture is maintained and inherited, but the ones that concern us here are those which maintain and transmit cellular characteristics during ontogeny. Often they are referred to as 'memory' systems, rather than 'inheritance' systems, because it is assumed that they do not transmit information from parents to their progeny, and consequently that they have no direct evolutionary significance. The ultimate origin of cell characteristics is in the sequence of DNA bases, but even if DNA sequence is kept completely constant, cells can have very different phenotypes, and these phenotypes can be transmitted for many cell generations.

In this chapter we first consider the general properties of the systems that have been called 'epigenetic inheritance systems' (EISs),[1] and then look at three different types of cellular EIS. For the time being we are concerned only with somatic cell lineages and the transmission of cellular phenotypes from one cell generation to the next. No change in DNA sequence is involved. In later chapters we develop the argument that cellular EISs are important not only in ontogeny, but also in the phylogeny of many unicellular and multicellular organisms, and that they have had far greater direct evolutionary significance than is generally acknowledged.

Epigenetic inheritance systems

Definitions of 'epigenesis' and 'epigenetics' vary.[2] According to Medawar and Medawar:

> In the modern usage 'epigenesis' stands for all the processes that go into implementation of the genetic instructions contained within the fertilized egg. 'Genetics proposes: epigenetics disposes.' (Medawar and Medawar 1983, p. 114)

A biological dictionary definition of epigenetic is:

> Pertaining to the interaction of genetic factors and the developmental processes through which the genotype is expressed in the phenotype. (Lincoln *et al.* 1982, p. 82)

A less formal, and more readily applied, definition is:

> Everything that leads to the phenotypic expression of the genetic information in an individual.

An *epigenetic inheritance system* (EIS) is a system that enables a particular functional state or structural element to be transmitted from one cell generation to the next, even when the stimulus that originally induced it is no longer present. In other words, EISs are the systems that enable the transmission of the various phenotypic expressions of the genetic information in an individual. They underlie cell memory. It is the EISs that ensure that when a fibroblast divides it gives rise to fibroblasts, and that when a kidney cell divides it gives rise to kidney cells. Although under some conditions the determined and differentiated states of cells can be reversed, normally they are remarkably stable, and are inherited stably. Yet when a cell lineage switches from one inherited functional or structural state to another, there is usually no change in DNA base sequence. The cells remain genetically equivalent, whatever their phenotype.

In 1958, soon after it was established that genetic information is encoded in the base sequence of DNA and that mutations are probably the result of changes in base sequences, Nanney pointed out that distinguishing between genetic and epigenetic inheritance is sometimes rather difficult. Phenotypic differences between cells can be perpetuated for many generations, even when there are no detectable genetic or environmental differences. These persistent inherited differences must be maintained through an epigenetic system. However, if persistent differences can be epigenetic as well as genetic, how is it possible to distinguish between the two types of inherited variation? Nanney suggested that the following criteria could help. First, although both genetic and epigenetic variations can show a range of stabilities, an EIS is usually less stable than the genetic system; EISs are more sensitive to environmental changes. Secondly, and more importantly, induced changes in EISs are probably more often directed than those in the genetic system: the variations are more predictable and repeatable. Finally, the variations resulting from changes in epigenetic systems are likely to show only a limited number of alternative states, because the alterations that are possible will be restricted by the genetic information in DNA base sequences.

Nanney realized that, even when taken together, the criteria he had suggested were too subjective to be really adequate. More recently, Meins (1983) proposed additional criteria for distinguishing between genetic and epigenetic changes. He stressed that epigenetic variations are always potentially reversible and, by definition, they cannot be transmitted through meiosis and passed to sexually produced progeny. The problem with this criterion is that although it is true that a high frequency of reversion following meiosis is a good indication that a variation is probably epigenetic, persistence or lack of persistence through meiosis is not a certain way of distinguishing a mutation from an epigenetic change. As we shall show in Chapter 6, there are many acquired variations that are not genetic, because no changes in DNA sequences are involved, yet they are transmitted from parents to sexually produced progeny.

So far, no one has provided an adequate set of criteria for distinguishing between inherited epigenetic and genetic variations. The reason for this is probably that the distinction between the two is not absolute. In order to see why, we first need to consider the nature of some of the epigenetic systems, and the way in which epigenetic information is transmitted in cell lineages. The study of cellular memory is still in its infancy, but the types of mechanism that might underlie it were clearly envisaged at least twenty years ago. In 1974 Cook suggested that information is carried not only in the primary structure of DNA, but also in its 'superstructure', or topology; DNA superstructure can be modified by interactions with proteins, and subsequently the altered topology can be inherited. He also pointed out the potentially heritable nature of the position of chromosomes in the nucleus, and the possible importance of self-sustaining feedback loops for transmitting states of gene activity. These ideas on the nature of cellular memory systems are very similar to those that have emerged from recent molecular studies.

The three types of cellular memory system that we describe in this chapter are: first, steady-state systems; second, structural inheritance systems; third, chromatin-marking systems. In all three, the variations that are inherited are not the direct result of differences in DNA base sequences. These epigenetic inheritance systems are therefore quite distinct from the usual genetic system.

Steady-state systems (or self-perpetuating metabolic patterns)

One of the first people to recognize that there are probably different types of inheritance system was Sewall Wright. He suggested that the persistence of alternative cellular states, 'cell heredity', could arise either as the result of changes in nuclear genes, or as a result of changes in the cytoplasmic constituents of the cell. One possibility he considered was the presence in the cytoplasm of plasmagenes—self-replicating, submicroscopic units. Another was the self-regulatory properties of cytoplasmic constituents:

Persistence may be based on interactions among constituents which make the cell in each of its states of differentiation a self-regulatory system as a whole, in a sense, a single gene, at a higher level of integration than the chromosomal genes. On this view the origin of a given differentiated state of the cell is to be sought in special local conditions that favor certain chains of gene-controlled reactions which cause the array of cytoplasmic constituents to pass the threshold from the previous stable state to the given one. (Wright 1945, pp. 298–299)

Throughout the late 1940s and 1950s, Hinshelwood and his colleagues argued that some of the persistent adaptive responses seen in bacterial cultures were caused by changes in metabolic states, rather than differential survival of pre-existing genetic variants.[3] A similar idea, but one with a more defined mechanism, was suggested in 1949 by Delbrück to explain the persistence through successive cell divisions of induced serotypes in ciliates, and similar alternative states of cell differentiation. His mechanism involved positive feedback: once initiated, a particular state is maintained by its own activity. The mechanism is illustrated in Fig. 4.1.

At the time that Delbrück proposed his mechanism, there were no well-studied examples of stable, self-maintaining, metabolic systems. However, a system of this general type was described in 1957 by Novick and Weiner. They showed that whereas when *E. coli* is cultured with a high concentration of an inducer of the *lac* operon all cells produce β-galactosidase, at lower concentrations there are two types of cell. Some cells are fully induced and synthesize β-galactosidase; they remain in this state and transmit it to daughter cells for as long as the inducer is present. Other, genetically equivalent cells do not synthesize β-galactosidase, and transmit this non-induced state to their daughter cells. The difference between the two types of cell is caused by chance fluctuations in the intracellular

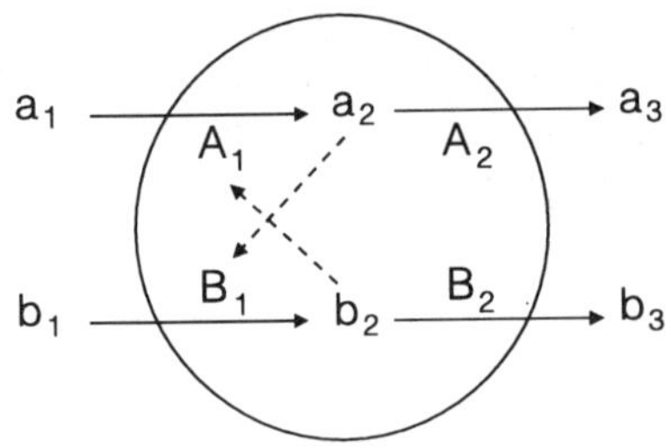

Fig. 4.1 Delbrück's (1949) model for the maintenance of alternative heritable phenotypes. Enzymes A_1 and A_2 catalyse the chain of reactions $a_1 \rightarrow a_2 \rightarrow a_3$, and B_1 and B_2 the chain $b_1 \rightarrow b_2 \rightarrow b_3$. Intermediate b_2 inhibits enzyme A_1, and a_2 inhibits B_1. Because of this inhibition, the chain of reactions that is functioning at any given time is perpetuated unless the relative concentrations of a_2 and b_2 are changed. For example, if a cell is carrying out pathway $a_1 \rightarrow a_2 \rightarrow a_3$, removal of a_2 would allow B_1 to become active and the cell would switch to the alternative steady state $b_1 \rightarrow b_2 \rightarrow b_3$.

concentration of the permease necessary to transport inducer into the cell. The permease gene is part of the *lac* operon, so it is induced by the same inducers as β-galactosidase. Consequently, in conditions of low inducer concentration, once a cell contains, by chance, a permease molecule, there is positive feedback: more inducer enters the cell and induces the formation of more permease, which enables more inducer to enter the cell, and so on until the cell is fully induced and synthesizes both permease and β-galactosidase at the maximal rate. This self-perpetuating system is shown schematically in Fig. 4.2. Notice that the system requires the continued presence of the inducer to maintain the distinction between the two functional states. Removal of the inducer from the medium results in all cells becoming non-induced.

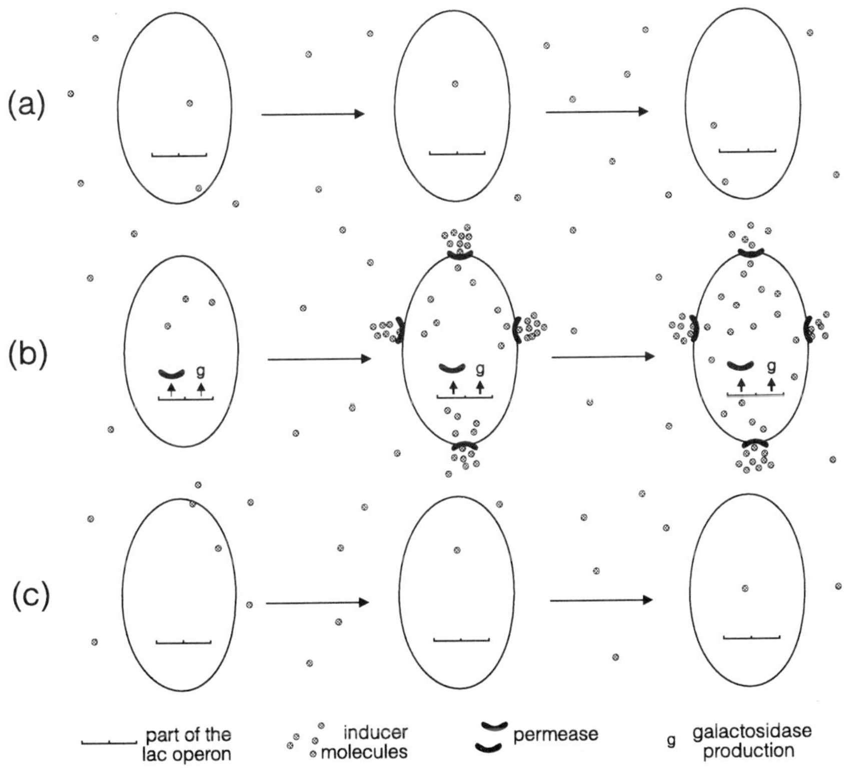

Fig. 4.2 Alternative states of induction of the *lac* operon in *E. coli* cultured with low concentrations of inducer. Because of random fluctuations in the level of internal inducer, cell (b) produces permease molecules and can therefore concentrate the inducer sufficiently to induce transcription of the *lac* operon; once induced, production of permease ensures further permease and β-galactosidase production. In cells (a) and (c), the inducer never reaches the critical concentration, so the operon remains uninduced.

Recent work with eukaryotes has shown that some genes are able to maintain their own activity, even in the absence of the inducing stimulus that initially triggered the change in their functional state.[4] For example, some homeobox-containing genes in *Drosophila* autoregulate their transcriptional activity through the action of their product on their enhancer. Similarly, some oncogenes regulate their own expression as well as that of other genes. A general scheme showing how such self-regulating genes may work is given in Fig. 4.3a. It shows that once a gene has been activated, it can maintain its state of activity, provided that its product is synthesized at a rate greater than that at which it is used or lost. When the rate of loss or use is greater than the rate of production, the system switches from stably active to stably inactive. Fig. 4.3b shows how such a gene should operate over time.[5]

It is not necessary for a gene to regulate its own activity in order for a functional state to be maintained: if two or more genes have certain regulatory interactions, then their states of activity can be perpetuated. An example of a regulatory network of this type is shown in Fig. 4.4. The primary stimulus causes the activation of gene *A*; gene *A*'s product acti-

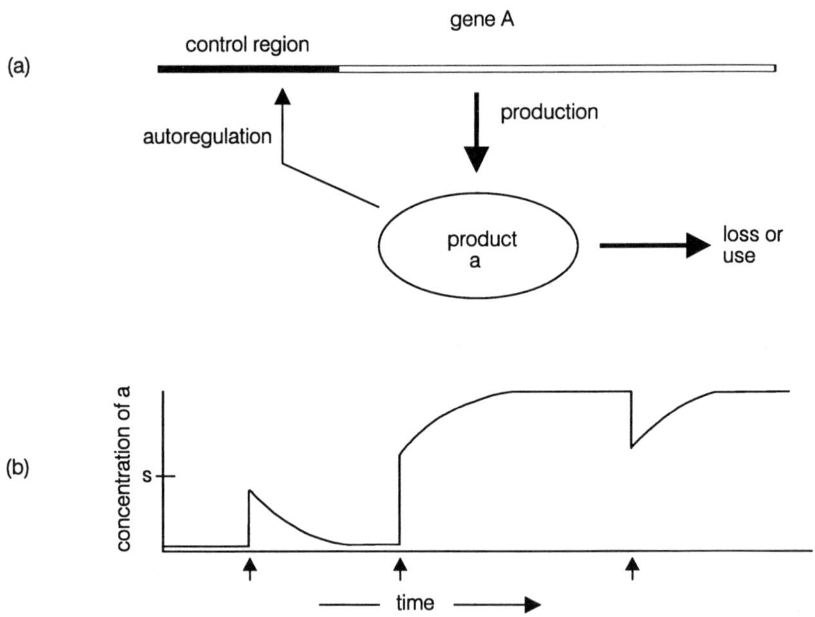

Fig. 4.3 (a) A steady-state system in which gene *A* is regulated by its own product **a**. (b) The expected behaviour of gene *A* over time. At the times indicated by the arrows, a stimulus causing changes in the concentration of **a** is introduced. Only when the concentration of **a** exceeds the threshold **s** is gene *A* turned on, but once turned on, its activity is maintained by positive feedback.

vates gene *B*, the product of which activates gene *A*. From the moment this state is established, both genes will remain active, even in the absence of the stimulus that originally activated *A*. This kind of regulation, as well as more complex regulatory networks, seems to be characteristic of several self-maintaining systems in eukaryotes. Systems of this type allow transient stimuli, such as various morphogens, to have far-reaching and permanent developmental effects. Once established, a functional state is stabilized by the regulatory network, and persists even in the absence of the original stimulus.

The property that is transmitted to daughter cells in these steady-state systems is the concentration of the regulatory proteins. The stability of the inheritance system therefore depends on the number of molecules of each type of protein, and on their distribution in the cytoplasm or in the nucleus. Provided that there is a more or less even distribution, and the number of molecules of each type of protein is not too low, the fidelity of these systems may be quite high. Formally, their behaviour is very similar to that of self-perpetuating cytoplasmic organelles such as chloroplasts and mitochondria. The transmission of functional states does not require any special

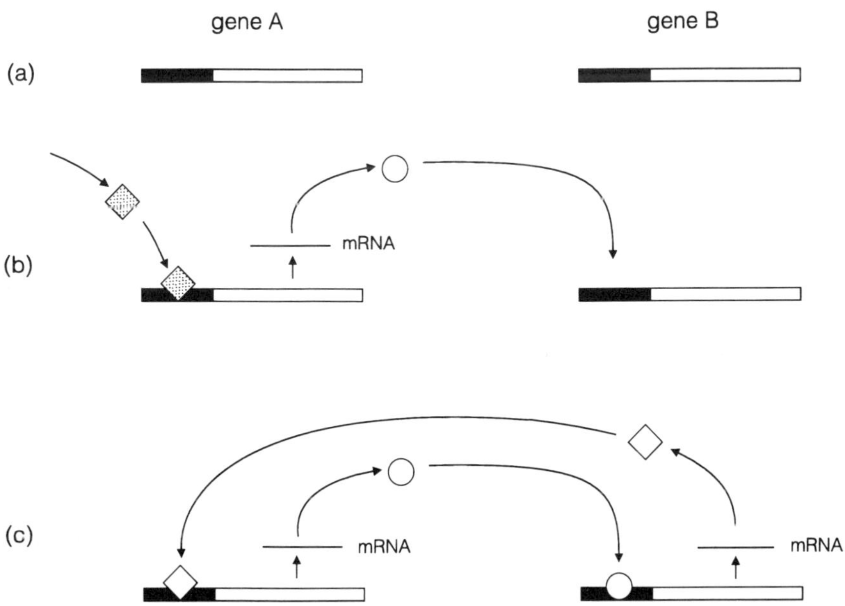

Fig. 4.4 A self-maintaining regulatory network involving two genes: control regions solid, transcription units open. (a) Neither gene is induced. (b) The presence of a transient inducer (shaded diamond) activates gene *A*. (c) The product of gene *A* (open circle) induces gene *B*; the product of gene *B* (open diamond) induces *A*. Once the state shown in (c) is established, both genes *A* and *B* remain active.

mechanisms: it is a simple consequence of the operation of the homeostatic system and of a more or less equal cell division.

The number of examples of self-maintaining regulatory loops that perpetuate differentiated cell states is increasing rapidly.[6] Steady-state systems are obviously important in the maintenance and transmission of information in cell lineages. In the types of system we have described, gene products have two functions: they participate in some process that leads to a characteristic phenotypic effect, and they have a regulatory role, controlling their own or each other's transcriptional activity. In our examples they were DNA-binding proteins, which exert a positive regulatory function by binding to specific control regions. One of the properties of this type of system is that a gene can have only two functional states: stably active, and stably inactive. Since 'intermediate' states of activity can also be inherited in cell lineages, there must be other ways in which information is transmitted between cell generations.

Structural inheritance systems

In 1968, in an article on cortical patterns in cellular morphogenesis, Nanney wrote that he wished:

> . . . to open the door to consideration of the possibility that essential biological information is encoded and transmitted by materials other than nucleic acids and by means other than linear templates. (Nanney 1968, p. 497)

The system Nanney had in mind was one in which existing supramolecular cell structures are used as templates for new similar structures. His ideas were based on experiments showing how the structural organization associated with the cell surface of *Paramecium*, *Tetrahymena*, and other ciliates is inherited.[7] The cell surfaces of these protozoa are covered with thousands of cilia arranged in fairly precise patterns. Each cilium is part of a larger cortical unit, which includes at least one basal body and various fibrous and membranous elements. These individual units are structurally asymmetrical, with a distinct anterior–posterior and left–right organization. Commonly, they are arranged in longitudinal rows, with all units within a row having the same symmetry and uniform orientation. When a basal body replicates, a new unit is formed anterior to the old one and it has the same orientation (Fig. 4.5). During cell division the longitudinal rows of ciliary units divide transversely.

Some ciliates show clonal variations in their ciliary patterns. For example, in several species, the average number of ciliary rows may differ in different clones. These differences are maintained through many generations, in spite of the genetic identity of the cells involved. Moreover, even when cells of different types go through the sexual process of conjugation,

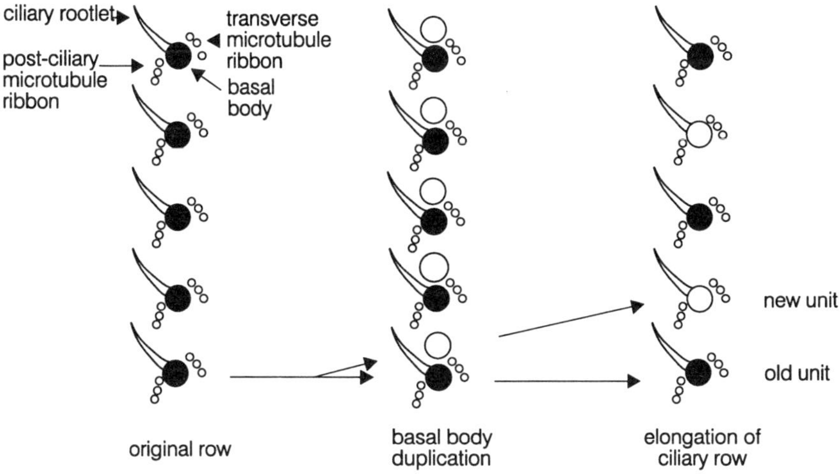

Fig. 4.5 The organization and multiplication of basal bodies along ciliary rows in *Paramecium*. (Redrawn from Jerka-Dziadosz and Beisson 1990.)

the inheritance of row number is not affected. Even more strikingly, cells in which the cortical pattern has been altered by environmental manipulation or microsurgery transmit the new pattern to their descendants. For example, in both *Paramecium tetraurelia* and *Tetrahymena thermophila*, when part of the cortex is rotated through 180° so that one or more of the ciliary rows have the opposite orientation, the rotated configuration persists and is stably inherited through many generations of asexual reproduction and through conjugation.[8] Clearly, induced variations in cell architecture, which do not involve changes in DNA sequence or in gene function, can be stably inherited.[9] Sonneborn (1964) called this type of inheritance, in which new cell structures are ordered and arranged under the influence of pre-existing cell structures, 'cytotaxis'. Grimes and Aufderheide (1991) preferred to talk about 'directed assembly', because it was a better reflection of their belief that the phenomema found in ciliates had implications for all eukaryotic cells.

Ciliates can perpetuate large-scale patterning, as well as patterns of ciliary rows. This is shown by *P. tetraurelia* and other species, where 'doublets' are formed when two cells remain stuck together following interference with cell division or conjugation. These cells have two sets of cortical structures, including two mouths and gullets. This new phenotype can be transmitted to daughter cells. It is clear that something other than direct templating must be involved in the inheritance of this type of cortical pattern. There must be some kind of organization that covers large regions of the cortex and is responsible for the perpetuation of these cellular structures.

The experiments with ciliates are very convincing because they show both that genotypically identical cells can perpetuate different supramolecular structures, and that a particular architecture is preserved even after the extensive genetic reorganization of the cells resulting from conjugation. Summarizing his studies on the inheritance of cortical patterns in *Tetrahymena*, Nanney wrote:

> A variety of pattern permutations can be established on a common genic basis, and these permutations have sufficient stability to be designated *hereditary* variants. The mechanisms of hereditary maintenance apparently do not involve genic differences—either nuclear or cytoplasmic, either structural or functional—but involve rather, a multidimensional information storage and transmission system whereby the pattern, in a sense, maintains itself. (Nanney 1968, p. 502; his italics)

Frankel (1983) has discussed the evolutionary implications of the inheritance of supramolecular structures in ciliates. He suggested that some cytotactic variations may have had evolutionary consequences. For example, one ciliate species, *Teutophrys trisulca*, possesses a single trunk but three anterior probosces. Each of these probosces is similar to the single proboscis found in another ciliate genus, *Dileptus*. *Teutophrys* is believed to be a distinct species, and it may have arisen as a result of a structural alteration in a *Dileptus*-like organism. The initial heritable variation is assumed to have been cortical—a developmental accident that produced a 'triplet' organization which was perpetuated by the structural inheritance system; later, perpetuation of the new phenotype was stabilized by genetic changes.

Is there any evidence that cytotaxis is not unique to ciliates? Does it have more general significance? In multicellular organisms cell division requires the duplication and organization of not only the nuclear material, but also of other cellular structures and cytoplasmic constituents. It seems unlikely that such a complex organization is produced solely by self-assembly, and that pre-existing structures do not influence or direct the formation of new structures. As Sonneborn wrote:

> . . . the integrity of nonrandom cell structure persists throughout growth and division which immediately suggests that the pre-existing structure plays a decisive role that may not be explicable by mere random self-assembly of genic products. (Sonneborn 1964, p. 924)

Structural inheritance may be responsible for the duplication of centrioles in animal cells. New centrioles are formed in association with old centrioles, suggesting a templating process similar to that found with protozoan cilia, but *de novo* production of centrioles can also occur.[10] Although the evidence is rather meagre, other examples of architectural continuity, which are probably instances of cytotaxis, have been found in somatic cell lineages. For example, genetically identical neuroblastoma cells in culture show a wide variety of neurite morphologies, but after cell division, 60% of sister pairs are identical twins or mirror images of each other. The close similarity is seen both in the number of neurites per cell, and in the fine details of neurite morphology. This architectural continuity persists for at least two cell divisions (Solomon 1979, 1981). Another experimental system in which cytotaxis has been demonstrated is 3T3 mouse fibroblast cells in tissue culture. After cell division, the pattern of migration, the cell shape, and the geometry of microfilament bundles in the cells are all transiently perpetuated (Albrecht-Buehler 1977).

This type of observation is not confined to cells in tissue culture. Structural continuity has been observed by Locke (1988) in the epidermis of caterpillars, where it is possible to compare structures in sister cells because they remain attached to each other by a narrow cytoplasmic bridge. Sister cells show similar or mirror-image patterns of nuclear and cytoplasmic structures such as the number, size and shape of nucleolar particles, and the number of actin bundles. Locke has called this transient inheritance of the three-dimensional cellular architecture 'somatic inheritance', and defined it as 'the inheritance of a pattern independently from the molecules or processes that form it'. He suggested that the sharing of common cytoplasm by the twin cells in the insect epidermis allows the nuclear skeleton and the cytoskeleton to reproduce themselves under conditions that maintain the initial pattern long enough for it to be recognized. Hjelm (1986) has argued that cytotaxis is a general property of living organisms, since all seem to use pre-existing structures for the reproduction of their cytoskeleton and cell surface elements. He suggested that sometimes heritable changes in cytotactic information may initiate carcinogenesis, because they lead to cellular abnormalities such as aberrant cell division and chromosome segregation.

After a detailed review of the transmission of cytoplasmic information, including the localized determinants present in the eggs of many metazoa, Grimes and Aufderheide concluded:

> The highly organized cytoplasm of a metazoan egg, therefore, *cannot* be solely the consequence of direct nuclear gene activity. Given the background of information from the Ciliophora, one would predict that structurally heritable information systems must be present in addition to direct nuclear (genic) control systems in the metazoa. The ciliated protozoa represent a group of organisms that have made exceptional use of the posttranslational, 'epigenetic' systems that contribute to the localization of gene products. . . . processes homologous, or at least analogous, to the directed assembly and directed patterning seen in ciliates are also functional in metazoa, and are of fundamental developmental significance. (Grimes and Aufderheide 1991, p. 67; their italics)

Structural inheritance may extend even beyond intracellular templating. In vertebrates there is evidence that the three-dimensional structure of the extracellular matrix in which cells are embedded is used by those cells as the template for the pattern of deposition of further matrix components. In this way the architecture of morphological structures is maintained, in spite of the turnover of their components. This type of structural inheritance is thought to be of major importance in morphogenesis.[11]

Chromatin-marking systems

The epigenetic inheritance systems for which we have most information are those that are based on the transmission of specific patterns of chromatin structure. They are systems that use the semi-conservative replication of DNA as the vehicle for carrying epigenetic information through cell division. The information is carried in chromatin as *chromatin marks*. Chromatin, the substance of chromosomes, is a complex of DNA and proteins. Chromatin marks are the non-DNA parts of the chromosomes, for example binding proteins or additional chemical groups attached to DNA bases, that affect the nature and stability of gene expression.

The structure and conformation of chromatin differs in the different cells of the body, in different chromosomes, in different chromosome regions, and at different times. The differences are associated with the types of genes the chromatin contains, and the activity of those genes. Some genes, known as housekeeping genes, are capable of transcriptional activity in all cells; their products are essential for intermediary metabolism, protein synthesis, and general cell maintenance. In contrast, the tissue-specific genes are active in only one or a few specialized cell types, and stage-specific genes are active only at specific stages of development. Some chromosome regions are never active, because they lack the DNA sequences essential for transcription. The chromatin conformations associated with these different types of genetic activity are not the same.

In order to simplify the terminology, we have suggested that when the chromatin structure at a locus is variable, the different chromatin conformations of a gene for which the DNA sequence remains unchanged should be referred to as the *gene's phenotypes* (Jablonka and Lamb 1989). The differing phenotypes of a gene reflect its accessibility to the factors necessary for transcription. A single gene may have several phenotypes, each associated with a different state of activity. It may be stably inactive and inaccessible to transcription factors under most natural conditions; alternatively, it may be inactive, but competent for activation by an appropriate stimulus at a later stage. There may be a range of stabilities for such transcriptional inactivity. When active, a gene can be either transiently active, with transcription depending on the continued presence of an inducing stimulus, or stably active, with transcription independent of the original stimulus. The transition from an active to an inactive state or *vice versa* is normally the result of a developmental or environmental stimulus, and often seems to be a multistage process, rather than a simple switch.

When gene phenotypes are inherited, the alternative phenotypes of a locus will be referred to as *epialleles*. Kermicle coined this term in 1978 to denote the different epigenetic forms which a gene acquires during gametogenesis in males and females.[12] We shall use the term rather more generally to refer to all heritable chromatin variants, not just sex-specific marks. Epiallelic variants arise as the result of the normal chromatin changes taking place during somatic differentiation. Epialleles can also arise through stochastic, non-directed processes (Holliday 1987). Theoretically the number of epialleles of a locus, and hence the heritable variability, is enormous: in a diploid multicellular organism the maximum number of epialleles is the number of cells in the body multiplied by two (the number of copies of each gene in the cell). Of course the actual number is much smaller: it is constrained by the DNA sequence of the locus, and by the fact many cells have identical epialleles. In different individual organisms with identical DNA sequences, the same cell type may have different epialleles. Consequently, both within individuals, and in populations of individuals, there is an abundance of heritable epigenetic variability.

In order to understand the mechanisms underlying the inheritance of epialleles, we need to know exactly what it is that is inherited. What features of chromatin characterize different epialleles? How is the organization of chromatin in chromosomes related to its structure and function, and how is this organization inherited?

1. *The structure of active and inactive chromatin*

In 1928, as a result of his cytogenetic studies, Heitz divided chromatin into two categories, heterochromatin and euchromatin.[13] Heterochromatin was the term used for chromosomes or chromosome regions that remain highly condensed during interphase, a period when the remaining chromatin, euchromatin, is diffuse. Later it was found that this type of heterochromatin, which is now called constitutive heterochromatin, contains little genetic information. Its DNA consists predominantly of repetitive sequences. A second type of heterochromatin, facultative heterochromatin, has been described. Facultative heterochromatin is chromatin that is condensed, but only in some cells and at certain stages of development. It contains normal genes, but when it is heterochromatic, these genes are inactive.

One of the classic examples of facultative heterochromatin, and one that has contributed a lot to our understanding of epigenetic inheritance, is the inactive X chromosome of female mammals. Early in development, one of the two X chromosomes in each cell becomes condensed and inactive. The inactive X chromosome can be seen in some interphase cells as a distinct structure, the Barr body, near the nuclear membrane. Functionally,

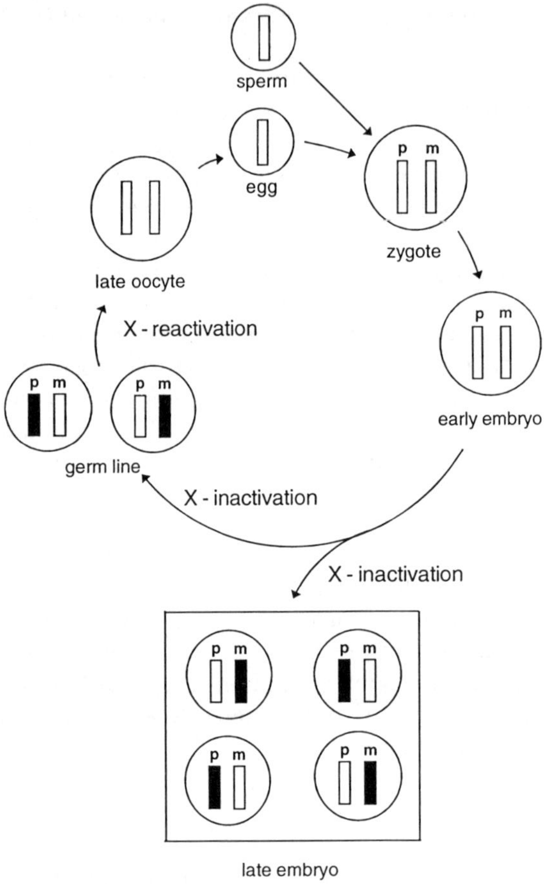

Fig. 4.6 The cycle of random X-inactivation in eutherian mammals. **p** is the X chromosome inherited from the father, **m** that inherited from the mother. Solid shading indicates inactivity.

although females have two X chromosomes, they have the same dose of X-linked genes as XY males because only one X chromosome is active. Figure 4.6 outlines the basic X inactivation cycle of eutherian mammals. Notice that both X chromosomes are active in the early embryo, and that the inactive X is reactivated in the germ line.

If autosomal genes are translocated into the X chromosome, they too may be inactivated, since inactivation can spread from the X for some distance into a translocation. The inheritance of the inactivated state of an autosomal segment is quite stable (Cattanach 1974). This effect is very similar to that known as 'position effect variegation' (PEV) which is found in *Drosophila* carrying chromosomal rearrangements. When flies have an inversion or translocation that has moved a euchromatic gene near to centromeric heterochromatin, the gene in some cells becomes heterochromatinized and inactive. This inactive state is clonally inherited. The result is a mosaic or variegated phenotype, with the gene being expressed in some patches of tissue but not in others. As with autosomal insertions into the mammalian X chromosomes, with PEV the chromatin structure of heterochromatin seems to spread to the neighbouring genes.[14]

Another early indication that there is a relationship between gene activity and chromatin structure came from studies of the polytene chromosomes found in cells of some dipteran insects, and the lampbrush chromosomes found in amphibian oocytes. In both of these specialized cell types, gene activity is associated with a more 'open' chromatin conformation. Amphibian oocytes often have a long and very active meiotic prophase in which they produce material needed to support early embryonic development. During this period, large, decondensed, chromatin loops can be seen extending from the main chromosomal axis, where most of the DNA remains in a highly condensed form. These chromatin loops—lampbrush loops—are the sites of active RNA synthesis. Comparable, extended, decondensed, transcriptionally active chromatin regions can be seen in the puffs of dipteran polytene chromosomes. Puffs are regions where the chromosome appears light and diffuse; they are formed when the many chromatin threads which make up a polytene chromosome decondense and become active. A stimulus-specific set of puffs is produced in response to stimuli such as the insect hormone ecdysone or a heat shock.

Although the early cytogenetic studies showed that there is a correlation between the degree of chromatin condensation and genetic activity, it is only recently that the techniques of molecular biology have begun to reveal in more detail how chromosomes are organized and how active and inactive chromatin differ.[15] One feature of chromatin now known to be strongly associated with gene activity is the time at which it replicates. DNA in inactive chromatin regions typically replicates later than that in active regions. For example, the inactive X chromosomes of female mammals are late replicating, as are most other types of transcriptionally inert heterochromatin; tissue-specific genes usually replicate late in tissues in which they are not expressed, and early in tissues in which they are expressed.[16]

The unit of replication seems to be the chromatin loop or domain.[17] Each loop or domain contains about 30–300 kb of DNA anchored at the ends to the nuclear matrix (Fig. 4.7). Domains are therefore larger than a typical single transcription unit, and since their DNA is anchored to the nuclear matrix, they are to some extent independent of each other. The autonomy of some looped domains has been confirmed by the discovery of 'boundary elements' located near their site of attachment to the nuclear matrix. These elements are only a few hundred base pairs long, but

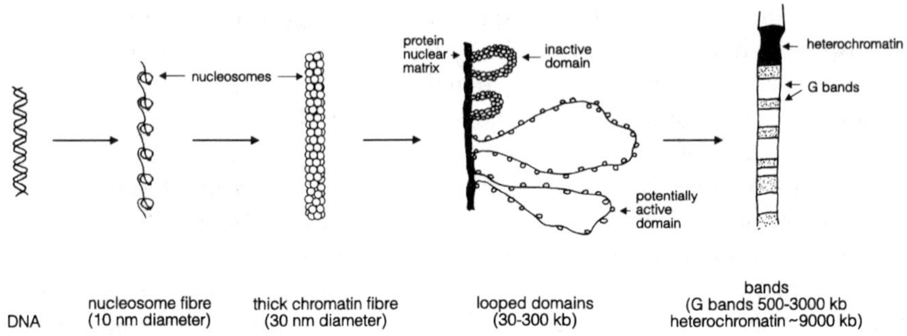

Fig. 4.7 The basic structure of mammalian chromatin.

insulate the genes within one looped domain from the regulatory influences of neighbouring domains. Although control sequences known as 'enhancers' can often act over large distances, an enhancer affects a gene's expression only if it is within the same domain. When separated from a gene by a boundary element, enhancers have no effect. The recent identification of DNA sequences that control the time of replication of whole domains lends further support to the reality and importance of these units of chromatin structure and function.[18]

Clearly there must be a difference between the organization of chromatin in an active, usually early-replicating domain, and that in an inactive, late-replicating domain. Many years ago, with her usual foresight, Barbara McClintock described an overall picture of tissue-specific gene expression in which some differentiation depends on transposition events, but in other cases it is brought about by changes in chromatin structure:

> Differential mitoses also produce the alterations that allow particular genes to be reactive. Other genes, although present, may remain inactive. This inactivity or suppression is considered to occur because the genes are 'covered' by other non-genic chromatin materials. Genic activity may be possible only when a physical change in this covering material allows the reactive components of the gene to be 'exposed' and thus capable of functioning. (McClintock 1951, p. 42).

We now know a little more about the nature of McClintock's 'covered' and 'exposed' states of the genic material. The genic material is DNA and the covering materials are methyl groups, proteins and RNA. We also know something about how the genic and non-genic components are organized. The basic unit of DNA packaging is the 10 nm diameter, 'beads-on-a-string' nucleosome fibre. Each nucleosome has slightly less than two turns of DNA wound round a histone core made up of two molecules of each of the histones H2A, H2B, H3 and H4. The DNA between the nucleosomes is known as 'linker' DNA. The next level of packaging is the thick (30 nm) fibre. The 10 nm diameter nucleosome fibre is compacted, probably by coiling into a cylindrical, solenoid-like structure. Another type of histone, H1, seems to play a major role in the establishment and maintenance of the thick fibre by holding together adjacent nucleosomes. Most non-expressed genetic information seems to be packaged in this form. Fig. 4.7 shows the way the nucleosome fibre is compacted to form the 30 nm fibre, and how the 30 nm fibre is organized into domains.

How else does the chromatin structure in active or potentially active domains differ from that in inactive regions? Some of the differences that have been detected are summarized in Table 4.1. The general picture that has emerged from biochemical studies is that a competent chromatin domain (i.e. chromatin of genes that are actually or potentially active) has a looser structure, which is detectable as an enhanced general sensitivity to DNA-digesting enzymes such as DNase-I or micrococcal nuclease. This looser structure is associated with changes in histones and in DNA

Table 4.1 Properties commonly associated with transcriptionally competent, or active, chromatin, and inactive, but functional, chromatin in mammals. (Based on Gross and Garrard 1987, Goldman 1988, Van Holde 1989, Garrard 1991, Wolffe 1991, Hansen and Ausio, 1992.)

Property	Competent chromatin	Inactive chromatin
Degree of condensation	Open, extended	Condensed
General DNase-I sensitivity	Sensitive	Less sensitive
Micrococcal nuclease sensitivity	Sensitive	Less sensitive
DNase-I hypersensitive sites	Yes	No
Supercoiling of DNA	More supercoiled	Less supercoiled
General level of methylation	Usually low	Usually high
Methylation sites in the control regions	Usually demethylated	Usually methylated
H1 histones	Not present, or differently bound	Present
H1 modification	Phosphorylated at amino terminus	Phosphorylated at carboxyl terminus
Modification of core histones	Highly acetylated	Deacetylated
HMG proteins 14 and 17	Abundant	Depleted
Time of DNA replication	Usually early	Usually late
Specific transcription-positive factors	Present	Usually absent

supercoiling.[19] DNA is still packaged into nucleosomes but, typically, the control sequences of genes within competent domains have nucleosome-free regions. These nucleosome-free regions are detected as DNase-I *hyper*sensitive sites (DH sites)—sites that are an order of magnitude more sensitive, and much shorter, than the general DNase-I sensitive regions just referred to. Although nucleosomes are absent, non-histone proteins, which may act as regulators of transcription, are commonly present at DH sites. Usually, DH sites appear before or at the same time as gene expression is induced, but the mechanisms leading to their formation are unknown.

Figure 4.8 shows how the DH sites in a region of the chicken lysozyme gene differ in different tissues and change during development. Lysozyme is an important egg-white protein which is produced in the hen oviduct in response to steroid hormones. It is also part of the body's defences against bacteria, and the gene is constitutively active in macrophages. Figure 4.8 shows that the distribution of DH sites is not the same in induced and non-induced oviduct genes, and that when the gene is active in the oviduct, the set of DH sites is different from that of the active gene in macrophages. This difference indicates that gene activity is regulated differently in the two cell types. In tissues such as the brain, where the gene is not expressed, many of the potential DH sites are organized in nucleosomes, so are not accessible to digestion by DNase-I. The various phenotypes of the lysozyme gene are therefore associated with its different activities in different cells and tissues.

Perhaps one of the most important aspects of the gene's phenotype, and one that is known to play a role both in the control of gene activity and in cell memory, is its methylation pattern. In many eukaryotes, some of the

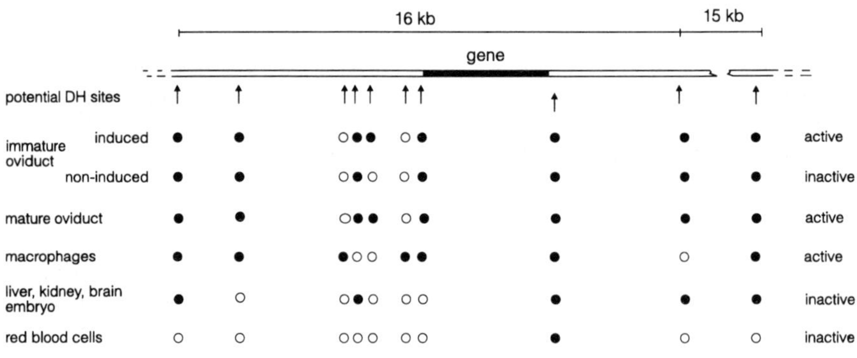

Fig. 4.8 The different phenotypes of the chicken lysozyme gene revealed by DNase-I hypersensitivity. Arrows indicate DH sites; potential DH sites are shown as open circles, filled circles show actual DH sites. Almost the whole of the region shown is within a region of general DNase-I sensitivity. (Based on Gross and Garrard 1988.)

cytosines in DNA can be modified by the enzymatic addition of a methyl group (Fig. 4.9). The methyl group does not change the coding properties of the base, but may influence gene expression. The evidence for this is substantial.[20] First, it has been found that highly methylated DNA is usually transcriptionally inactive, whereas DNA depleted of methyl groups is more often active. Second, there is a correlation between the pattern of methylation at specific sites, particularly in promoter and enhancer regions, and the state of activity of the gene. The promoter regions of housekeeping genes are unmethylated, whereas those of tissue-specific genes are usually methylated, except in the cells where they are expressed (Figure 4.10). Third, experimental manipulation of methylation, such as treatment with the demethylating agent 5-azacytidine, may change gene activity. Fourth, methylation-deficient mutants in mice die in mid-gastrulation.

Although there is good evidence of a relationship between cytosine

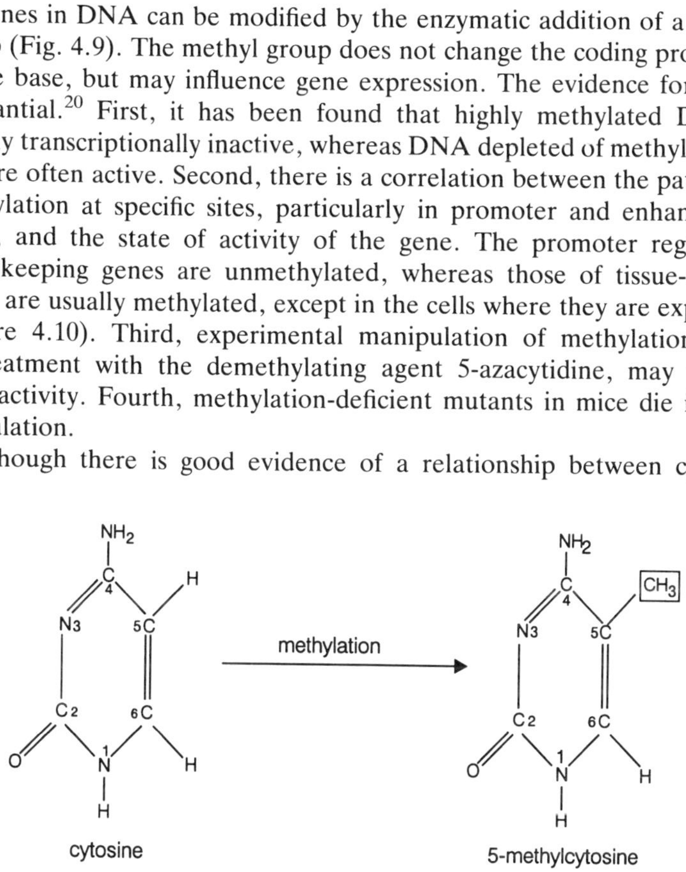

Fig. 4.9 The formation of 5-methylcytosine by enzymatic methylation of a cytosine in DNA.

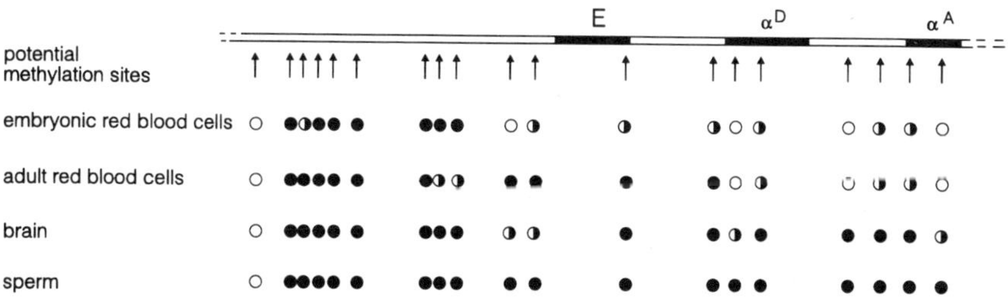

Fig. 4.10 Methylation patterns in the chicken α-globin gene cluster. Embryonic red blood cells express mainly the E-globin gene, whereas adult cells express only α^{A}- and α^{D}-globins. Filled circles: methylated sites; open circles: sites with little methylation; half-filled circles show partially methylated sites. (Based on Haigh *et al.* 1982.)

methylation and the regulation of gene expression, observations in many experimental systems suggest that methylation probably stabilizes, rather than establishes, the functional state of genes: high levels of methylation stabilize the inactive state, low levels stabilize the potentially active state. In some way methylation must affect DNA-protein interactions, but it is not clear at which level or levels of chromatin structure—at that of the nucleosomal fibre, the thick fibre, or the domain—it is most important. The extent of methylation affects all the different levels of chromatin organization (Levine *et al.* 1991, 1992). Proteins that bind to DNA sequences containing methylated cytosines have been identified, and seem to contribute to the control of transcription, but the precise details of the relationship between DNA methylation and chromatin structure have yet to be resolved. Although the methylation system is not fully understood, the way in which it acts as a memory system is clear, and certainly plays a major role in cellular inheritance in many organisms. However, since some organisms such as yeast lack DNA methylation, and in others such as *Drosophila* it is barely detectable, methylation can be only one of several mechanisms controlling competence for transcriptional activity.

The nucleosomal fibre, which is twisted into a thick fibre that then forms looped domains, seems to be the characteristic organization of eukaryotic chromatin, but in higher vertebrates, and probably most other eukaryotes, there is a further level of chromatin organization associated with differences in gene expression. A variety of staining techniques have shown that in addition to chromosomes having regions of constitutive heterochromatin (C bands), the euchromatin and facultative heterochromatin regions are also organized into bands (Fig. 4.7).[21] These bands, which are clearly visible only in late prophase and metaphase, contain 500–3000 kb of DNA, and include clusters of looped domains. The names given to the bands (light and dark G bands, R bands, Q bands, replication bands) depend on the techniques used to reveal them, but the different types of bands are in fact related to each other. Bands have characteristic properties that seem to reflect fundamental differences in chromatin organization in different regions. Some of the features of dark G bands, light G bands and C bands are summarized in Table 4.2, which shows that they differ in their base composition, the types of highly repetitive DNA sequences they contain, their general sensitivity to nucleases, and their time of replication. The location of a gene or a domain within a chromosome—whether it is in a dark or light band—may affect its function, since the potential for transcriptional activity is not the same in all types of band. The functional significance of grouping genes and domains into bands is not clear. Banding patterns are stably inherited in cell lineages, and do not seem to differ between tissues, yet the correlations shown in Table 4.2 suggest that they must have some role in the overall control of gene activity. We look at this again in Chapter 7.

Table 4.2 Some of the properties of different types of chromosomal bands in vertebrates. (Based on information in Holmquist 1988, John 1988, Bickmore and Sumner 1989, Sumner 1990.)

	C bands[a]	Positive G bands (dark)[b]	Negative G bands (light)[b]
Location	Constitutive heterochromatin	Euchromatin	Euchromatin
Timing of replication	Usually replicate very late in S-phase	Usually replicate late in S-phase	Usually replicate early in S-phase
Condensation	Condensed from a very early stage in the cell cycle	Condense early in the cell cycle	Condense late in the cell cycle
Type of genetic information	Very few genes; most DNA is non-coding	Few genes; some claim that most genes in these bands are tissue-specific	Most genes; both housekeeping and tissue-specific genes
Type of repeated DNA sequence	Tandem repeats of long and/or short sequences (satellite DNA)	Rich in long interspersed repetitive sequences (LINES)	Rich in short interspersed repetitive sequences (SINES)
Sensitivity to endonucleases	DNase-I insensitive in *in situ* nick translation[c]	DNase-I insensitive in *in situ* nick translation[c]	DNase-I sensitive in *in situ* nick translation[c]
DNA sequence composition	Some satellite sequences C and G rich, some A and T rich	A and T rich	G and C rich
Meiotic chromatin structure	Non-chromomeric	Pachytene chromomeres	Interchromomeric regions

[a] C banding—Giemsa banding after treatment with hot alkali and then warm saline.
[b] G banding—Giemsa banding after incubation in warm salt solution or trypsin.
[c] *In situ* nick translation is a technique used to detect the pattern of general DNase-I sensitivity in chromosome preparations.

2. *The inheritance of chromatin structure in cell lineages*

Once a cell is determined (i.e. is committed to a particular fate) or differentiated (i.e. has assumed its ultimate phenotype), this epigenetic state can be transmitted to its daughters.[22] Somehow the epigenetic information in chromatin is passed to progeny cells. Exactly how this is brought about is far from clear. One intriguing idea is that once established, the timing of DNA replication in a chromatin region is self-perpetuating and governs many other aspects of chromatin behaviour (Riggs 1990, Riggs and Pfeifer 1992). The argument is this: during the early part of S-phase (the period of DNA synthesis), proteins that are necessary for the assembly of newly replicated DNA into an active chromatin conformation are produced. Since chromosome regions with an active conformation are preferentially replicated early in the next cell cycle, early DNA replication is self-perpetuating. Similarly, late replication is also self-perpetuating: DNA that is late replicating is inactive because the proteins needed for its assembly into an active chromatin conformation are no longer available late in S-phase. Since it has an inactive conformation, this DNA will be late replicating in the next cell cycle. Once again, the time of replication and the consequent state of activity are maintained. Figure 4.11 illustrates this system. It is, of course, a type of steady-state system, albeit one that is not acting at the level of individual genes.

Although their precise role in the control of gene activity remains to be elucidated, the way in which DNA methylation patterns can be stably inherited is clear. Indeed, at present it is the only biochemically well-characterized mechanism of epigenetic inheritance. The inheritance of methylation patterns in eukaryotes is based on the fact that methylation occurs in CpG doublets or CpNpG triplets.[23] Methylation is therefore

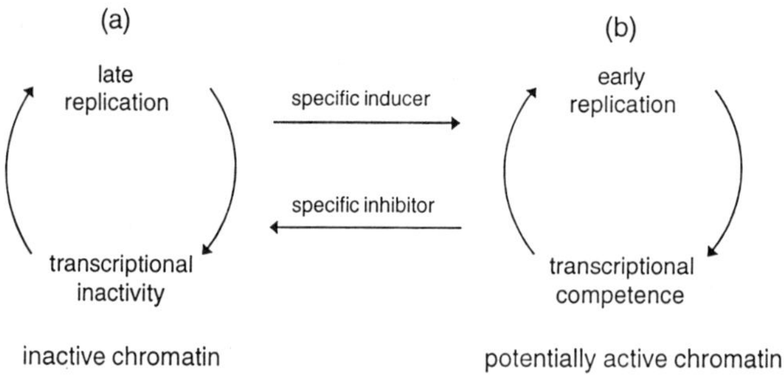

Fig. 4.11 Self-sustaining loops in which (a), the inactive state, is maintained by late replication, and late replication is maintained by the inactive state; (b), the competent state is maintained by early replication, and early replication maintains a competent state. Transitions between (a) and (b) are brought about by inducers or inhibitors that override the control of the steady-state. (Based on Riggs and Pfeifer 1992.)

symmetrical on the two DNA strands. Complementary base pairing means that a CpG (or CpNpG) on one strand is partnered by the same sequence, but in the opposite direction, on the other strand. After replication, the parental strand is methylated, but initially the new strand is unmethylated. An enzyme, methyltransferase, recognizes this asymmetrical state, and preferentially methylates the CpG of the new strand (Fig. 4.12a). The accuracy with which methylation patterns are transmitted has been estimated to be over 99% for some somatic cells in culture, although other studies suggest a lower fidelity.[24]

Mechanisms rather similar to those responsible for transmitting methylation patterns are thought to underlie the inheritance of DNA-protein interactions,[25] although far less is known about these systems. Many of the DNA sequences at the binding sites for specific transcription factors are, like methylation sites, symmetrical on the two DNA strands, suggesting that protein subunits may bind symmetrically (Latchman 1990). According to most models for the transmission of DNA-protein interactions, after DNA replication each parent strand retains its bound protein subunits. The semi-bound sites on the new DNA molecules are then preferential sites for the assembly of free protein subunits, or for the activity of a 'protein-transferase' which restores the original structure (Fig. 4.12b). As with the mechanism for the inheritance of methylation patterns, these

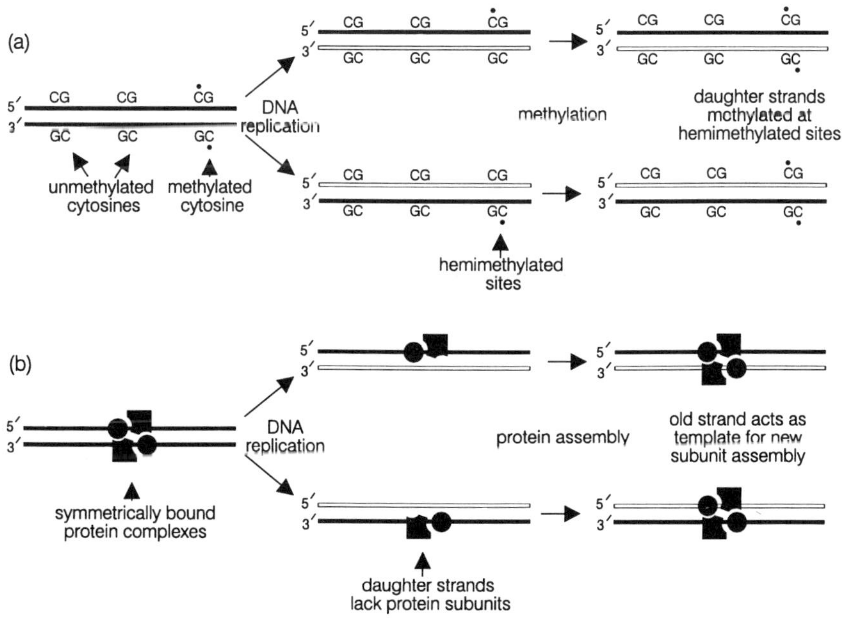

Fig. 4.12 Possible mechanisms for the inheritance of (a) methylation patterns; (b) protein–DNA interactions.

hypothetical mechanisms for the inheritance of DNA-protein complexes exploit the semi-conservative replication of DNA to perpetuate particular chromatin structures.[26]

The transmission of epigenetic information, like the transmission of genetic information, is not error free. Holliday (1987) has termed randomly produced modifications of methylation patterns and other heritable alterations in chromatin structure 'epimutations'. Epimutations are similar to classical mutations in that they are random with respect to the stimulus that induced them (if there was one), and with respect to the chromatin region which is modified. In the fluctuation test, epimutations should give the same type of distribution as is given by random mutations, with a variance much greater than the mean (Chapter 3, p. 59). The process of epimutation produces different epialleles of a gene. Epialleles can therefore be produced by stochastic processes, as well as by developmentally or environmentally induced changes. They can be either random or directed.

With chromatin-marking systems, the potential for inherited epigenetic changes is enormous. In contrast to the steady-state EIS, where only stably active or stably inactive states can be perpetuated, intermediate levels of transcriptional competence can be transmitted. Since the chromatin-marking maintenance mechanisms are independent of the functional state of the gene, and the gene product is not required to have a specific regulatory role, the system is potentially much more flexible. It is able to change both during development and during evolution.

Multicellular organisms, which have many different stably determined and differentiated cell types, clearly need the type of cellular memory provided by the chromatin-marking systems. Tissue-specific genes frequently have to remain permanently inactive. Through the chromatin-marking systems, large regions of chromosome can be stably inactivated. The classic example of regional control is the inactive X chromosome of female mammals, but there is increasing evidence that entire domains and bands act as heritable units of regulation.[27]

Genetic systems and EISs: a comparison

We have described three types of EIS, each of which seems to have distinct properties that are different from the DNA inheritance system. But in fact the different EISs overlap. For example, if the mechanism outlined in Fig. 4.11 is correct, chromatin marks that determine the time of replication are inherited by means of a steady-state system. Similarly, if the inheritance of chromosomal proteins is through a templating mechanism such as is shown in Fig. 4.12b, then it is a form of structural inheritance. There is not always a clear distinction between different types of epigenetic inheritance: the systems are interrelated.[28]

There is further problem in categorizing different inheritance systems. As we discussed at the beginning of this chapter, in the early days of molecular biology, Nanney realized that there were difficulties in distinguishing between genetic and epigenetic variations. What is the situation today? Can genetic variation and epigenetic variation be distinguished? In many ways, the vast expansion of knowledge during the thirty years since Nanney highlighted the problem has made distinguishing between the two types of variation even more difficult. It might seem that the difference is obvious: genetic variations involve changes in DNA sequence, epigenetic variations do not. However, this distinction does not accommodate very comfortably developmental changes such as those in the immune system. Differentiation in the immune system involves changes in DNA sequences, yet it is clearly an *epi*genetic process. Similarly, in *Drosophila* the chorion genes are amplified during egg production; although the number of copies of a DNA sequence is increased, this too is an epigenetic rather than a genetic change. As we showed in the last chapter, there are many epigenetic variations in DNA sequence.

If epigenetic changes involve DNA changes, does it mean that epigenetic variations cannot be distinguished from genetic changes? We think that the answer to this question is that although each has characteristic features, it is not possible to make a rigid distinction between the mechanisms underlying genetic and epigenetic changes. What we can distinguish between is different inheritance systems—the DNA inheritance system and the non-DNA system. Some epigenetic changes involve the DNA system, but most do not. In subsequent chapters we concentrate on epigenetic variations that are not direct consequences of DNA sequence changes, so in Table 4.3, where we have summarized the differences between genetic and epigenetic variations, the epigenetic variations considered are those that do not involve changes in DNA sequence.

One important difference between the epigenetic and genetic systems is that the range of stabilities of EISs is greater than that of the genetic system: the frequency with which epialleles arise, and the frequency of their reversion, are typically far higher than for classical mutations, but some epialleles have levels of stability comparable with those of mutations. For example, both *in vivo* and in cell culture, the inactive X chromosome in female mammals remains inactive through thousands of cell divisions: the stability of transmission is similar to that of DNA replication.

Further evidence for the stability of epialleles has come from studies of certain mammalian cell lines in which mutant phenotypes can be induced remarkably easily. The reason for the apparently high mutability was found to be that the cells were heterozygous for active and inactive alleles. Therefore, any new mutations in the active locus had a phenotypic effect, even if it was 'recessive'.[29] The most significant finding for the present argument was that although the inactive allele in these cell lines was very

Table 4.3 A comparison of heritable epigenetic and genetic variations

Property	Epigenetic variation	Genetic variation
Type of variation	Usually does not involve change in DNA sequence; involves change in chromatin structure, gene activity, or architectural organization of cellular structures	Involves change in DNA sequence
Origin of variation	(a) Random changes due to imperfections of the copying system or to non-directed effects of environmental factors (epimutations)	(a) Usually random
	(b) Directed variations due to a specific environmental agent inducing specific and predictable heritable changes	(b) Some DNA changes are believed to be directed
Unit of variation	The activity of the gene; in structural inheritance, the organization of gene products	DNA bases in DNA base sequences
Frequency of 'forward' variation	Very wide range: up to 100% per locus	More limited range: $<10^{-4}$ per locus[a]
Frequency of 'backward' variation	Very wide range: up to 100% per locus	More limited range: usually very low
Locus and tissue specificity	May be highly specific; the probability of a specific change could be 100% for the appropriate gene in the relevant cell type at the appropriate stage of development; coordinated changes in several loci probably common	The probability of a particular change varies, but is always extremely low; several genes may change, but only rarely in a co-ordinated way
Adaptiveness of the induced response	The modification may have non-random, although not necessarily adaptive, biological significance	Usually no connection between the molecular event and its potential biological significance
Transmission through the germ line[b]	Depends on the nature of reprogramming processes and on the efficiency of repair and cell selection	Depends on the efficiency of repair and cell selection processes
Unit of inheritance	The gene's phenotype, cell activity, or cellular architecture; the inheritance of acquired characters is probably common	The genotype; direct inheritance of acquired characters is probably rare

[a] An exception to this is the acquisition or loss of cytoplasmic organelles or plasmids, which can occur in all cells.
[b] See Chapter 6.

stable, its inactivity was not the result of a deletion or change in DNA sequence. It was caused by an epigenetic modification, which could be reversed by treatment with the demethylating agent 5-azacytidine: it was an epiallele.

In contrast to the high stability of such epialleles, the stability of some genes is now known to be much lower than was at one time assumed. Usually, the rate of spontaneous mutation is between 10^{-5} and 10^{-9} per locus per generation, and that of induced mutations about 10^{-4}. However, genetic changes such as those mediated by transposable elements may occur at a very high rate, and with some genetic backgrounds, or under some stressful conditions, the majority of cells may become mutant. Obviously, lack of persistence does not preclude a genetic basis for a variation. Although it is still true that, in general, epigenetic variants are less stable than genetic variants, this criterion alone cannot distinguish between the two inheritance systems in any specific case.

A more reliable criterion for distinguishing between the genetic and epigenetic inheritance systems is the greater predictability and specificity of epigenetic changes. Many epigenetic changes are directed. A directed epigenetic variation occurs whenever a stimulus causes a heritable change in a specific gene or genes, in a particular cell type, at a particular stage of development, and the stimulus has no consistent effect on other genes, or on the same gene in other cell types or at other developmental stages. In the case of cortical inheritance, a directed epigenetic variation occurs if the stimulus produces a predictable and heritable change in the architectural organization of cells. However, the criterion of predictability is also inadequate for any specific case. Many epigenetic variations are epimutations, i.e. variations that are random with respect to the inducing agent (if there is one), and with respect to the gene that is modified. For example, 5-azacytidine causes a genome-wide reduction in the level of DNA methylation: any gene with activity influenced by DNA methylation may be reactivated. The probability of reactivation varies from 0.1 to 30%, depending on the gene and the cell type involved. Yet in spite of the high frequency of reactivation, the response to 5-azacytidine is general, not specific (Jones 1985). Similarly, epimutations arising from accidental mechanical alterations in cortical structures are also non-specific. Not only are some epimutations non-specific, some genetic variations may be quite specific: as we discussed in Chapter 3, not all DNA sequence changes are random. Nevertheless, it is still true that directed epigenetic changes are much more common than directed genetic changes, and a directed change usually indicates epigenetic variation.

Another difference between the epigenetic and genetic inheritance systems is that with EISs an acquired change in cell phenotype may often be directly heritable, whereas with the genetic system, a change in phenotype can be transmitted only if it is transferred to the genotype by causing a concordant change in DNA sequence. This probably requires complex mechanisms such as those discussed in Chapter 3. Although such mechanisms seem to exist, it is likely that they are restricted to rather special environmental conditions, and to certain classes of DNA sequence.

One marked difference between the epigenetic and genetic systems is that epigenetic variations, unlike mutations, are often adaptive. After all, normal development depends on the ability to acquire and transmit epigenetic variations. However, not all acquired variations are adaptive. Commonly those produced by unusual or extreme environmental conditions are not. Phenocopies, the environmentally induced phenotypes that mimic genetic mutants, are sometimes very specific, particularly when they are produced by chemical agents (Lambert *et al.* 1989), but like the mutations they mimic, most phenocopies are detrimental. The epigenetic changes underlying phenocopies, like those underlying adaptive changes, can be transmitted in cell lineages. For example, exposure of early embryos of *Drosophila* to ether vapour induces a *bithorax* phenocopy which is seen in the adult as a fly with four wings; the determined state induced by the ether treatment is transmitted through somatic cell divisions occurring long after the ether has been withdrawn (Capdevila and Garcia-Bellido 1974). A similar long-term retention of a specific induced variation has been found with the globin genes: in culture, DH sites induced by sodium chloride are transmitted to daughter cells for over twenty generations after the removal of the inducing stimulus (Groudine and Weintraub 1982).

Another excellent example of stable inheritance of a cellular phenocopy is found in Meins's (1983, 1985) studies of induced cytokinin autotrophy in cultured tobacco cells. These cells can be either cytokinin requiring (C^-) or cytokinin habituated (C^+), depending on their tissue of origin. In culture, C^- and C^+ cells normally maintain their phenotypes for many cell generations, but when C^- cells are grown on medium with progressively decreasing concentrations of cytokinin, C^+ cells appear. The C^+ phenotype occurs at frequencies greater than 10^{-3}, so it is unlikely that it is the result of mutation. Once established, the C^+ phenotype is extremely stable. Since there is evidence suggesting that cytokinins trigger cytokinin production, Meins suggested that the variations induced in these tobacco leaf cells are maintained and inherited through a steady-state system with an autoregulatory mechanism based on positive feedback.

As we discussed at the beginning of this chapter, Meins argued that the best criterion for distinguishing between genetic and epigenetic variations is the persistence or non-persistence of a variation through meiosis. This distinction is useful, but only in the negative sense: a high frequency of reversion following meiosis does indeed point to an epigenetic variation, but persistence through meiosis does not mean that a variation is the result of a change in DNA sequence. For example, some cytotactic changes in ciliates persist after meiosis and conjugation. Similarly, Pillus and Rine (1989) found that genetically identical yeast cells can exist in phenotypically distinct states of gene activity that are not only mitotically stable, but are also retained following meiosis. The initiation of these alternative states is apparently due to random fluctuations in some intracellular constituents. The mechanism that sustains them through cell divisions is not known, but either a steady-state or chromatin-marking EIS could be responsible. Whatever the underlying mechanism, it is clear that in this case epigenetic variations are transmitted through meiosis. In Chapters 5 and 6 we look at some of the many other cases in which epigenetic information is transmitted through meiosis. Persistence or lack of persistence through meiosis is not a satisfactory criterion for distinguishing between epigenetic and genetic variations.

There is one other feature of epigenetic systems which distinguishes them from genetic systems: epigenetic variations, particularly directed epigenetic variations, can occur coordinately at several loci; several loci can acquire new epialleles at the same time. This is what happens during determination and development. The evolution of developmental systems through selection for the ability to respond to specific stimuli has resulted in cells and organisms having a repertoire of coordinated inducible activities that can be inherited. Of course, as Nanney pointed out long ago, the range of epigenetic variation is restricted by the DNA base sequence. However, this is not a very formidable restriction, especially in a complex system with many genes. Even if the DNA sequence of the genome remains constant, the number of combinations of different epialleles is enormous. A piece of poetry in which the poet does not invent new words is also restricted by the number of words in the language, but the number of possible combinations of words allowed by the laws of grammar still allows remarkable feats of creativity.

Summary

In this chapter we have discussed three types of EIS that operate between cell generations, either in unicellular organisms or in the somatic cell lineages of multicellular organisms. Information can be transmitted through inheritance systems based on self-maintaining metabolic patterns, on structural templating, or on chromatin structure, as well as through the genetic system based on DNA sequence. Although there are characteristic differences between genetic and epigenetic variations, there is no single criterion by which a hereditary variation can be identified as being genetic (i.e. inherited through DNA) or epigenetic (i.e. inherited through non-DNA systems). Epigenetic changes are part of normal development in multicellular organisms, and highly specific coordinated responses to internal and external stimuli are transmitted from one somatic cell generation to the next through the EISs. In the next two chapters we argue that EISs also operate between generations of multicellular organisms, and responses to external and internal stimuli may be transmitted from parents to offspring.

Notes

1. The abbreviation EIS for an epigenetic inheritance system was suggested by Maynard Smith (1990).
2. A useful discussion of these terms is given by Hall (1992*b*, Chapter 6).
3. See, for example, Hinshelwood (1953), Dean and Hinshelwood (1963). According to Stent and Calendar (1978), the advocacy of this Lamarckian type of adaptation in bacteria by such an influential British scientist as Hinshelwood was responsible for the delayed development of bacterial genetics in Great Britain. Rubin (1990), however, believes that Hinshelwood's ideas were 'a prophetic foreshadowing of views that are currently coalescing about the adaptive behavior of animal cells'.
4. See Serfling (1989) for a brief review.
5. Bussey and Fieldes (1974) proposed a model similar to that shown in Fig. 4.3b to explain the stable inheritance of environmentally induced changes in flax and other plants that were discussed in Chapter 3. Meins (1989*a*) also described a steady-state model to explain cytokinin habituation.
6. Blau (1992) gives more examples. Not all steady-state systems operate at the level of transcription: even when a gene is permanently active, post-transcriptional regulation involving a positive feedback loop can perpetuate alternative states of cell activity. For example, inherited post-transcriptional regulation occurs with *sex-lethal*, a gene that plays a key role in sex determination in *Drosophila*. The way the primary RNA product of this gene is spliced is sex-specific: initially the type of splicing is determined by transient events that depend on the sex-chromosome constitution, but once determined, the type of splicing is stably maintained through positive autoregulation by the gene's protein products (Bell *et al.* 1991).
7. For more details of this work, see Nanney (1985) and Frankel (1989, Chapter 4).
8. Details of these experiments can be found in Beisson and Sonneborn (1965), and Ng and Frankel (1977).
9. The recent observation that centrioles (and probably also basal bodies) contain DNA does not mean that structural inheritance is mediated directly by DNA, because it is very unlikely that surgical reorganization of the cortex is accompanied by a corresponding change in the DNA of basal bodies. For a fuller discussion of this work, see Frankel (1990).
10. See Alberts *et al.* (1989, pp. 651–652).
11. For evidence for the role of the extracellular matrix in the pattern of deposition of new matrix components, see Markson *et al.* (1991*a,b*). A fuller discussion of the role of this type of structural inheritance in morphogenesis can be found in Ettinger and Doljanski (1992).
12. Jorgensen (1993) has used epialleles in a more general sense, to refer to all heritable directed changes arising in development. He included directed changes in DNA sequence, such as those discussed in Chapter 3, as well as heritable marks that do not involve DNA sequence changes.

13. John (1988) gives a comprehensive discussion of heterochromatin.
14. A comprehensive review of position effects is given by Wilson *et al.* (1990).
15. Reviewed by Weintraub (1985), Gross and Garrard (1987), Goldman (1988), Van Holde (1989), Zlatanova (1990), Wolffe (1991), Fedor (1992), Hansen and Ausio (1992), and Kornberg and Lorch (1992).
16. Although a switch to transcriptional activity is often accompanied by a switch to early replication, not all early replicating genes are transcriptionally active. The relationship between early replication and activity is in fact quite complex. For example, in the trophectoderm of female mammals, it is the inactive X that is early replicating. Some other inactive genes also replicate early in the cell cycle (see Chapter 5).
17. For a concise review of some of the evidence that looped domains and replicons are related, see Kitsberg *et al.* (1991).
18. Eissenberg and Elgin (1991) and Kellum and Schedl (1991) have reviewed the evidence for the existence of boundary elements. Lowrey *et al.* (1992) have characterized a region in the human globin locus that seems to affect the activity of the whole domain.
19. For a review of the role of histones and nucleosomes in transcription, see Hayes and Wolffe (1992).
20. Reviewed in Riggs (1989), Adams (1990), Lewis and Bird (1991), Razin and Cedar (1991), Hergersberg (1991), and Bird (1992).
21. Reviewed by Holmquist (1988), Bickmore and Sumner (1989), Manuelidis (1990), and Sumner (1990). It has been suggested that the apparent absence of bands in some vertebrates may be a consequence of technical problems rather than a difference in chromosome organization. Organisms may show some types of banding but not others.
22. For general reviews see Conklin and Groudine (1984), Riggs (1989), and Holliday (1990*a*).
23. 'CpG' stands for a DNA site in which C (cytosine) is followed by G (guanine); p denotes the phosphate group, so the C is in the 5′ position relative to G. 'CpNpG' is a DNA site at which C and G are separated by N, which stands for any base.
24. For a fuller discussion of the stability of methylation, see Jablonka and Lamb (1989) and references therein.
25. See Alberts *et al.* (1977), Weintraub (1979), Brown (1984), Weintraub (1985), and Alberts *et al.* (1989, p. 576).
26. As indicated earlier, the methylation system also involves DNA binding proteins. Selker (1990*b*) has proposed a plausible model for the control and inheritance of states of chromatin activity which incorporates both DNA methylation patterns and DNA binding proteins.
27. Rivier and Rine (1992) describe many of the classic examples of large-scale repression in the genome. Selig *et al.* (1992) analysed regional control around the cystic fibrosis gene and showed that the gene is part of a much larger (500 kb) unit with a characteristic pattern of replication. Regional units containing several genes seem to be a general feature of mammalian chromosome organization (Holmquist 1992).
28. Blau has argued that most of the heritable control of gene expression occurring during differentiation is achieved though steady-state systems. She makes a distinction between actively maintained steady-state systems, and the passive type of control of gene expression seen in the methylated inactive X chromo-

some (Blau and Baltimore 1991, Blau 1992). However, this is a false dichotomy, because *all* control systems depend on the active synthesis of either structural proteins or the enzymes that determine the heritable state of activity of genes or chromosome domains. Even apparently permanent decisions such as X-inactivation cannot be regarded as being passively maintained.

29. Reviewed by Holliday (1987) and Harris (1989).

5

Genomic imprinting

> My son, hear the instruction of thy father, and forsake not the law of thy mother:
> For they shall be an ornament of grace unto thy head, and chains about thy neck.
>
> Proverbs 1. 8–9

The inheritance systems described in the last chapter were concerned with the faithful transmission of epigenetic information either in somatic cell lineages, or from one generation of unicellular organisms to the next. It is usually taken for granted that in multicellular organisms, unless reproduction occurs vegetatively, such systems do not operate between generations. Normally, epigenetic information is erased during gametogenesis. Consequently, the only ways in which information can be transmitted from parents to offspring are through DNA base sequences, or through the cytoplasm of the egg, or, in higher animals, through learning. Yet in spite of the widespread assumption that these are the only mechanisms of information transfer between generations, it has been known for many years that epigenetic information *is* sometimes transmitted from parents to their offspring. In this chapter and the next we shall review the evidence for this, and consider some of the ways in which it may occur. Obviously, if EISs operate between parents and their offspring, they must have relevance to the problem of the inheritance of acquired characters.

The first recognition that epigenetic information from one generation can affect gene activity in the next came from studies of what is now called *genomic imprinting*. Usually the parental origin of a gene or chromosome is of no importance: the way that genetic information is transmitted and expressed is the same, regardless of whether it was inherited from the mother or the father. But this is not always so. In an increasing number of cases it is being found that the genetic contributions of the male and female parents are not functionally equivalent: the expression and transmission of a gene, a chromosome region, a whole chromosome, or a whole set of chromosomes depends on the sex of the parent from which it was inherited. The processes that establish the differences between the paternally and maternally derived genetic material are known as imprinting.

The first person to use the term 'imprint' in connection with the parent-dependent behaviour of chromosomes was Crouse, in 1960. She was studying the peculiar chromosome behaviour of the dipteran fly *Sciara*. In this insect, X chromosomes inherited from the father are regularly eliminated from all somatic cells. In addition, during gametogenesis in the male, the entire paternally-derived set of chromosomes is eliminated. The male transmits to his offspring only the set of chromosomes he inherited from his mother. Crouse pointed out that the selective elimination of paternally-derived chromosomes must mean that they had acquired some imprints of their parental origin. In some way the chromosomes from the male parent must be marked or tagged differently from those of the female parent and, in the cells of the offspring, this difference is recognized and responded to.

Similar types of chromosome behaviour have been found in other insects.[1] In the somatic cells of male mealy-bugs and some other coccids, the whole set of chromosomes inherited from the father becomes heterochromatic in the early embryo and is eliminated completely during spermatogenesis. In the most specialized coccids, the armoured scale insects, instead of just becoming heterochromatic, the paternal set of chromosomes is eliminated from somatic cells, as well as from germ cells. It is clear that in these insects the haploid chromosome set inherited from the father must differ in some way from that inherited from the mother. The maternal and paternal chromosome sets are preconditioned to behave differently in the embryo. They must carry different imprints.

Sciara and the coccids are somewhat obscure and unusual insects, so it might be thought that imprinting is just another peculiarity of these groups, and has no general importance. But imprinting is more widespread than this, and does seem to have more general significance. Unfortunately, it is not easy to assess how common it is, because the effects of imprinting can easily be confused with maternal effects, or be dismissed as cases of incomplete penetrance or variability of gene expression. In very few organisms has there been any systematic search for evidence of imprinting.[2] In spite of this, imprinting has been found in many different groups of plants and animals. Examples of probable cases of genomic imprinting in non-mammalian organisms are summarized in Table 5.1. The table does not include the many cases of parent-dependent differences in gene or chromosome activity that have been found in interspecific crosses. These are discussed in Chapter 9.

Imprinting in mammals[3]

Although imprinting was first recognized in insects, and subsequently has been found sporadically throughout the plant and animal kingdoms, much of the information about imprinting has been provided by studies of mammals. It is seen most clearly in non-random X chromosome inactivation. In female mammals, one of the two X chromosomes in almost all somatic cells is inactive. In eutherians, in most cells inactivation is the

Table 5.1 Non-mammalian groups in which imprinting has been found

Type of organism	Species	Events showing the effects of imprinting	References
Unicellular alga	*Chlamydomonas reinhardii*	Loss of paternally derived chloroplast DNA	Sager and Kitchin (1975)
Fission yeast	*Schizosaccharomyces pombe*	Mating type conversion patterns involve imprinting of one strand of the DNA	Klar (1990)
Dipteran insect (fruit fly)	*Drosophila melanogaster*	Phenotype of inversion Uab^1 depends on whether inherited from a male or female	Kuhn and Packert (1988)
		Extent of position effect variegation	Spofford (1976)
		Time during development when viability effects are seen	Biémont (1991)
Dipteran insect (hessian fly)	*Mayetiola destructor*	Elimination of the paternally derived chromosomes in spermatogenesis	Stuart and Hatchett (1988)
Dipteran insect (midge)	*Sciara coprophila*	Elimination of paternally derived chromosomes	Crouse (1960)
Coccid (mealy-bug)	*Planococcus citri*	Heterochromatinization of paternal set of chromosomes	Brown and Nelson-Rees (1961)
Water fern	*Marsilea vestita*	Selective segregation of old and new strands of the paternal DNA in some regions	Tourte *et al.* (1980)
Rye	*Secale cereale*	B chromosome transmission (depends on the B chromosomes of the maternal parent)	Puertas *et al.* (1990)
Maize	*Zea mays*	Aleurone mottling due to R locus depends on transmitting parent	Kermicle (1970)
Axolotl	*Ambystoma mexicanum*	Paternal gene for DNA-ligase I activity non-functional in the early embryo	Signoret and David (1986)

result of a random process: in some cells the X chromosome inherited from the father is inactive, in others the inactive X is that inherited from the mother. An exception is found in the extraembryonic tissues of rodents. Here it is the paternally-derived X chromosomes that are preferentially inactivated (Fig. 5.1a). They are also heterochromatic and replicate asynchronously.[4] Clearly, in order for this preferential inactivation to occur, maternal and paternal X chromosomes must be imprinted in a way that in some embryonic tissues makes the chances of their inactivation unequal. Non-random X-inactivation is also found in marsupials. In this group, the paternal X chromosome is preferentially inactivated in almost all somatic cells (Fig. 5.1b).[5]

Imprinting in mammals is not limited to the sex chromosomes. Evidence from several different sources shows that a substantial part of the mammalian genome carries imprints of its parental origin. Moreover, normal development occurs only if the chromosomes carry the appropriate imprints from each parent. Perhaps the most striking and clear-cut evidence of this comes from nuclear transplantation experiments in the mouse.[6] In newly fertilized mouse eggs, the maternal and paternal pronuclei differ from each other in both position and appearance. Consequently, it is possible to remove a nucleus and replace it with one of the same or the opposite sex. Using this and similar techniques, it has been found that embryos having both haploid sets of chromosomes from parents of the same sex do not complete development. Androgenetic embryos, which have two paternally derived sets of chromosomes, have relatively normal extraembryonic membranes, but the embryo itself is poorly developed. In contrast, gynogenetic embryos, produced from two female pronuclei, have poor development of the extraembryonic tissues, but the embryo itself is more normal. It seems that parental chromosome contributions from both male and female are necessary, and they have complementary roles. Maternal chromosomes are more involved with the development of the embryo itself, whereas the paternal set is essential for the development of extraembryonic tissues. Of course, nuclear transplantation experiments like those described for the mouse have not been carried out in humans, but mistakes in human development sometimes produce tissues with two paternal or two maternal sets of chromosomes. The abnormalities seen in such tissues are consistent with the different roles of the paternal and maternal chromosomes found in the mouse.

Additional evidence for the existence of functional differences between maternally and paternally derived autosomes comes from mice that inherit both copies of a chromosome or a chromosome region from a single parent.[7] Frequently such mice do not complete development. When they do, the phenotype of the mice that inherit both autosomes from their mother may differ from that of those inheriting both from their father. For example, newborn mice having two copies of part of chromosome 2 from

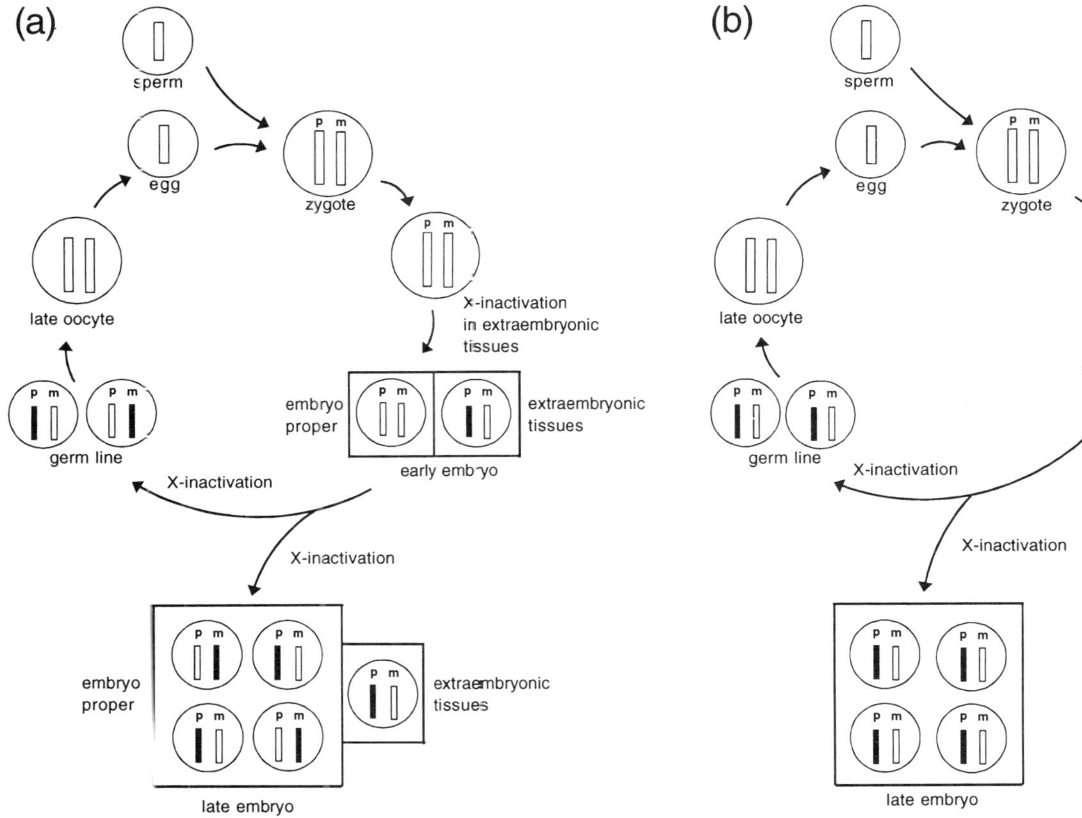

Fig. 5.1 Non-random X-inactivation (a) in the extraembryonic membranes of eutherian mammals; (b) in marsupials. The solid chromosome is the inactive X chromosome.

their mother are described as 'flat-sided, arch-backed, and hypokinetic' and usually die within 24 h; the reciprocal genetic type, with two copies of the chromosome from the male parent, shows the opposite phenotype, with short square bodies, broad, flat backs, and hyperkinetic behaviour. It has been estimated that about a third of the mouse chromosomes have regions that must be inherited from both parents in order for development to be normal (Searle and Beechey 1990).

In both mice and men, the effect of being heterozygous for a deletion of part of a chromosome sometimes depends on the parent from which the deleted chromosome was inherited. There is an intriguing situation in humans, whereby two very different genetic disorders, Prader-Willi syndrome and Angelman syndrome, both seem to be associated with a deletion of one particular region of chromosome 15.[8] Prader-Willi syndrome seems to be the result of the transmission of the deletion by the father, Angelman by the mother. Some cases of Prader-Willi syndrome are not associated with a deletion; in these cases the two copies of chromosome 15 are both inherited from the mother. Prader-Willi syndrome therefore seems to be caused by the absence of a paternally-derived region of chromosome 15. Similarly, Angelman syndrome is caused by lack of maternally-derived genes. Paternally- and maternally-derived chromosomes are not equivalent. The non-equivalence of maternal and paternal chromosomes is also clear from studies of some types of tumours. The chromosomes, or parts of chromosomes, that are lost in these tumours are always those from one particular parent (Reik and Surani 1989). The implication is that the maternal and paternal copies of the chromosomes responsible are not the same, and carry different imprints.

Imprinting can also show itself in more subtle ways. The severity of the effects of some dominant traits in the mouse depends on the sex of the parent transmitting the gene.[9] There are also some human conditions, such as the fragile-X syndrome, in which variability in expression is thought to be in part a consequence of parental imprinting. Hall (1990) has suggested that many human genetic traits that show variable expressivity and incomplete penetrance may be influenced by imprinting.[10] She has described the patterns of inheritance that should be seen in pedigrees for traits for which expression depends on the sex of the parent transmitting the gene. Examples of these patterns are shown in Fig. 5.2. Obviously, with such complex patterns of transmission, many traits showing the effects of imprinting are likely to have been missed.

The first direct evidence of an effect of parental imprinting on the expression of individual genes, rather than whole chromosomes or regions of chromosome, came from work on transgenic mice. Transgenic mice are mice that have had foreign genes inserted into their genomes. It has been found that whether or not a transgene is expressed sometimes depends on

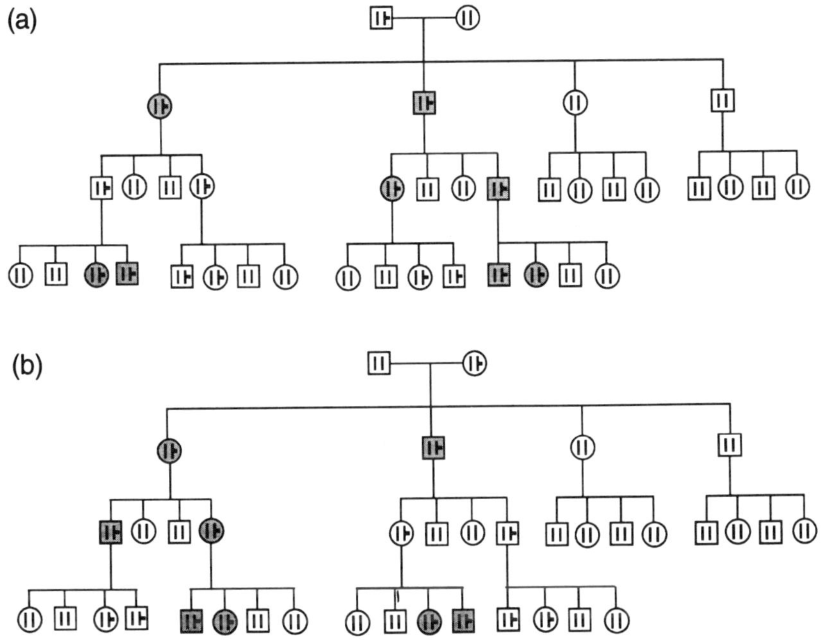

Fig. 5.2 Idealized pedigrees for an inherited trait the manifestation of which is determined by genomic imprinting. (a) The gene is expressed only when transmitted by the male parent. (b) The gene is expressed only when transmitted by the female parent. Shaded symbols represent those individuals manifesting the trait; ▮ is the chromosome carrying the abnormal allele. (Based on J.G. Hall 1990.)

the sex of the parent from which it is inherited. For example, Swain and her colleagues (1987) found that a transgene carrying the oncogene c-*myc* was expressed only when it was inherited from a male parent. When a daughter of this male passed the gene to her offspring, they did not express it, but if the daughter's sons inherited the gene, some of their offspring could express it.

The development of new molecular techniques has made it easier to study and characterize the imprinting of normal endogenous genes in the mouse.[11] So far, the number of genes studied is small, but some generalizations seem to be emerging. The genes for which information has been obtained are all involved in growth and development: *Igf2* (*insulin-like growth factor II*) and its receptor *Igf2r* (which is also the receptor for manose 6-phosphate) are expressed at high levels in the early embryo; *H19* is a gene which codes for an RNA product of uncertain function which is abundant in the developing embryo, and *Tme* (*T-associated maternal effect* mutation)[12] and *Fu* (*Fused*) are both associated with developmental defects. *Fu* has a visible phenotypic effect in the adult—a short and deformed tail. There seems to be no bias as to whether it is the maternally or paternally derived allele that is active in the developing embryo: with *Igf2r*, *H19* and *Tme* the maternal allele is active, whereas for *Igf2* it is the paternal allele, and with *Fu* there is a decrease in the gene's penetrance when it is transmitted through the female. The imprinted genes are clustered in two groups: *Igf2r*, *Fu* and *Tme* are found in a region of chromosome 17 known as the *t*-region, which is genetically extremely complex. *Igf2* and *H19* lie fairly close to each other on chromosome 7. In spite of their proximity, the activity of the two genes is not the same: the maternal *H19* allele is active, but it is the paternal *Igf2* allele that is expressed.

For the imprinted regions of chromosome 7 and 17, the time of replication of the paternally- and maternally-derived regions differ: the paternally-derived regions replicate earlier (Kitsberg *et al.* 1993). The same is true for the X chromosomes in eutherian trophectoderm: the paternal X replicates before the maternal. It also seems to be true for the imprinted region of human chromosome 15 that is involved in Prader-Willi and Angelman syndromes: the paternal chromosome is the first to replicate. In none of these cases is the maternal chromosome particularly *late* replicating: it simply does not replicate as early as the paternal chromosome. The sizes of the asynchronously replicated regions are quite large, at least 150 kb for chromosome 7, and 500 kb for chromosome 17. This means that the regions of asynchronous replication are at least as big as a looped domain, and may even approach the size of a replication band. Since the two imprinted genes in chromosome 7 behave differently, with one being active when derived from the mother and the other being active when derived from the father, it is obvious that there is no correlation between the time of replication and transcriptional activity. Moreover, the replication asynchrony of the imprinted regions is not always associated with a difference in the transcriptional activity of the two homologues. The insulin II and tyrosine hydroxylase genes are located within the differentially replicated region of chromosome 7, but both parental alleles are transcribed equally. Similarly, in chromosome 17, not all genes in the differentially replicated region show differential activity. By the criterion of differential gene expression, only a restricted region is imprinted; by the criterion of differential replication, the region is much larger.

These studies of imprinted normal genes suggest that at the level of replication-timing there is regional control of imprinting. Although regional control could be maintained by a steady-state inheritance system based on the time of replication (Fig. 4.11, p. 100), it is much more likely that replication asynchrony and the associated chromatin structure are maintained through a chromatin-marking EIS, based on DNA-protein and protein-protein interactions and on methylation.

The origin and maintenance of imprints

The diversity of groups in which genomic imprinting has been found suggests that the phenomenon may be quite common. Whether or not the mechanisms underlying it are the same in all groups is unknown. Imprinting must involve some kind of chromatin modifications that are capable of influencing gene expression, that persist during cell division, and that can be reversed. The differences between homologous, but differently expressed, chromosomes must be erased before or during the production of gametes, so that the two chromosomes transmitted by a parent are equivalent and carry the imprints characteristic of the individual's sex. For example, in the case of imprinting involving chromosome elimination such as is found in *Sciara* and the lecanoid coccids, the maternally derived chromosome set retained by the male must acquire a male imprint in the germ line before it is passed to his offspring. Similarly, in the germ line of females, both the set of chromosomes inherited from the father and the set inherited from the mother must become equivalent and acquire female-specific imprints.

In theory, imprinting could take place at any stage in the production of gametes—either before, during or after meiosis, or it might be a process involving several stages. We favour the idea that the removal of old imprints and the establishment of new ones is a gradual process. It probably begins in the early embryo with the removal of most parental imprints. It ends in the germ line, where the final erasure of old imprints and formation of new ones starts either before meiosis or in early meiotic prophase, and continues in the haploid stage. Old imprints probably have to be eliminated by the time that meiotic pairing begins because, in order for chromosomes to pair properly, the two homologues must have a similar chromatin conformation (see Chapter 9, p. 254). This is seen most readily with the mammalian X chromosomes, where the inactive X is reactivated before the onset of meiosis. If old imprints are not erased before pachytene, it is probable that the conformational differences between homologous chromosomes would impede normal pairing. This in turn could lead to sterility.[13]

The reason for suggesting that new imprints are established gradually during gametogenesis is that we believe that the most plausible origin of imprinting is in the differences in chromatin structure that are inevitable consequences of the different ways in which the chromosomes are packaged in the sperm and the egg. The restructuring of chromatin during gametogenesis involves a complex series of changes. Those that occur during spermatogenesis are better understood than those that occur during oogenesis, and are particularly dramatic. In many different groups, somatic histones are replaced by testis-specific histone variants (Grimes, 1986). In some groups, at a later stage in spermatogenesis, these testis-specific variants are themselves replaced by other proteins. For example, in many mammals they are replaced by basic 'transitional' proteins in the mid-spermatid stage, and these in turn are replaced by protamines in late spermatids. These changes are thought to facilitate the condensation and packaging of chromatin into the head of the sperm. It is clear, however, that at least in mammals, some restructuring of chromatin occurs long before the haploid stage. In the rat there are changes in the histones of spermatocytes even before meiosis begins (Meistrich and Brock 1987), and DNase-I sensitivity is different at the pachytene-pairing stage (Rao and Rao 1987), showing that chromatin structure is changed well before the gametes are formed.

There are no detailed studies of histone replacement and sensitivity to endonucleases during oogenesis comparable to those made for spermatogenesis, but very marked changes in chromatin structure can be seen in some groups.[14] In mammals, reactivation of the inactive X chromosome occurs before the onset of meiosis, and in birds and other groups in which the female is the heterogametic sex, the Y chromosome, which is heterochromatic in somatic cells, is reactivated. Even more dramatic changes are seen in those species of amphibia that produce lampbrush chromosomes in meiotic prophase. The diffuse appearance of lampbrush chromosomes is thought to be associated with the intense synthetic activity that takes place in oocytes.

As a result of the chromatin modifications that have occurred during gametogenesis, at fertilization the paternal and maternal chromosome sets have different conformations. In mammals the overall difference between the two chromosome sets is retained in the fertilized egg. As the sperm nucleus is transformed into the male pronucleus, the chromatin becomes more dispersed and basic nucleoproteins are replaced by proteins similar to those found in the egg pronucleus. Decondensation of the chromosomes continues in the male pronucleus (Longo 1987). Cytologically, the paternal chromosomes in the mammalian zygote are more elongated than the maternal complement, but this difference disappears after the first cleavage division.

Until quite recently, only gross changes in chromatin conformation could be followed through development, but techniques have now been developed that make it possible to follow changes in the methylation patterns of specific genes (Kafri *et al.* 1992). Figure 5.3 illustrates the types of change seen. It shows that although some genes remain unmethylated throughout development, for most of the genes studied, methylation patterns change. The patterns seen in the sperm and egg are sometimes quite different, but all differences are erased in the pre-blastocyst embryo, when every one of the sites studied becomes demethylated. Most sites are re-methylated during the post-blastocyst stage, but in the primordial germ cells sites remain unmethylated for far longer. Eventually, they become re-

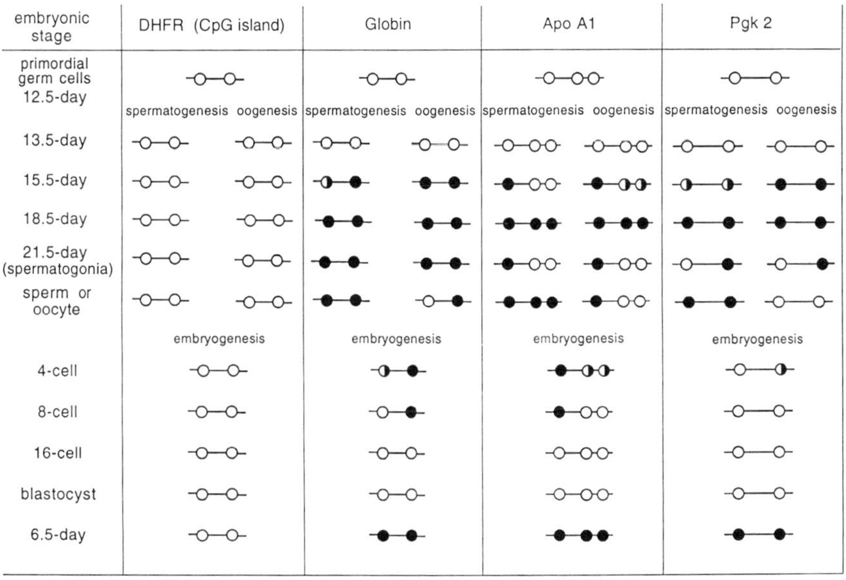

Fig. 5.3 Changes in the methylation of specific CpG sites in the DHFR, globin, Apo A1 and Pgk2 genes during development of the mouse. Open symbols: ‹10% methylated; half-filled: 20–50% methylated; solid: 50–100% methylated. (Based on Kafri *et al.* 1992.)

methylated, and as the germ cells mature, gradual, site-specific demethylation establishes the different patterns seen in sperm and egg. The lower level of methylation in the oocytes compared with that in sperm may reflect the intense gene activity that occurs in oocytes.

Since there are such dramatic changes in methylation and the appearance of chromatin early in embryogenesis, it is at first sight surprising that any imprints remain. There seems no reason why, following the replacement of the proteins and changes in methylation, the paternal and maternal chromosomes should not become equivalent in every sense. Probably, at the majority of sites they do, so most imprints are erased in early development. Undoubtedly, though, some differences remain. The most likely reason for this is that since the maternal and paternal chromosomes start from different states, their affinity for various DNA binding proteins is not the same. Any differences between the chromosomes, such as differences in methylation patterns, may make a particular site on one chromosome more likely to bind a regulatory protein than the equivalent site on the other chromosome. This difference in bound proteins may lead to further methylation differences, and so on. Thus, two alleles that start out with different epigenetic marks in the fertilized egg may change their marks during early development, but still not acquire identical marks.[15] Whether or not an initial difference is retained will depend not only on the nature of that difference, but also on the availability of binding proteins in the cytoplasm. In some cases, such as that of preferential inactivation of the paternal X chromosomes in marsupials, the initial epigenetic differences between the parental chromosomes are retained in most tissues throughout development. In other cases, differences seem to be retained in some tissues but not in others. For example, the maternal *Igf2* gene is silent in embryos, but both the maternal and paternal alleles of the gene are active in some tissues of the adult central nervous system (DeChiara *et al.* 1991).

Methylation differences have been found in all imprinted genes that have been studied at the molecular level.[16] Some of these differences are known to be established in the parental germ line. The idea that methylation is important in determining the differences in the activities of maternal and paternal alleles has been strengthened by studies of mice that are deficient in methylase activity (Li *et al.* 1993). In such mutant mice the differential expression of imprinted loci is abolished.

Differences between the maternally and paternally derived chromosomes in the embryo do not always result in easily recognizable or detectable differences in methylation patterns in the adult, even when other aspects of the genes' phenotypes show that they retain a memory of their parental origin. The paternal X chromosome of marsupials provides a good example. It is generally inactive, and there are obvious visible differences between the states of condensation of the two X chromosomes, and between the times at which they replicate their DNA. Yet no differences in the methylation of regulatory sequences have been found (Kaslow and Migeon 1987). This suggests that although differential methylation may be involved in much of the imprinting seen in mammals, methylation differences are not responsible for the maintenance of all imprints present in adults.

Even if differential DNA methylation is found to be important in most imprinting in mammals, it is unlikely that it is involved in all cases of imprinting. Male mealy-bugs have a higher level of methylation than females, so methylation may be associated with the imprinting that enables the entire paternal set of chromosomes to become heterochromatic. However, many insects seem to have very little methylated DNA at all (see Table 8.2, p. 210).[17] Methylation is therefore unlikely to play a major role in the imprinting mechanisms of many species in this group. Presumably, a different chromatin-marking system underlies the parental origin effects found in insects and other groups with little or no cytosine methylation.

The imprinting of transgenes

Historically, interest in imprinting was given a boost when studies of transgenes made it possible to investigate the molecular basis of parental origin effects. The methylation-phenotype and expression of a transgene were sometimes found to be different when inherited from a male than when inherited from a female.[18] For example, in the study of the c-*myc* transgene mentioned earlier, the transgene was undermethylated and expressed when it was inherited from the father, whereas when inherited from the mother it was more heavily methylated and not expressed (Swain *et al.* 1987). In general, however, there is no correlation between transgene expression and methylation: some transgenes are always expressed, even though their methylation state depends on the parent from which they were inherited; others show parental origin differences in methylation, but are never expressed.

Although transgenes have been invaluable in establishing the relationship between imprinting and chromatin modifications such as methylation, they have properties that make it necessary to be cautious in interpreting the results obtained by using them.[19] First of all, the animals studied are usually hemizygous for the transgene, and the lack of a pairing partner in meiosis could make chromatin restructuring atypical. Second, transgenes are frequently present in multiple copies; they integrate as a block of head to tail tandem repeats, and there can be several hundred copies at a single site in the mouse. Third, the expression of transgenes is known to be influenced by their site of insertion, and the sites of insertion may not be random. Fourth, a very high proportion of transgenes (10–30%) show imprinting,[20] even though the regions into which they integrate often show no parent-dependent methylation differences; transgene imprinting seems to be a more local event than the imprinting of endogenous genes. Finally, with the exception of a single transgene which is also unstable, all the imprinted transgenes that have been studied are more methylated when inherited from the mother than when inherited from the father. This is not what is seen with normal genes, and is contrary to what one might predict from the fact that, in general, there is less methylation in the female germ line than in the male (Monk and Grant 1990).

Is there any explanation for these unusual features of transgene behaviour and their methylation phenotypes? Sapienza (1990) has proposed a model of chromosomal imprinting which he believes provides an explanation. According to this model, imprinting is the result of the activity of sex-linked modifier genes that produce binding proteins affecting chromatin condensation. In the germ line of female mammals, both X chromosomes are active and produce a lot of binding protein. Chromosomes therefore become heterochromatinized. Males, with their single X chromosome, produce less protein from their sex-linked modifiers, so the conformation of their chromatin is more open. Methylation stabilizes the germ-line chromatin conformations, so the heterochromatinized transgenes of females become more highly methylated than those of males. The problems with this explanation are: first, so far none of the known modifiers of imprinting is sex-linked; and second, some normal genes that show imprinting are active when male-derived, others when female-derived. If the hypothesis Sapienza proposed is correct, normal genes, like transgenes, should also be more highly methylated and be more likely to be inactive when female-derived.

An alternative explanation for the high levels of methylation in transgenes transmitted by females is that it is the result of protective mechanisms directed at silencing or inactivating foreign DNA. Frequently, foreign DNA in eukaryotic cells becomes methylated and inactivated.[21] Exactly how foreign sequences are recognized as targets for methylation is not known, but one way may be by a process comparable to the 'ripping' that occurs in fungi. We described this briefly in Chapter 3: in some fungi, duplicated DNA sequences are excised through intrachromosomal recombination, or they are inactivated by methylation and mutation.[22] These processes happen in each of the two haploid nuclei present in the cytoplasm at the heterokaryon stage, and seem to be based on some kind of self-checking mechanism that requires pairing of the duplicated sequences. When pairing occurs, both of the paired sequences are methylated and sometimes mutated. There is an interesting parallel between this and the behaviour of transgenes in plants. Transgenes that are copies of the genes already present in the plant interact with their homologue in a way that results in both the natural gene and the transgene becoming suppressed by methylation. A type of ripping process seems to take place.[23] Kricker and colleagues suggested that a similar process may operate in mammals.[24]

Although similar, the processes through which repeats in mammals and plants are inactivated are not quite the same as those found in the fungus *Neurospora*. Inactivation seems to involve DNA modification rather than mutation. For this reason the term RIGS (*R*epeat *I*nduced *G*ene *S*ilencing) has been suggested for this type of inactivation.[25] This term will be used here for the long-term repression of foreign and parasitic DNA through methylation or proteins binding to the DNA. RIGS processes are probably defence mechanisms, protecting the genome by inactivating viral DNA sequences and transposons. They also affect transgenes because transgenes are normally present in multiple copies and their sequence composition is unfamiliar. The extent of transgene methylation is roughly proportional to the number of repeats,[26] as might be expected if multiple copies of DNA are silenced by a RIGS mechanism.

If RIGS does occur in mammals, it is unlikely that it takes place premeiotically or at the haploid stage as in fungi. In fungi the DNA sequences that are ripped are multiple copies of genes that are normally present as single copies, but many mammalian genes are normally present as multiple copies. Therefore they would be in danger of being ripped if the process occurred at the haploid stage. RIGS is much more likely to occur during meiosis, when foreign duplicated sequences above a certain length would be recognizable because of irregularities in meiotic chromosome pairing.

Although the existence of RIGS processes that inactivate foreign DNA may explain why transgenes are commonly methylated, it does not explain why they are sometimes differently methylated in males and females. The more highly methylated state of female-transmitted transgenes implies that RIGS is more effective in females. If it is, the explanation may lie in the greater gene activity that occurs in oogenesis compared with spermatogenesis. During the lengthy period of transcriptional activity in oogenesis, the opportunities for foreign viral DNA or mobile elements to integrate into the genome are far greater than in the short and relatively inactive period of spermatogenesis. Moreover, since in oogenesis inserted sequences would often impair gene activity, the consequences of DNA insertion are likely to be more serious than in the male. Selection to eliminate or inactivate newly integrated DNA must be far greater in females than in males, where rapid meiosis may be more selectively advantageous. If, as the data from fungi suggest, gene inactivation is a slow process, for males the possible advantages of RIGS may be outweighed by the disadvantage of the additional time needed to complete meiosis. This may be why effective RIGS in mammals is more efficient in females.[27]

According to the argument just outlined, methylation of female-derived transgenes is a consequence of RIGS processes that occur during meiosis. If, as in fungi, inactivation is progressive, the hypothesis explains why several generations are sometimes necessary to inactivate a block of transgenes. It also explains why a transgene's methylation phenotype is independent of that of neighbouring sequences: there is no reason why the extent of methylation of transgenes should be related to that of neighbouring genes.

Although the hypothesis is speculative, there is evidence from studies of the molecular basis of some human genetic disorders, such as the fragile-X syndrome, that repeated sequences are treated differently in the male and female germ line.[28] The fragile-X syndrome is a relatively common disorder involving mental retardation. It gets its name from a fragile site on the X chromosome: the X frequently breaks at this particular site and, in certain conditions in which the rest of the X is condensed, the fragile site appears cytologically as a gap. Apparently the gap is caused by a failure of the chromosome to condense. The pattern of inheritance of the fragile-X syndrome is unusual. In order for the mutant allele to be expressed either cytologically or phenotypically, it must first be passed through a female. Even then, not all of the progeny that inherit the fragile-X chromosome show the clinical phenotype. Manifestation of the disease is associated with an increase in the number of copies of a simple trinucleotide tandem repeat, CCG: normal people have 6–60 copies, an individual who transmits the disease has 60–200 copies, and an affected person has over 600.

The length of the repeated sequence in both affected and carrier individuals is meiotically and mitotically unstable, and the repeat tends to be progressively amplified as it is passed through the generations. However, the amplification of repeats in this case occurs only in the germ line of females. It seems, therefore, that repeats can be treated differently in male and female gametogenesis.

Whatever the explanation of the amplification of repeats in the fragile-X syndrome, the molecular studies of the disease have highlighted differences between events in the male and female germ lines. The same is true of studies of transgenes, which are often more highly methylated when female transmitted. Although both the fragile-X and possession of transgenes are abnormal genetic conditions, they, like normal genes, show that epigenetic changes in parents can be transmitted to their offspring.

Modifying the effects of imprinting

Chromatin marks established during gametogenesis in the parents must interact with DNA binding proteins in the offspring. Since the types and proportions of the various proteins involved in chromatin remodelling are unlikely to be the same in all strains, it is not surprising to find that both the establishment and manifestation of imprints are strain-dependent. In the parents, allelic differences may affect the formation of imprints, and in the offspring they will modulate imprinted gene expression in different ways. The phenotypic effect of allelic variations in modifiers may be seen as differences in the penetrance and expressivity of mutant alleles. In both humans and mice, some dominant mutations that show parental origin effects also show variable penetrance and expressivity.[29] A simple scheme illustrating how 'modifiers' involved in the production of DNA binding proteins could affect the phenotypic expression of imprinted alleles is shown in Fig. 5.4.

The way in which genetic background influences epigenetic modifications and gene expression has been investigated in the mouse using both natural genes and transgenes. Using classical genetic techniques, Ruvinsky and Agulnik (1990) studied the inheritance and manifestation of the *Fused* gene. *Fused* is dominant, but shows incomplete penetrance: penetrance is lower when *Fu* is inherited from the mother than when inherited from the father. Three different dominant modifiers affect *Fu* expression. One, the t^{12} haplotype, is on the same chromosome as *Fu*. It lowers the penetrance of the maternal allele, acting before fertilization, presumably during meiosis. The second, which is unlinked to *Fu*, also suppresses the penetrance of female-derived *Fu* alleles, but acts both before and after fertilization. The third modifier is also unlinked, but acts only after fertilization. In all three cases, the susceptibility of the *Fu* allele to the suppressing effect

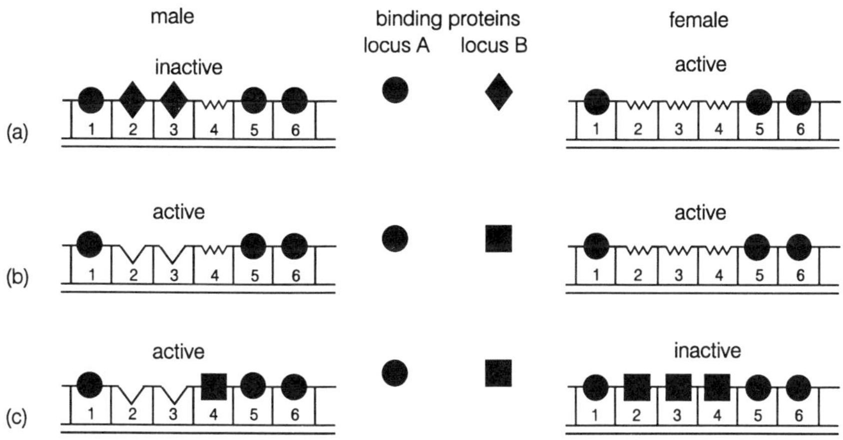

Fig. 5.4 The interaction of imprinted alleles and modifier genes. The left and right parts of the figure show the chromatin marks (imprints) transmitted by male and female gametes; the different shapes of the numbered blocks represent different imprints. Protein binding-factors produced by embryo genes *A* and *B* are shown as different solid shapes. Imprinted genes are active only if sites 2 and 3 remain unbound. (a) The paternal gene is inactive and the maternal active as a result of their different imprints. (b) Imprints are the same as in (a), but both maternal and paternal genes have the same activity because a different allele for one of the binding factors is present. (c) The binding factor alleles are the same as in (b), but the inherited marks are different, so effects of imprinting are again apparent, although in this case the paternal gene is active and maternal inactive.

of these modifiers depends on the parental origin of the *Fu* mutation: when female-derived, the allele is much more susceptible.

The mouse *Tme* (*T-associated maternal effect*) locus is on the same chromosome as *Fused* and also shows imprinting in most laboratory strains. The maternally derived allele is active, the paternal inactive. Consequently, when a female transmits a deletion of the locus to her offspring, they do not survive because they have no active copy of the gene; when the male transmits the deletion, the offspring are viable because they have the active maternal allele. Forejt and Gregorová (1992) found a related species of mouse (*Mus mus musculus*) in which both male- and female-derived alleles of *Tme* are active, suggesting that the gene is not differentially imprinted. By using strains from this species, they were able to identify a modifier gene responsible for the imprinting of *Tme* in the laboratory mouse strains. The modifier gene acts before fertilization. During spermatogenesis the allele carried by laboratory strains leads to *Tme* acquiring an imprint that causes its inactivation in the offspring. The allele present in *M. m. musculus* does not produce the imprinting that causes *Tme* to be inactive.

Strain-dependent variations have also been found in the imprinting of transgenes in mice. In one study, all offspring inheriting a quail *troponin* transgene from their mother had heavy methylation of the transgene, whereas those inheriting it from their father had variable patterns of methylation. In the latter case, the methylation phenotypes depended both on the methylation pattern of the transgene established in the father, and on the genetic background contributed by the non-transgenic mother (Sapienza *et al.* 1989). For one transgene, pHRD, the genetic basis of the strain effect has been identified. This transgene became highly methylated when bred onto a C57BL/6 background, but progressively less methylated on the DBA/2 background.[30] Genetic analysis identified a strain-specific modifier of methylation on chromosome 4 (Engler *et al.* 1991).

A very complex interplay between parental imprinting and genetic background has been found in studies of the mouse transgene known as TKZ751 (Allen *et al.* 1990). The level of methylation and expression of this particular transgene depends on its site of insertion, on the modifiers present, and on the parental origin of the modifiers. On the background provided by the BALB/c strain, the transgene becomes inactive and heavily methylated, providing the BALB/c modifiers are maternally transmitted. In contrast, on a DBA/2 background the transgene's expression is enhanced and methylation is reduced. The changes were found to be cumulative over successive generations, and resulted in complete methylation of the transgene after three generations of selection for low expression, and almost complete demethylation after three generations of selection for high expression (Fig. 5.5). Obviously, passage through the germ line did not erase the epigenetic information acquired previously, for had it done so such cumulative effects would not have been seen. Significantly, after the locus had become fully methylated and inactive, it remained that way, even in the presence of low methylation modifiers from the DBA/2 background. The fully methylated state had become irreversible. A similar permanent change in methylation phenotype occurred in an HBsAg (hepatitis B surface antigen) transgene: it became highly methylated and irreversibly repressed after passage through female gametogenesis (Hadchouel *et al.* 1987). The initially variably imprinted gene acquired a permanent gene phenotype; it became a new epiallelic form.

These last two cases of variability in the imprinting of transgenes are especially interesting and important. As far as we know, most imprints in normal genes last for only one generation; indeed from the point of view of the germ cell, imprints last only a single life cycle. The two cases of imprinted transgenes just described show that some epigenetic changes can be more permanent. Not only can information acquired by germ-line cells be transmitted to the next generation, but through selection in successive generations the information can become fixed and independent of sex. Although in both cases the genes showing this were transgenes and might

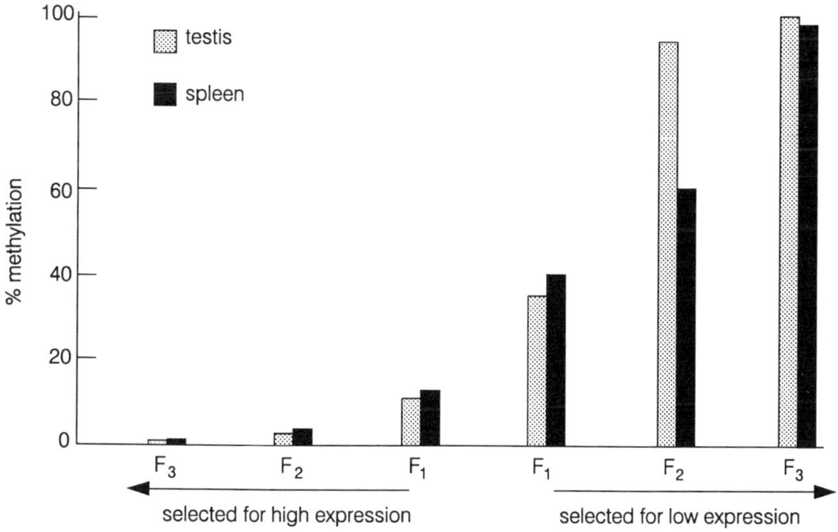

Fig. 5.5 Changes in the methylation of a transgene following selection for high and low expression. (Redrawn from Allen *et al.* 1990.)

therefore be regarded as atypical, as we show in the next chapter, permanent inherited changes in marks are not peculiar to transgenes. Normal genes show the same type of behaviour.

Summary

Evidence of parental imprinting has been found in many different groups of animals and plants, but exactly how parental imprints are established, how they are erased, and how they exert their influence is not yet known. However, studies of imprinting are beginning to produce some intriguing and exciting results. The different types of modifications that chromatin undergoes during male and female gametogenesis suggest that imprinting should be considered neither as an 'active' process, nor as a 'passive' process. Rather, it is the result of a difference between the male and female chromatin contributions to the zygote that is sometimes retained in embryogenesis and, when retained, sometimes has a phenotypic effect. It is clear that in mammals complementary imprints are necessary for normal development.

Work with imprinted transgenes suggests that their imprinting may be rather different from that of normal genes. There may be special defence mechanisms in mammalian germ-line cells that enable them to detect and inactivate foreign DNA. If so, the differences between male-transmitted and female-transmitted transgenes may be the result of a difference between the sexes in their ability to carry out these processes.

Does the work on imprinting have anything to tell us about the inheritance of acquired characters? It certainly shows that epigenetic information acquired in the germ line of one generation can be transmitted to the next. However, it also shows that this information from the previous generation usually is not permanent: it lasts for only a single generation, being initiated in the germ line of one generation and finally erased in the germ line of the next. The erasure of imprints suggests that there has been strong selection against carrying the remnants of an individual's epigenetic history into the next generation. If so, can epigenetically acquired information play any role in adaptive evolution? If all epigenetic information is transient, it seems unlikely that it can. But some of the results obtained in studies of imprinting suggest that not all epigenetic changes are ephemeral. The studies of methylated transgenes have shown that some changes can persist through several generations. If the same is true for normal genes, changes such as these may be important in evolution. In the next chapter we show that persistent epigenetic variations of this type are not unique to transgenes. Both random and induced epigenetic variations in normal genes can be transmitted through many generations.

Notes

1. Reviewed by Brown and Chandra (1977).
2. One organism in which parental imprinting has been sought and not found is the nematode *Caenorhabditis elegans* (Haack and Hodgkin 1991), although even here, as the authors acknowledge, a small amount of imprinting could occur without detection. Quite possibly imprinting has been looked for without success in other species, but, as so often happens when negative results are obtained, the studies have not been reported.
3. Imprinting in mammals is receiving a lot of attention because it may be important in understanding a number of human diseases, including some forms of cancer. General reviews of the subject and more details about most of the work considered in this section can be found in Monk (1988), Solter (1988), J.G. Hall (1990), and Reik (1992).
4. The first tissue of eutherian mammals to show X-inactivation is the trophectoderm where initially the inactivated paternal X is very early replicating, but later becomes late replicating (Takagi and Sasaki 1975).
5. See Sharman (1971) and Cooper *et al.* (1971, 1977).
6. See McGrath and Solter (1984), Surani *et al.* (1984).
7. Reviewed by Cattanach (1986).
8. J.G. Hall (1990) gives a comprehensive review of the effects of genomic imprinting in various genetic diseases, including Angelman and Prader-Willi syndromes.
9. Discussed in Bander *et al.* (1989), Ruvinsky and Agulnik (1990).

10. The expression of an allele often depends on the environment and on the effects of non-allelic genes. These non-allelic genes that can modify to a greater or lesser extent the expression or penetrance (likelihood of having detectable phenotypic effects) of the allele under consideration are called 'modifiers'. Until very recently, terms such as 'modifier', 'incomplete penetrance' and 'variable expressivity' have often been used as a fig leaf to mask our ignorance of the mechanisms underlying differences in phenotypes.
11. Details of the genes described can be found in DeChiara *et al.* (1991) for *Igf2*, Barlow *et al.* (1991) for *Igf2r*, Bartolomei *et al.* (1991) for *H19*, Forejt and Gregorová (1992) for *Tme*, and Ruvinsky and Agulnik (1990) for *Fu*.
12. It is not certain that the *Tme* gene is distinct from the *Igf2r* gene.
13. The evidence that homologous chromosomes must have a similar chromatin conformation for normal pairing, and the functional and evolutionary problems associated with inadequate pairing, are discussed by Jablonka and Lamb (1988, 1990*a*) and by Hulten and Hall (1990).
14. The changes in the conformation of sex chromosomes that occur during development are described more fully in Jablonka and Lamb (1990*a*).
15. For examples of the way in which endogenous imprinted genes change their pattern of methylation marks during early development, see Brandeis *et al.* (1993) and Stöger *et al.* (1993).
16. See Brandeis *et al.* (1993), Dittrich *et al.* (1993), and Stöger *et al.* (1993).
17. Scarbrough *et al.* (1984) found evidence of methylation in mealy-bugs, but in *Drosophila* the level of methylation is barely detectable (Urieli-Shoval *et al.* 1982; Achwal *et al.* 1984; Patel and Gopinathan 1987), and Gerbi (1986) failed to detect methylation in *Sciara*.
18. See Sapienza *et al.* (1987), Reik *et al.* (1987), Hadchouel *et al.* (1987), and Swain *et al.* (1987).
19. Wilson *et al.* (1990) give a concise account of some of the general properties of transgenes. Reik *et al.* (1990) summarized data on the properties of imprinted transgenes.
20. Reik *et al.* (1990) suggested that 10–20% of transgenes are normally imprinted, whereas Sasaki *et al.* (1991) suggested that as many as 30% show methylation imprinting.
21. Doerfler (1991) reviews literature on the methylation of foreign DNA.
22. Reviewed by Selker (1990*a*).
23. Reviewed in Jorgensen (1990), Dooner *et al.* (1991), and Matzke and Matzke (1993).
24. Kricker *et al.* (1992) compared the frequency of CpG dinucleotides in unique and repeated DNA sequences in mammals. When cytosine is methylated, it is much more likely to be deaminated to thymine than when unmethylated. Consequently, DNA sequences that have a history of being highly methylated are deficient in CpG dinucleotides, because many cytosines have been replaced by thymines. Kricker *et al.* found that CpG depletion is much more substantial in various types of repeated mammalian sequences than in unique sequences, provided that the repeated sequences are identical for lengths greater than about 130 base pairs. They suggested that the duplicated sequences were targets for methylation because they paired, and this methylation then led to higher rates of sequence divergence through deamination of cytosines to thymines.
25. Signer, reported by Martin and Jones (1992).

26. Mehtali (1990). See also Linn *et al.* (1990) and Renckens *et al.* (1992) who have found that inactivation by methylation of transgenes in *Petunia* plants is more likely when the gene is present in more than one copy.
27. Barlow (1993) also suggests that imprinting in mammals may be associated with a female-specific defence system operating in oocytes.
28. For reviews of the fragile-X syndrome, see Laird (1989), Richards and Sutherland (1992*a,b*).
29. See Surani *et al.* (1990), Allen *et al.* (1990), and Reik (1992).
30. C57BL/6, DBA, and BALB/c are different commonly used strains of laboratory mice.

6

The inheritance of directed epigenetic variations

> The fathers have eaten a sour grape, and the children's teeth are set on edge.
>
> Jeremiah 31.29

With classical imprinting, parents transmit to their offspring chromatin marks carrying the stamp of the parent's sex. The mark is usually transient, and is reversed in the next generation if it passes through the germ line of the opposite sex. Imprinting is therefore a rather special case of the transmission of epigenetic information between generations. A few cases in which a mark is transmitted in a sex-independent fashion and in a more permanent manner have been found with imprinted transgenes, but transgenes are in many ways atypical and differ from normal genes. To show that epigenetic inheritance can occur between generations and can have direct evolutionary significance, similar sex-independent and more permanent transmission must be found with normal genes.

We have seen in Chapter 4 that for both cortical inheritance in ciliates and epigenetic inheritance in yeast, the processes of meiosis do not erase all of the epigenetic information established and transmitted during previous asexual generations. It could be argued, however, that in these cases epigenetic information is retained simply because the organisms are unicellular and lack specialized gametes. Do the processes of differentiation into specialized sperm and eggs in multicellular organisms erase all prior epigenetic information? It seems that usually they do. The dramatic chromatin restructuring that goes on during gametogenesis resets the genome and ensures that most previous epigenetic information is erased, and at the same time establishes sex-specific marks. In spite of this, however, some epigenetic information persists. Comparisons of methylation levels in germ cells and embryonic cells show that the methylation status of some DNA sequences is essentially the same in both.[1] Even when radical changes in methylation take place in gametogenesis, some traces of the past—some aspect of the former phenotype of the gene—can be retained. For example, the chromatin structure of housekeeping genes in chicken sperm differs from that of the same genes in somatic tissues and spermatogonia: the overall methylation level in sperm is higher. However, this is not so for all DNA sequences of these genes: previously DNase-I hypersensitive regions are preferentially *under*methylated. These undermethylated regions seem to be 'footprints' of the past DNase-I hypersensitive state; they may also be 'blueprints', which are interpreted by the early embryo and enable it rapidly to reconstitute the DNAse-I sensitive conformation. Therefore, although the structure of chromatin is altered in the sperm, for these housekeeping genes, a memory of the past conformation is retained, encoded as special marks.

Is there any reason for thinking that changes in the epigenetic state of one generation can influence subsequent generations? In this chapter we review the evidence that epigenetic variations (other than those involved in imprinting) are transmitted between generations of eukaryotic organisms, that sometimes they are transmitted through the germ line, and occasionally they are transmitted for many generations. This evidence is summarized in Table 6.1. We also discuss the kind of systems in which epigenetic inheritance is likely to be important, and the types of genes and environmental conditions in which directed heritable epigenetic variations are

Table 6.1 Examples of what may be transgenerational epigenetic inheritance in eukaryotes

Organism	Type of heritable variation	Type of EIS	References
Paramecium (protozoan)	Induced responses to temperature, salt, and arsenic (Dauermodifikationen)	Not known	Jollos (1921)
Paramecium, Tetrahymena, Stylonychia, Paraurostyla, Euplotes (protozoa)	Accidental or induced variations in cortex structure	Structural	Sonneborn (1964), Nanney (1968, 1985), Nelsen *et al.* (1989), Ng (1990)
Difflugia corona (protozoan)	'Teeth' structure[a]	Structural	Jennings (1937)
Saccharomyces (yeast)	Utilization of melibiose[a]	Not known, but probably steady-state	Spiegelman *et al.* (1945)
	Transcriptional state of *HML*-α locus	Not known, but protein marks suggested	Pillus and Rine (1989)
Aspergillus nidulans (fungus)	Fluffy phenotype induced by 5-azacytidine	Chromatin marks —methylation pattern	Tamame *et al.* (1988), Tamame and Santos (1989)
Coprinus cinereus (fungus)	Methylation pattern at a centromere-linked locus	Chromatin marks —methylation pattern	Zolan and Pukkila (1986)

Table 6.1 (*continued*)

Organism	Type of heritable variation	Type of EIS	References
Pisum sativum (pea)	Induced response to temperature	Not known	Highkin (1958*a,b*)
Arabidopsis thaliana (thale cress)	Expression of a transgene	Not known, possibly chromatin marks	Scheid *et al.* (1991)
Petunia hybrida (petunia)	Changes in methylation and expression of a transgene	Chromatin marks —methylation pattern	Meyer *et al.* (1992)
Phaseolus vulgaris (bean)	Induced response to temperature	Not known	Moss and Mullett (1982)
Oryza sativa (rice)	Induced dwarfism and reduced level of methylation	Chromatin marks —methylation pattern	Sano *et al.* (1989, 1990)
Zea mays (maize)	Transposition of *Spm*, *Ac* and *Mu* transposable elements	Chromatin marks —methylation pattern	Fedoroff *et al.* (1989), Dennis and Brettell (1990), Martienssen *et al.* (1990)
	Paramutation at the R locus[b]	Chromatin marks —methylation pattern	Brink (1973), Dooner *et al.* (1991)
	Paramutation at the B locus	Chromatin marks —proteins	Patterson *et al.* (1993)
Nicotiana tabacum (tobacco)	Requirement of leaf cells for cytokinin	Not known, but steady-state suggested	Meins (1985, 1989*a,b*)
	Expression of genes in T-DNA	Chromatin marks —methylation pattern	Matzke and Matzke (1990)
Lolium perenne (perrenial ryegrass)	Colchicine-induced variation	Not known	Francis and Jones (1989)
Triticum (wheat)	Cytosine methylation of glutenin genes	Chromatin marks —methylation pattern	Flavell and O'Dell (1990)
Triticale (cereal crop)	Induced growth-phenotype and reduced methylation level	Chromatin marks —methylation pattern	Heslop-Harrison (1990)
Many species of angiosperm	Developmental phase[a]	Not known	Brink (1962)

Table 6.1 (*continued*)

Organism	Type of heritable variation	Type of EIS	References
Stenostomun incaudatum (flatworm)	Induced resistance to lead acetate[a]	Not known	Sonneborn (1930)
Caenorhabditis elegans (nematode)	Induced resistance to high temperature	Not known	Brun (1965)
Asplanchna sieboldi (rotifer)	Size	Not known	Badino and Robotti (1975)
Daphnia pulex (water flea)	Electrophoretic mobility of G6PD	Not known	Ruvinsky *et al.* (1983*a*,*b*)
Pieris brassicae (butterfly)	Induced resistance to LSD and changes in diapause	Not known	Vuillaume and Berkaloff (1974)
Myzus persicae (aphid)	Insecticide resistance	Chromatin marks —methylation pattern	Field *et al.* (1989)
Drosophila melanogaster (fruit fly)	Penetrance of an induced bithorax phenocopy	Not known	Ho *et al.* (1983)
	Modifying ability of Y chromosome	Chromatin marks —proteins	Dorn *et al.* (1993)
Carausius morosus (stick-insect)	Induced change in food preference	Not known	Sladden and Hewer (1938)
Philodina citrina, *Euchlanis triquetra* (rotifers), *Drosophila melanogaster* (fruit fly), *Poecilia reticulata* (fish)	Lansing effects—various characters show cumulative progressive changes with parental age[b]	Not known	Lansing (1954), Lints (1978), Beardmore and Shami (1985), Jablonka and Lamb (1990*b*)
Mus musculus (mouse)	Expression of *Fused* gene[b]	Not known	Belyaev *et al.* (1983), Ruvinsky (1988)
	Expression and methylation of transgenes	Chromatin marks —methylation pattern	Hadchouel *et al.* (1987), Allen *et al.* (1990), Engler *et al.* (1991)
	Methylation of endogenous sequences	Chromatin marks —methylation pattern	Sasaki *et al.* (1991)

Table 6.1 (*continued*)

Organism	Type of heritable variation	Type of EIS	References
	Haemoglobin level	Not known	Kahn (1982)
Rats, mice, guinea pigs, rabbits	Drug and hormone induced changes in endocrine function	Not known	Campbell and Perkins (1988)
Homo sapiens (man)	Methylation of endogenous sequences	Chromatin marks —methylation pattern	Silva and White (1988)

[a] The variation is propagated only in asexual reproduction.
[b] Similar observations have been made in other organisms: see Lints (1978) for a review of Lansing effects, Brink (1973) for paramutation, Ruvinsky (1988) for an account of characters in foxes which behave similarly to *Fused* in the mouse.

likely to occur. We end by presenting a model of epigenetic inheritance which shows how some types of acquired characters could be transmitted to the next generation. The EIS on which most of our discussion is based is the chromatin-marking EIS, since this seems to be the predominant system for transmitting epigenetic information in the cell lineages of multicellular organisms.[2] If transmitted between sexual generations, permanent chromatin marks (epialleles) should behave like chromosomal variations in DNA base sequence, and segregate in a Mendelian fashion at meiosis, although if marks are less stable, Mendelian proportions may not be found.

Transmission of epigenetic variations to the next generation

The old genetics literature has many descriptions of peculiar patterns of inheritance that are not readily explained in classical Mendelian terms. Some of these patterns are of exactly the type expected for inherited epigenetic variations.[3] The hereditary variations are not transient, like some somatic modifications of gene expression, nor are they reversed in a predictable way like imprints. However they are not as stable as classical mutations caused by changes in DNA base sequence. This type of variation was studied by Jollos in the 1920s. He called them 'Dauermodifikationen'—'enduring' or 'lingering' modifications.[4] In his studies of *Paramecium*, Jollos found that exposure to high temperatures, high salt concentrations, or arsenic induced a specific change in resistance to these stimuli. The modifications in resistance were inherited: they persisted through many generations of asexual reproduction, long after the inducing stimulus was removed. Gradually they faded away, until eventually, after hundreds of generations, they disappeared. Typically, such modifications were lost quickly after sexual reproduction, although this was not always so.

Jollos's experiments with *Paramecium* were repeated and the results confirmed by later workers,[5] but the mechanisms underlying them remain unknown. Jollos also studied the inheritance of heat-induced changes in *Drosophila*. With this organism his results were less clear-cut and more controversial, but as discussed in Chapter 3, they certainly did not fit comfortably with the idea that all induced changes are the result of random mutation.

Examples of lingering modifications have subsequently been found in many different organisms. They seem to be particularly common in plants. One case (illustrated in Fig. 6.1) was described by Highkin (1958*a,b*) who grew plants from two pure lines of peas in a constant temperature regime instead of the normal fluctuating one. The constant regime was detrimental, causing a reduction in vigour. The effects on vigour were cumulative, eventually levelling off after five generations of growth at the constant temperature. The interesting result was that, when returned to the normal fluctuating environment, the plants still produced progeny that were inferior to those of the controls. This carry-over effect could have a conventional explanation: it could have been caused by the production of inferior

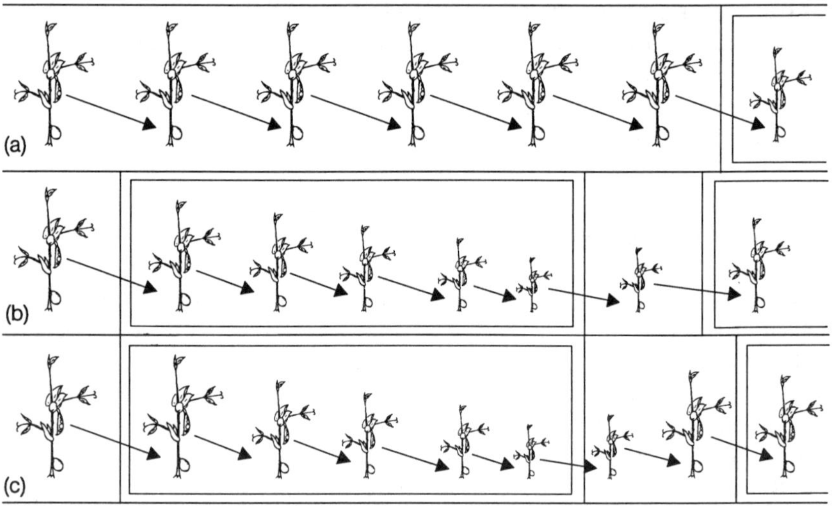

Fig. 6.1 The persistent effect on pea growth of ancestral exposure to a new environment. (a) Control plants transferred to the detrimental conditions for one generation. (b) Plants grown in the detrimental conditions for five generations, followed by one generation in the normal environment before returning to detrimental conditions. (c) As in (b) but with two generations of recovery in the normal environment. Boxed areas show periods in detrimental conditions. (Based on Highkin 1958*a,b*.)

seeds in the detrimental conditions. This explanation seems to be supported by the finding that when seeds from ancestors grown in detrimental conditions were grown for *two* generations in normal conditions, their offspring hardly differed from control plants. However, the plants were not in fact the same: when their seeds were returned to the detrimental constant environment, they were much less vigorous than seeds of plants with ancestors that had never experienced the detrimental conditions. This difference persisted for at least three generations. As Highkin pointed out, his results showed a marked resemblance to the lingering modifications described by Jollos.

A similar type of phenomenon—an effect of the temperature treatment in previous generations on the present sexually-produced generation—has been found by Moss and Mullet (1982) in the bean *Phaseolus vulgaris*. After beans were grown in a 27 °C day/22 °C night regime for four generations, they were more vigorous, and this effect was seen whatever the temperature regime in the final generations.

Francis and Jones (1989) found an extremely persistent heritable change after treating one-week old seedlings of inbred *Lolium perenne* with a low concentration of colchicine. The treatment induced non-random changes in quantitative characters such as tiller number, fresh weight, and dry weight. These changes persisted for seven years of vegetative propagation, and were also transmitted sexually when the plants were self-fertilized. The nature of the colchicine-induced changes is at present unknown.

A very different, and very good, example of inherited induced variations comes from the work of Meins and his colleagues on tobacco plants regenerated from cells grown in culture.[6] In Chapter 4, we described how tobacco cells can exist in different epigenetic states with regard to their need for an exogenous source of the hormone cytokinin. Cultures of leaf cells need cytokinin for rapid growth—they are stably C^-; stem-cortex tissues do not—they are stably cytokinin-independent, or C^+. A third tissue, pith parenchyma, initially requires cytokinin, but in cultures containing cytokinin, some cells lose their requirement and become habituated; they change from C^- to C^+. The rate of change in these inducible cells is greater than 10^{-3} per cell generation—100–1000 times faster than the normal somatic mutation rate. Other pith-parenchyma cells are non-inducible and remain C^- under normal conditions. As we noted in Chapter 4, these epigenetic differences are stable in cell culture, but usually they are not permanent: when the cells are used to regenerate whole plants, the tissues of the regenerated plants have the same cytokinin phenotypes as those of plants grown from seeds (Fig. 6.2). The epigenetic state of the cells must therefore be reset during regeneration. However, there are exceptions to this. Leaf cells that are normally non-inducible and stably C^- can be habituated by subculturing on media containing successively lower concentrations of cytokinin. Under these conditions, C^+ variants arise with

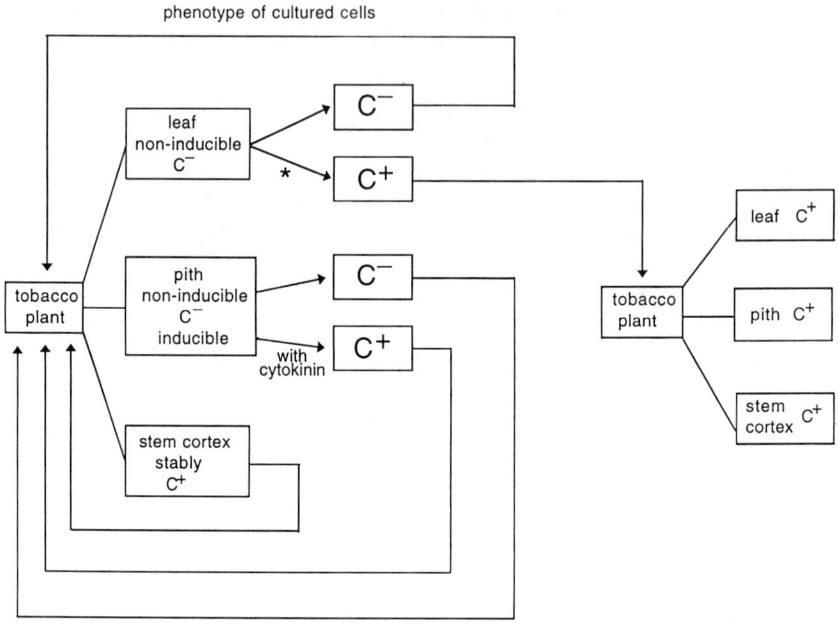

Fig. 6.2 Inheritance of cytokinin phenotype in tobacco plants. ★: cultures treated by lowering the cytokinin concentration of the medium. See text for explanation. (Based on Meins 1989*a* and personal communication.)

a frequency of between 1 in 100 and 1 in 1000 per generation. What is interesting is that in this case the leaves of plants regenerated from the variants do not revert to the normal, C^-, cytokinin-requiring state—they retain the C^+ state that was induced in culture. The acquired C^+ state persists when the plants are propagated asexually, and when selfed, C^+ and C^- phenotypes segregate, indicating that the regenerated plants are 'heterozygous'. In crosses between C^- and C^+ plants, the C^+ phenotype segregates in the F_2, behaving like a single Mendelian dominant factor. The molecular basis of this induced heritable change is not known, but the data suggest that a locus has been modified epigenetically, and the induced epiallele is transmitted through the germ line. The frequency with which the change arises (10^{-2} to 10^{-3}), and its directed nature, makes it very unlikely that it is a conventional mutational change.

The idea that genes can be transmitted through the germ line in different states of potential activity has been used to explain several puzzling patterns of inheritance in both mammals and plants. Many years ago Zuckerkandl and Pauling suggested that some genes may be 'dormant'.[7] A dormant gene is one that is inactive because there is no tissue in which intracellular conditions are suitable for its expression. Such a gene can be dormant for many generations, only being activated when environmental conditions are altered in ways that result in changes in intracellular conditions. Zuckerkandl and Pauling assumed that the production and changes in state of dormant genes were the result of chromosome rearrangements or of transposition events. They did not consider the possibility that they could be due to non-genetic mechanisms. Other workers did, however.

In the USSR, Belyaev and Ruvinsky and their colleagues took up the concept of dormant genes.[8] They suggested that the frequent and reversible changes in the expression of some mammalian genes, particularly those seen under the conditions of stress resulting from domestication, were the result of non-genetic changes that activated dormant genes. Their most detailed studies have been of the *Fused* (*Fu*) gene in the mouse. In both heterozygotes and homozygotes, *Fu* causes developmental defects including a short and deformed tail. However, the number of progeny from crosses between heterozygotes and wild type is not the 1:1 ratio of Fused to wild type that is expected for a dominant trait. The number of normal progeny is greater than expected, and this is not because wild type mice are more viable. The explanation is found in the observation that some phenotypically normal F_1 progeny produce Fused offspring when mated with wild type mice. Such results are commonly attributed to 'incomplete penetrance'—the *Fu* allele does not show itself in all individuals because of the segregation of modifiers which affect its expression. However, the pattern of inheritance of Fused and the switches between wild type and mutant phenotypes are not easily explained in terms of the segregation of modifiers. Rather, Belyaev and his colleagues suggested, the *Fu* gene switches from an active to a dormant state. Figure 6.3 shows the high frequency with which these changes occurred.

Two other results from crosses with *Fused* are worth noting. First, the gene shows classical imprinting, with penetrance depending on whether it is transmitted by the male or female parent. Second, if hydrocortisone is injected into male mice during the period of spermiogenesis when chromatin

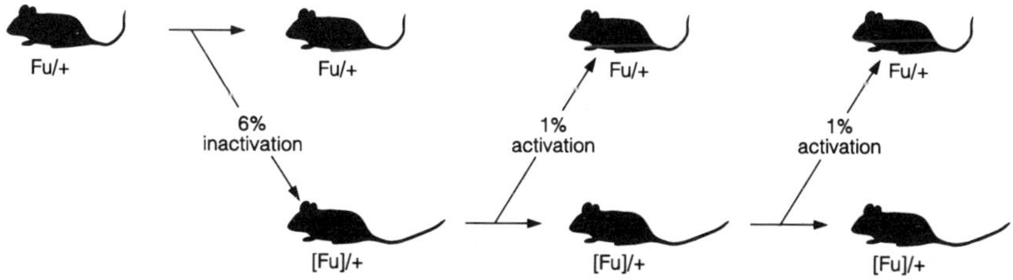

Fig. 6.3 The inheritance of alternative states of *Fused* in the mouse. When the symbol *Fu* is in square brackets, the gene is dormant. (After Belyaev *et al.* 1981*a*.)

is restructured, there is a significant decrease in the penetrance of *Fu* in the offspring. These two observations make the indirect evidence suggesting that variations in the penetrance of *Fu* are associated with heritable differences in chromatin structure quite strong. Belyaev and his colleagues summarized the situation in this way:

> The extremely frequent transition of the *Fu* gene from one state to another makes it difficult to explain in terms of the classical mutation theory. It is also difficult to attribute this manifestation-nonmanifestation of the *Fu* gene to the influence of a set of modifier genes. The whole body of evidence obtained refutes this possibility. The results of genetic analysis of animals having the *Fu* gene, active or inactive, in homologous chromosomes indicate that either state is independently inherited. . . . The reasonable inference is that the state of the *Fu* gene is related to intrinsic properties of the chromosome bearing it, but not to the genotypic environment, although this does not rule out the effect of the latter on its transition probability from one state to another. (Belyaev *et al.* 1981*a*, p. 110)

A similar case, also described by Belyaev and his group, involves *Star*, a gene determining a piebald pattern in domesticated silver foxes. This trait is inherited as an 'autosomal semi-dominant' but, like *Fused*, phenotypically Star is under-represented in the progeny of crosses. Again, as with *Fused*, this cannot readily be explained by differential viability or by the segregation of modifiers causing 'incomplete penetrance'. Belyaev and his colleagues suggested that during the course of domestication there were transitions of the *Star* gene from a dormant to an active state (see Chapter 9, p. 231). The Star phenotype is rarely found in wild strains, but more than 1% of foxes selected for tameness showed Star. The altered state usually persisted in the progeny of those showing the Star phenotype, but it could revert to the dormant state.

Many cases of 'incomplete penetrance' may turn out to be the result of differences in the inherited epigenetic state of the gene. This point has been made by both Sapienza (1989) and J. G. Hall (1990), who suggested that parental imprinting may underlie some of the incomplete penetrance seen in human pedigrees. However, the cases Belyaev and his colleagues investigated belong to a broader category than this, since the inherited epigenetic state was not determined solely by parental sex. The mechanisms responsible for the switches between the dormant and active states of the genes they studied are not known. They suggested, with some supporting evidence, that the transition is caused by hormonal changes affecting chromatin organization. If so, the persistence of an altered state through several generations means that chromatin changes induced by hormone action are not totally erased during sexual reproduction. There is other evidence pointing to the same conclusion: Campbell and Perkins (1988) have summarized and discussed a large amount of experimental data showing that hormonal stresses can have effects that carry-over into the next generation. They believe that many results are best explained by assuming that hormones can affect DNA sequence in a directed way, but the data are explained at least as well by assuming hormones produce heritable changes in chromatin structure.

The Russian workers who studied Fused and Star likened their findings to a phenomenon found in plants that Brink (1960) called paramutation.[9] The most well-studied system showing paramutation involves the *R* locus in maize, which affects pigment intensity in various parts of the plant. It shows some very strange patterns of inheritance. Under the influence of the allele on the homologous chromosome, some *R* alleles change with very high frequency into variant alleles. This specific and directed change happens in somatic tissues, but is frequently transmitted to sexually produced progeny. The new state may persist, but it can also revert to its former state. An allele that shows this unusual behaviour is called 'paramutable', and the allele that induces such a transformation 'paramutagenic'. The change in the paramutable allele is progressive, that is, it produces increasingly more extreme phenotypes during several passages through heterozygotes. The sensitivity of the *R* locus to paramutation is altered by the presence of nearby heterochromatin. The *R* locus also shows genomic imprinting.

Although more than thirty years ago Brink (1960) speculated that paramutation might involve changes in chromatin structure, the nature of the changes in the *R* locus remained a mystery for many years. It is only recently that some of the molecular events that take place are beginning to be uncovered. Changes in *R* are accompanied by changes in methylation: as the paramutable allele becomes progressively altered through successive generations, it becomes progressively more methylated.[10] It seems that, as Brink predicted, inherited variations are associated with heritable alterations in chromatin structure.

The modifications of chromatin structure associated with another paramutable locus in maize, the *B* locus, are not changes in methylation (Patterson *et al.* 1993). But neither are they changes in DNA sequence. Patterson and his colleagues concluded that heritable changes in DNA-protein complexes were responsible for the paramutable behaviour of the *B* locus.

Transmission of altered DNA-protein complexes has also been found in *Drosophila*. Changes in the proteins bound to the DNA of Y chromosomes, brought about by an autosomal 'imprintor' gene, have been faithfully transmitted for at least eleven generations in the absence of the original imprintor gene (Dorn *et al.* 1993). This work is discussed in more detail in Chapter 7 (p. 180). For present purposes it is only necessary to note that it provides another excellent example of how new chromatin marks are stably inherited through the germ line of sexually-reproducing organisms.

It is clear from the cases that we have discussed so far that epigenetic changes can be carried through meiosis and gametogenesis for a substantial number of sexual generations. It is difficult to estimate how general this is, for there have been very few experiments designed specifically to look at the question. Ruvinsky and his colleagues realized that one of the best places to look for inherited epigenetic changes is in the clones of genetically similar individuals produced through ameiotic parthenogenesis. In such organisms, any possible ambiguities resulting from the segregation of undetectable modifiers is avoided. Accordingly, they studied spontaneous and induced changes in the electrophoretic mobility of glucose 6-phosphate dehydrogenase (G6PD) in clones of the water flea *Daphnia pulex*.[11] Some clones were unstable, with two variants of G6PD, a normal form (N) and a slow form (S), being found in different individuals. The spontaneous rate of transition between the two forms was between 10^{-1} and 5×10^{-1}. This is so high it effectively rules out classical mutation as the cause of the change, even though the variants were heritable. Evidence for heritability came from the effectiveness of selection for one of the variants within clones.

A rapid change from the N to the S form could be induced in individuals of some clones by brief exposure to glucose. The clones in which the S form was inducible were the same ones that showed spontaneous transitions from N to S. This makes it likely that the same mechanism underlies both. The one suggested by Ruvinsky and his colleagues was similar to that which they thought was responsible for heritable transitions in the activity of *Fu* in the mouse. They suggested that *Daphnia* has a regulator gene that modifies G6PD proteins in a way that affects their electrophoretic mobility. This gene can exist in a heritable state of activity or inactivity. The differences between clones reflects variation in the stability of the inactivation. It is interesting that in natural populations, the S variant was common only in stressful, fluctuating, environmental conditions. Clones that had a high frequency of transition from N to S also showed a high frequency of transition from parthenogenesis to sexual reproduction. According to Ruvinsky and his colleagues, in stressful fluctuating environments, natural selection favours the ability to make heritable changes in gene activity and to alternate efficiently between sexual and asexual reproduction. Both the ability to switch to sexual reproduction, and the appearance of the S variant, are consequences of the activation of dormant genes. The ease with which dormant genes are activated is under genetic control.

Heritable variation within lineages has also been found in aphid clones. Although some of the reported cases may have been the result of contamination or mutation,[12] in others variability is associated with heritable changes in gene expression. In the peach potato aphid, insecticide resistance is sometimes the result of DNA amplification. Some resistant clones that carry a chromosomal translocation lose their resistance in the absence of selection by the insecticide, although their amplified sequences are retained. The changes in resistance are correlated with changes in methylation of the amplified locus: resistance is associated with increased methylation, and sensitivity with loss of methylation (Field *et al.* 1989). Such variation within clones is reminiscent of the variation within somatic cell lineages that characterizes position effect variegation (PEV) in *Drosophila*. In both cases, a chromosome rearrangement is associated with alternative phenotypes that can be transmitted to the next clonal generation.

Clonal plants also show very extensive heritable variation (Silander 1985). The origin of this variation is unknown, but it is difficult to explain in straightforward genetic terms. Often clonal plants show extreme phenotypic plasticity. Probably both genetic and epigenetic inheritance contributes to the variation found in these plants.

Some strains of sexual organisms also show more phenotypic variability than is expected from the genetic differences between individuals. For example, the differences between sublines of mice maintained for many generations by brother-sister mating is far greater than it should be according to classical population genetics theory.[13] To produce so much divergence, the mutation rate would have to be two or three orders of magnitude greater than that usually found. Several explanations for the discrepancy between theory and observation have been proposed. For example, Grüneberg (1970) suggested that some variability is the result of the action of latent viruses, rather than nuclear genes. It is also possible that the mutation rate may have been high for reasons similar to those that McClintock suggested led to stress-induced transposition in plants. The stress in inbred lines would be caused by internal conditions, generally by high homozygosity. More recently, Holliday (1987) suggested that the variability in inbred lines could be the result of methylation defects—of epimutations. Fitch and Atchley (1985*a,b*) argued that the most plausible explanation of the data they collected was that during the early stages of inbreeding, there had been selection for heterozygosity. This would explain not only the rapid divergence rate in inbred lines, but also two other oddities in their data: first, there were only two different alleles for most loci, and second, no new mutants were found in the inbred lines. However, as Fitch and Atchley pointed out, there are difficulties with this hypothesis, and the data can be explained equally well by a conversion or switching process operating between two variants. The type and frequency of change they describe is not unlike that expected for an epigenetic system in which alternative states are inherited.

Epigenetic inheritance may also explain certain progressive changes found in strains believed to lack genetic variability. For example, Brun (1965) found that the self-fertilizing nematode *Caenorhabditis elegans* showed a gradual adaptation to high temperature over several generations. He thought it unlikely that this was the result of selection, since there could be little genetic variation present. He suggested that changes in the cytoplasm were responsible for the adaptation process. Ho and her colleagues (1983) gave a similar interpretation of the progressive increase in penetrance of a bithorax phenocopy in inbred lines of *Drosophila*. The phenocopy was induced by treating embryos with ether in each generation. The change occurred in the absence of artificial selection, and in lines which, even if not homozygous, were unlikely to have contained much genetic variation. In cases like these, where genetic variation is low, it seems far more likely that the basis of the change is epigenetic rather than genetic.

Lansing effects and parental age effects

There is another category of heritable changes which is well-documented in the old literature, but so far has not received an adequate explanation. This category includes parental-age effects, in which the age of the parent influences the phenotype of the offspring, and Lansing effects.[14] Lansing effects were named after A.I. Lansing who drew attention to the way in which parental age affected the longevity of the next generations. Working with parthenogenetic rotifers, he found that the progeny of old parents do not live as long as those of young parents, and, rather surprisingly, this effect is cumulative. It is also reversible.[15] Lines of rotifers maintained through old parents eventually die out, but the cumulative detrimental effects can be reversed if the age at reproduction is changed. Similar cumulative heritable effects of the age of parents on the longevity of offspring have subsequently been found in several species of insect, and in nematodes and mammals. It is certainly not a universal finding, but Lansing effects have been reported sufficiently often for one to feel confident that the phenomenon is real.

We have argued previously (Jablonka and Lamb 1990*b*) that the reversibility of Lansing effects makes it highly unlikely that they are caused by the accumulation of detrimental mutations in the germ line of the parent. However, reversibility is not unexpected if Lansing effects are caused by age-related epigenetic modifications of germ-line chromatin. Evidence supporting this interpretation comes from three types of observation showing that ageing is accompanied by changes in chromatin structure. First, some somatic cells *in vitro* show a progressive decrease in the level of methylation as cultures age, and when cultures are treated with a demethylating agent, culture lifespan is reduced. Second, methylation levels in some somatic cells *in vivo* decrease with age, and do so faster in short-lived species. Third, studies of the stability of X-inactivation in the mouse have shown that genes that were formerly inactive become reactivated as the animal ages. If, as these observations suggest, heritable chromatin marks in somatic cells are affected by age, it is not unreasonable to suggest that sometimes chromatin in germ-line cells is also affected. Provided that any changes are not totally erased during gametogenesis, they could lead to Lansing effects as well as less specific types of parental-age effects.

Parental-age effects are seen in many different characters, and in many different organisms. Probably the majority are caused by mutational changes in the parents' germ lines, or by ageing changes in the cytoplasm of the egg. However, not all cases can readily be explained in this way. In particular, when the father's age influences gene expression, explanations in conventional terms seem implausible. Explanations based on inherited age-related changes in marks seem much more appropriate for many paternal-age effects, although there is no direct evidence that the explanations are correct.

Transmission of methylation patterns between generations

So far, most of our evidence for the transmission of epigenetic information from one generation to the next has been based on the patterns of inheritance of phenotypes. The traits we described do not show orthodox Mendelian inheritance, but the molecular basis of the odd inheritance patterns is not always clear. However, molecular studies are providing increasing evidence showing that chromatin marks are passed from generation to generation. In several different species of animal and plant, it has been found that methylation patterns are transmitted through meiosis and segregate with the DNA sequences on which they are imposed, sometimes having phenotypic effects, and sometimes not.

Inbred lines have reduced genetic variability, yet as indicated earlier, they show substantial heritable variation. It is therefore of interest to find that even when individuals have identical DNA sequences at a particular locus, the chromatin marks associated with these sequences are not always identical. In an inbred line of wheat, Flavell and O'Dell (1990) found that genes coding for a high molecular weight glutenin protein had seven methylation variants—seven epialleles. Some of these patterns of methylation were stably inherited both somatically and between generations. In crosses between epiallelic variants, the F_1 individuals showed the patterns of both parents, and in the F_2 generation the patterns segregated so that both the F_1 and parental patterns were found. In other words, the methylation patterns showed the same behaviour as Mendelian alleles, and segregated with the DNA sequences on which they were imposed. Occasionally a new methylation pattern appeared, showing that the gene had a high epimutability.

Similar behaviour of chromosome marks has been found by Zolan and Pukkila (1986) in the fungus *Coprinus cinereus*. The methylation patterns of identical DNA sequences at a centromere-linked locus are inherited. In tetrad analysis, crosses between strains with identical DNA sequences but different methylation patterns show a 2:2 segregation of methylation patterns, just like that obtained with classical Mendelian alleles.

Further evidence for the transmission of epigenetic information through the germ line comes from studies of methylation in human and mouse tissues. In some human tissues, the pattern of methylation at two allelic sites differs (Silva and White 1988). This variation, which is tissue specific, is inherited in a Mendelian fashion for at least three generations. However, the allele-specific methylation patterns are not preserved in the sperm, where the methylation pattern of the locus is uniform and sperm-specific. The methylation variants are therefore not transmitted directly. Nevertheless, some blueprint of the methylation patterns which enables them to be reconstructed in the next generation must be established during gametogenesis. Silva and White suggested that the basis of this blueprint could be DNA binding proteins that segregate with the chromosomes at meiosis. Comparable epialleles have been found in the mouse by Sasaki and his colleagues (1991), who suggested that this type of variation could be ubiquitous in mammalian genomes.

The cases just described show that methylation marks can be transmitted from parents to progeny, without obvious phenotypic effects. However, there are several experiments that suggest that there is a link between changed methylation marks and changed heritable characters. One example is found in rice, where Sano and co-workers (1989,1990) found that exposure of germinating seeds to a demethylating agent (5-azacytidine or 5-azadeoxycytidine) induced dwarfism, as well as demethylating DNA. The dwarf phenotype was inherited for the three generations they studied, and the low level of methylation induced by the treatment segregated with the dwarf phenotype. Sano and his colleagues suggested that there is a direct and causal relationship between the inheritance of height and the inheritance of methylation level. A comparable heritable effect of 5-azacytidine on both DNA methylation and phenotypic traits has been found by Heslop-Harrison (1990) in a stable variety of Triticale, a wheat x rye hybrid. Seeds were treated with the demethylating agent and the first and second generations were examined. The treatment caused DNA demethylation and also heritable phenotypic changes such as increases in stature and number of tillers, and changes in the time of ripening. The effects were inherited for at least two generations.

A highly specific new heritable phenotype called 'fluffy' has been found by Tamame and coworkers (1988, 1989) following treatment of the fungus *Aspergillus* with 5-azacytidine. Although *Aspergillus* is an organism with a very low level of methylation, some sites are methylated, so the explanation that demethylation of a particular locus caused the fluffy phenotype is not implausible.

Perhaps the most clear-cut demonstration of a relationship between heritable differences in methylation and heritable differences in gene activity has come from studies of maize transposable elements. These elements are able to move around the genome and affect the expression of the genes into which they insert. In her pioneering studies, McClintock found that during development some transposable elements make regular, heritable but reversible, transitions between active and inactive states. It is now clear that many of these transitions are associated with changes in the methylation level of the elements: a genetic property, the transposability of an element, is correlated with its level of methylation. This has been shown very clearly by Fedoroff and her co-workers in their studies of the *Spm* (*Suppressor-mutator*) elements.[16] *Spm* is a transposable element that encodes the proteins necessary for its transposition and for the regulation of its expression. It can exist in three interconvertible heritable states: active, inactive (or *cryptic*), and *programmable*. *Programmable* elements shift between active and inactive states in a regular way during development. The three different states are associated with differences in the extent of methylation in a cytosine-rich region around the transcription start site (Fig. 6.4). The methylation level of the upstream control region (UCR) is associated with transcriptional activity and transposability: active elements have low methylation, inactive high. Methylation of the control region downstream from the transcriptional start site (DCR) determines the heritability of the element's activity and its developmental expression programme: active elements are hypomethylated, *cryptic* elements are highly methylated, and inactive *programmable* elements are partially methylated.

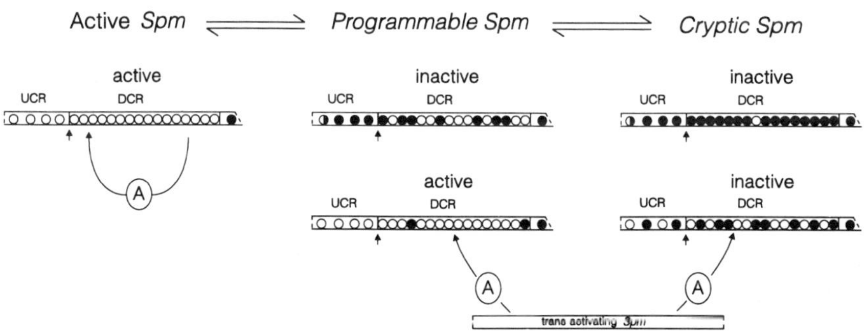

Fig. 6.4 Fedoroff *et al.*'s model of the different states of the *Spm* element. UCR: upstream control region; DCR: downstream control region; solid symbols: methylated sites; half-filled symbols: partially methylated sites; hollow symbols: unmethylated sites; A: the autoregulatory gene product. The active element autoregulates its own activity. The programmable element is transcribed when an active element provides gene product A, which reduces methylation and activates transcription. Product A also reduces the level of methylation of cryptic elements, but not enough to lead to transcription. (Redrawn from Fedoroff *et al.* 1989.)

The likelihood that a *programmable* inactive element will become heritably active or vice versa depends on several factors, all of which have been shown to affect methylation levels. First it depends on position in the plant. When transmitted through tiller gametes, the likelihood of reactivation in the following generation is increased, whereas transmission through gametes produced on the main stalk tends to stabilize the inactive state. In general a plant's ability to inactivate an active element increases with the distance from the base of the plant. Second, change in a programmable element depends on the presence or absence of other active *Spm* elements. An active element can reactivate a *programmable* inactive element and increase the probability that the new activity state will be transmitted to the next generation. The presence of an active element also increases by two to three orders of magnitude the likelihood of reactivation of a *cryptic* element in the same genome. An active *Spm* element thus behaves like a modifier of inactive elements. At the molecular level, it has been shown that an active *Spm* element causes a reduction in methylation of both UCR and DCR sequences in inactive elements, presumably as a result of the activity of its regulatory gene product. The third factor influencing the activity state of *programmable* elements is their parental origin: activation is most likely if the element is transmitted through female gametes.

The switch from an active to inactive state can be a gradual process. Breeding from plants in which there was a low frequency of reactivation of a programmable element produced progeny in which the element was even more inactive. Eventually, after several generations of selection, the element no longer showed differential expression during development. It had become a *cryptic* element. This progressive change to a stably inactive element was paralleled by a progressive increase in methylation.

From the evidence just outlined, it is clear that heritable changes in the genetic behaviour of *Spm* are associated with heritable changes in the epigenetic state of the element. Studies of other transposable elements in maize and other plants have led to similar conclusions.[17] Fedoroff and her colleagues concluded an analysis of *Spm* regulation and transmission by stating:

> Perhaps the most striking observation that has emerged from the analysis of the *Spm* element's developmental control mechanism is that epigenetic changes in the present generation can influence the expression pattern of the element in the next generation. (Fedoroff *et al.* 1989, p. 143)

Many of the observations on transposable elements in maize have parallels in other systems. For example, in transgenic tobacco plants, one transposon can affect the expression and methylation of another. Active, unmethylated, genes on a T-DNA sometimes become inactivated and heavily methylated when cells are transformed with a second T-DNA.[18] Whether or not the second T-DNA suppresses expression in the first depends on its site of insertion. If plants with a suppressed T-DNA are crossed so that the two T-DNAs segregate, the first T-DNA is reactivated and demethylated in those progeny that lack the second. However, in the first generation of offspring demethylation is not complete, and individuals have two populations of cells, some with active and demethylated T-DNA, and others in which it is methylated and inactive. Plant transgenes also show changes in methylation and behaviour that are associated with endogenous factors and environmental conditions (Meyer *et al.* 1992). Petunia plants that carry a single copy of a maize gene giving them a salmon-red phenotype produce some plants with white or variegated flowers. In most cases the new phenotypes are correlated with changes in the methylation of the transgene's promoter. These changes can occur before or during meiosis in the parent plant, or at any time during the development of the progeny. The extent of methylation and likelihood of producing the modified phenotypes depend both on growth conditions and the age of the plant when it produces seeds.

Some general conclusions

Can any general conclusions be drawn from this survey of some of the evidence for epigenetic inheritance? The first and most important conclusion is that not all of the random and directed epigenetic information acquired in one generation is erased during the production of the next. Even with sexual reproduction, marks—in the cases investigated DNA methylation patterns[19]—can be transmitted and are perpetuated through several generations. As yet we do not know exactly what happens during gametogenesis—how marks are altered, what determines which marks are altered, whether marks that are altered usually, or rarely, leave 'footprints' of their previous nature, and so on. We need to know more about the processes occurring in the germ line in order to be able to assess the extent and persistence of inherited epigenetic variations. The stability of inherited marks seems to vary: some persist for only a few generations, whereas others seem very stable. We consider this in more detail in Chapter 8 when we look at models of 'lingering modifications' and consider the evolutionary implications of epigenetic inheritance.

Another conclusion that seems to be justified by the evidence assembled in Table 6.1 is that a high proportion of the cases of epigenetic inheritance in multicellular organisms have been found in plants. This may not be a coincidence. If an epigenetic mark is to be transmitted to the next generation, it must be present in the germ line. Organisms with late or non-existent germ-line segregation can pass on variations that occur in somatic cells because these somatic cells can become germ-line cells. When a new epigenetic variation is advantageous at the tissue level, selection may result in cells with the new variation coming to dominate the tissue. The chances that the cells will become germ cells and transmit the new variant to the next generation increase. Therefore, in plants and other organisms that lack a segregated germ line, new epigenetic variants occurring in somatic lineages may be inherited. In contrast, organisms such as mammals, in which the germ line segregates early in development, can transmit to the next generation only those new variations that occur either before germ line segregation, or in the germ line itself. Consequently, most cases of the inheritance of epigenetic variations in somatic characters are likely to be found in organisms with late or no sequestration of the germ line. This argument is, of course, exactly the same as that considered in Chapter 2 with respect to the selection and transmission of somatic mutations. Both genetic and epigenetic variants are most likely to be selected and transmitted to the next generation in plants and in those animals that lack a distinct germ line.

Many of the more detailed studies of epigenetic inheritance have shown that there are interactions between the marked gene and other genes in the genome. Effects on penetrance are common. This is not surprising, because as we discussed in the last chapter with respect to imprinted genes, gene expression depends on interactions between the regulatory elements of the gene and trans-acting factors. Epigenetic marks altering the extent or specificity of gene expression should behave like mutations in the regulatory elements. Some marks will alter the binding affinity of trans-acting factors, others will not. The trans-acting products of some alleles will bind to a new, differently marked sequence, whereas those of other alleles may have radically changed affinities, and consequently have pronounced effects on gene expression. Selection will always be for a combination of the mark and the trans-acting factors produced by 'modifier' genes. Many of the so-called 'modifiers' may be genes coding for proteins that regulate chromatin structure. They could modify the expression of several different marked genes.

Why are there not more cases of the inheritance of acquired epigenetic variations?

One of the objections to the idea that the inheritance of epigenetic variations is important in evolution is that in spite of much searching in the first half of the twentieth century, very few examples of this type of inheritance were found. The examples we give in Table 6.1 are certainly a rather limited rag-bag of old and new observations, and it is tempting to dismiss them as mere curiosities. But this would be a mistake. There are several good reasons why data showing epigenetic inheritance between generations are limited. First of all, most ideas about EISs and the experimental attempts to understand epigenetic inheritance, are recent. Indeed, as Table 6.1 indicates, many of the studies providing conclusive evidence of the inheritance of epigenetic information have been made in the last few years. It is likely that many more examples will be found as the techniques of molecular biology expand and are applied to the problem. A second reason why there is a paucity of examples is that many instances of epigenetic inheritance may have been misinterpreted. Some new variants thought to be caused by conventional DNA sequence changes may turn out to be heritable epigenetic variations. We noted in Chapter 4 that when heritable changes in cultured mammalian cells were studied in greater detail, some 'mutations' were found to be epimutations. The same may be true for variants that are transmitted from an organism to its offspring.

Perhaps the main reason why epigenetic inheritance has rarely been observed is that people have looked for the wrong type of effect. It is often implicitly assumed that the inheritance of epigenetic variations is the same as the inheritance of acquired characters, so that new inherited epigenetic variations will be seen as 'acquired characters'. But this is not so. Epigenetic variations can be random, and if they are, they will not produce an 'acquired character'. Even directed epigenetic variations will not produce acquired characters unless they have distinct, recognizable, phenotypic effects. Many directed changes in marks will have no obvious phenotypic effect at all. Rather than directly changing a gene's expression, they may affect the way that it is influenced by other factors. As is clear from the work on *Spm* in maize and transgenes in the mouse (Chapter 5), the mark (in these cases methylation patterns) on a gene is a kind of label that changes the way the gene is 'interpreted' by trans-acting factors. Only rarely will a changed mark be sufficient to alter gene activity. The more subtle effects of an altered mark may be difficult to analyse because they appear as quantitative variations in a character. The character will often show incomplete penetrance and variable expressivity.

Looking for the wrong type of effect cannot, however, be the major reason for the lack of examples of inherited epigenetic variations. The cases we have reviewed show that some environmentally-induced epigenetic changes do result in distinct acquired characters that can be inherited. Yet these cases are few. Since people were looking for instances of the inheritance of acquired characters for many years, why did they not find more? It may be that usually they were looking in the wrong place. Most people were looking for the inheritance of somatic adaptations. It was the giraffe's neck, the colour of salamanders, the degeneration of eyes in cave dwellers, and so on, that interested them. But obviously the type of character one should be looking at is a character that can be changed in cell lineages that can contribute to gametes. In many animals this means looking at characters induced either early in development, or at characters acquired in, and affecting, the germ-cell lineage itself. Characters acquired in somatic lineages, even when adaptive, cannot be inherited, unless the somatic cells can become germ-line cells.[20]

A further reason for the failure to find inherited acquired characters is that sometimes the acquired character being sought was an adaptation to a stimulus to which it was impossible for the organism to adapt. For example, adaptation to high concentrations of DDT is physiologically impossible. The organisms die. Only organisms with pre-existing adaptive mutations can survive in such conditions. There is no chance of acquiring resistance if it is not already present. In Chapter 3 we discussed how overlooking this type of limitation led to the misleading generalization that all mutations in bacteria are random. It has probably also impeded searches for inherited epigenetic variations.

These considerations lead to a further question: in what type of character are we most likely to find evidence of the inheritance of acquired characters? Consider an organism with a segregated germ line. In theory, all loci in the germ line can be epigenetically modified in a random manner (i.e. acquire an epimutation), and all loci can acquire directed variations, which may or may not be adaptive. However, in two types of gene epigenetic modifications will have particular significance: epigenetic variations in germ-line-specific genes are rather special, as are those in the genes involved in housekeeping functions, which are important in all cell types. Unlike variations in other genes, variations in these genes, which are expressed in the germ line, may have immediate phenotypic effects, and they can be selected at the level of the germ-cell lineage. In other words, these types of epigenetic variations can be selected in germ-cell lineages in the same way as Buss and Klekowski described for mutations (Chapter 2). Moreover, variations in genes that are expressed and selected in the germ line are unlikely to have deleterious pleiotropic effects in the next generation: if the loci are concerned with housekeeping functions, the variations should be as good in the somatic cells of the offspring as in the germ cells of the parent; if they are germ-line-specific, then by definition they are expressed only in the germ line of the offspring, so unless the epigenetic change has altered their specificity, they are likely to be neutral in the soma and adaptive in the germ line of the offspring. Consequently, in organisms with a segregated germ line, epigenetic variations in loci that are expressed in germ-line cells will form a disproportionately high fraction of the adaptive variations in the offspring. In organisms without a sequestered germ-line, the spectrum of genes likely to be involved in acquired epigenetic variations is much wider.[21]

Resetting the epigenetic state in the germ line

With asexual reproduction, the transmission of marks from generation to generation is no different from that in cell lineages. As in somatic cell lineages, once a mark is acquired, the fidelity with which it is transmitted

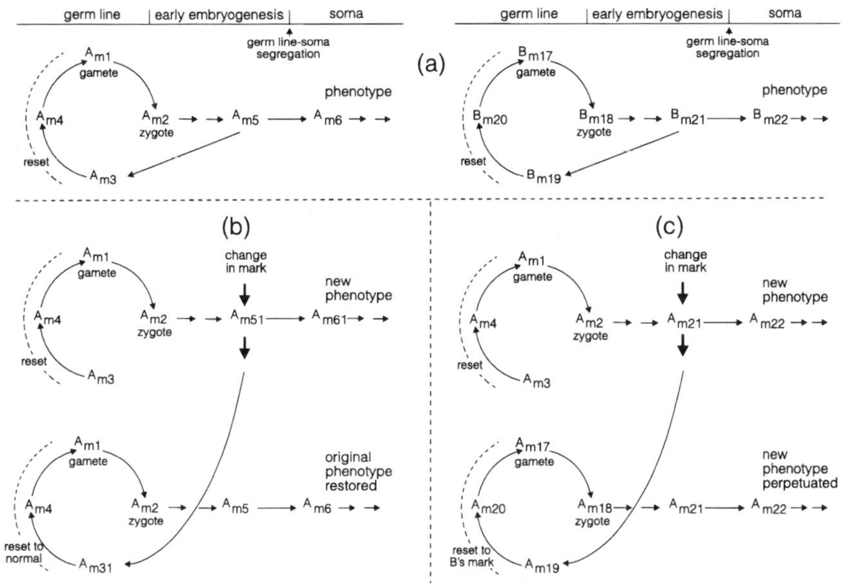

Fig. 6.5 Temporary and permanent changes in inherited marks. Each gene carries a chromatin mark (m) that changes during development. The different numbers represent different marks. (a) Changes in the marks on genes *A* and *B* during normal development showing how developmental cycles of marks are perpetuated. (b) An altered mark on gene *A* which does not lead to an inherited change in phenotype. The normal mark *m5* is altered to *m51*, and this produces a change in phenotype. The transition of this to the germ-line state results in a new mark *m31*. However, *m31* is sufficiently similar to *m3* for it to be reset to the normal mark *m4*. The normal cycle is restored and there is thus no memory of the epigenetic change. (c) An altered mark on *A* which results in an inherited variation. The normal mark *m5* is changed to *m21*, a mark that mimics the one normally found on *B*, and produces a phenotypic change. In the germ line, as for *m21* on *B*, *m21* on *A* is reset to *m20*. From this point on, marks on *A* are changed and inherited like those on *B*, and the new phenotype is perpetuated. (Based on Jablonka *et al.* 1992.)

can be quite high. However, in sexually reproducing organisms, newly acquired marks may be reset to the original germ-line state during gamete production.[22] Yet in spite of the chromatin changes in germ-line cells, the evidence we have presented in this chapter shows that some new epigenetic marks are transmitted through sexual reproduction, and permanent heritable changes can be effected by epigenetic modifications in the previous generation. In Fig. 6.5 we show the kind of route by which we think this may happen. In the top part of the figure, we show how the marks on two genes might change during development, and be reset to the original state in the germ line. We assume that genes *A* and *B* carry different marks, and respond to, and are changed by, different developmental inducers. In the bottom half of Fig. 6.5 we show what could happen if the mark on gene *A* is altered, either because of a random change, or because a new environment produced an abnormal inducing stimulus. A new phenotype is produced, but whether or not this affects more than one generation depends on what happens in the germ line. In Fig. 6.5b the new mark on *A* is sufficiently like the old one for it to be recognized and reset in the germ line to the normal germ-line mark. The new variation will therefore not be inherited. Figure 6.5c shows a different fate for a new mark on *A*. This new mark is such that it mimics the normal mark on *B*. Consequently, in the germ line, the mark will be reset in the same way as that on *B*. A heritable change in the mark on *A* has occurred, and the new phenotype will be perpetuated. In the figure we have shown the change in mark taking place in early embryogenesis, but it could equally well occur at any stage prior to germ-line segregation, and have the same effect.

Outcomes of changed marks other than those shown in Fig. 6.5 are possible. One likely possibility would be a domino effect in which a new mark is not recognized by any of the existing programmes that change marks, and consequently undergoes a series of essentially random changes. Ultimately, this would lead to a new stable state or, more probably, to the death of the lineage.

Genes seem to vary in the extent to which they are able to retain a memory of their past history. It is noticeable that several of the examples of inherited epigenetic variations described in this chapter involve genes that also show evidence of imprinting. This is true for the *R* locus in maize and for *Fu* in the mouse. Some transgenes in the mouse at first showed parental imprinting, but in later generations their marks were transmitted stably, regardless of sex. It seems that genes that can be marked in ways that make them retain a memory of parental sex are also able to acquire the more permanent marks that survive transmission through gametogenesis in both sexes. In other words, genes that can acquire the 'stubborn' marks that are not erased and lead to a difference between maternal and paternal alleles, may also be able to carry the even more stubborn marks that result in epigenetic variations being transmitted for several generations.

Why do some genes acquire stubborn marks? Is the phenomenon of any importance in development and evolution? Jorgensen (1993) argued that when plant genes such as the *R* locus, the *Spm* element, and various transgenes, carry epigenetic information between generations, it is merely an aberrant manifestation of normal developmental processes. In natural conditions the changed epialleles would be eliminated. If this is so, the role of germinal epigenetic inheritance in evolution is minimal. In contrast to Jorgensen, we believe that the DNA sequences able to carry stubborn chromatin marks, and the enzymatic machinery that enables the transmission of epigenetic variations between generations, has been positively selected. For some types of gene, and in some types of environment, the ability to acquire transmissible marks is an advantage. We discuss this further in Chapter 8.

In this chapter we have focused on transmission through the chromatin-marking EIS. During sexual reproduction the inheritance patterns of stubborn chromatin marks will be the same as that of chromosomes and the alleles they carry—epialleles show Mendelian inheritance. With the steady-state and structural EISs, patterns of inheritance are different. Most sexual organisms have dissimilar-sized male and female gametes. Since the female contributes most to the zygote, variations in the steady-state and structural EISs will be transmitted largely through the female: they will show maternal inheritance. Given the diversity in the size, formation, and development of eggs, making general models for the behaviour of epigenetic variations in the steady-state and structural inheritance systems is more difficult.

Summary

In spite of claims to the contrary, there is evidence that epigenetic information acquired in one generation can be inherited by subsequent generations. Studies of methylation patterns, the only epigenetic inheritance system that can be easily studied at present, have shown that differences in methylation patterns can be inherited, even through sexual reproduction. In several cases, these heritable differences are known to be associated with differences in gene activity. Chromatin marks are therefore one mechanism by which memories of cellular activity in one generation can be transmitted to the next.

At present it is difficult to know how common is the inheritance of acquired modifications, because many changes are likely to have only small phenotypic effects, seen mainly as quantitative, rather than qualitative, changes. Some of the variability between individuals in both clonal and sexual populations probably results from epiallelic differences. In sexual organisms, adaptive 'acquired characters' are most likely to be found in germ-line-specific characters, or in functions that affect all cell types. They should be common in organisms such as plants, which lack a distinct or early segregated germ line.

Although in this chapter we have concentrated on induced epigenetic changes in a single gene, the response to a new environmental stimulus is likely to be epigenetic modification of many loci. The evolutionary opportunities following changed environmental conditions are therefore enormous. The importance of coordinated changes in many loci in response to new environmental challenges are examined in Chapters 8 and 9.

Notes

1. Groudine and Conklin (1985) studied changes in chicken chromatin, and Kafri *et al.* (1992) studied changes in the mouse.
2. In addition to the empirical evidence that chromosomes carry epigenetic marks, it was recognized long ago that there are theoretical reasons for thinking that EISs associated with chromosomes may be important in both cell heredity and inheritance between generations. Chromatin marking influences gene expression at the level of transcription, which is an early, and therefore economical, stage at which to stop the production of gene products (Nanney 1960). Furthermore, as Mather (1961) pointed out, mitosis ensures the regular and precise transmission of any controls carried by, or associated with, the chromosomes. Since transmission of cytoplasmic elements is far less regular than the transmission of chromosomes, controls based on cytoplasmic elements are likely to be transmitted with much lower fidelity.
3. Rubin (1990) has commented on the neglect of some of this early work on 'enduring modifications', emphasizing how important the type of variations described in it may be for understanding adaptation at the cell level.
4. See Jollos (1921). Hämmerling (1929) gives many more examples of enduring modifications in a variety of organisms.
5. See Beale (1954).
6. See Meins (1985), but also Meins and Foster (1986) and Meins (1989*b*).
7. Zuckerkandl and Pauling (1962) made their suggestion about dormant genes in a discussion of the role of gene duplication in evolution. They postulated that modified, unfavourable, or unwanted duplicate genes could be preserved in a permanently dormant condition if no existing tissue provided conditions suitable for their expression. Such genes could be reactivated by a change in the intracellular environment brought about by adaptations to changed external conditions. They suggested that reactivation of dormant genes could be the cause of the explosive evolution that occurred after periods of major environmental change.
8. Much of this work was originally published in Russian, but Belyaev *et al.* (1981*a*, 1983) give descriptions of the work on Fused in the mouse, and Belyaev (1981*b*) summarizes work on Star in foxes. More general accounts of the interesting work carried out by this group can be found in Belyaev (1979) and Ruvinsky (1988).
9. Brink, who in the 1960s first called attention to paramutation, defined it in the following way: 'Paramutation is an interaction between alleles that leads to directed, heritable change at the locus with high frequency, and sometimes invariably, within the time span of a generation' (Brink 1973, p. 129). The Mendelian rule that, in a heterozygote, the passage of different determinants through meiosis does not alter their properties, is violated.
10. Unpublished observations by Alleman and Kermicle, cited by Dooner *et al.* (1991).
11. Ruvinsky *et al.* (1983*a,b,* 1986). Since *Daphnia* populations are known to have higher levels of genetic variability than is expected from their low levels of sexual reproduction (Mort 1991), the phenomena described by Ruvinsky *et al.* in these papers may be common.
12. See Blackman (1979) and Mackenzie (1992) for evaluations of these experiments. They are discussed further in Chapter 9, p. 237.

13. See Grüneberg (1970), Fitch and Atchley (1985*a,b*), Johnson *et al.* (1985), and Green *et al.* (1985).
14. For an extensive review of both Lansing effects and parental-age effects, see Lints (1978).
15. Lansing's results are summarized in Lansing (1954). Lints (1988) has correctly pointed out that Lansing's results are commonly misrepresented, but for the present discussion this is not important. The relevant point is only that parental age has cumulative and reversible effects on the longevity of offspring.
16. Details of the control of the *Spm* element are given in Banks *et al.* (1988), Banks and Fedoroff (1989), Fedoroff (1989), and Fedoroff *et al.* (1989).
17. The *Ac* element (Chomet *et al.* 1987, Kunze *et al.* 1988) and the *Mu* element (Chandler and Walbot 1986, Martienssen *et al.* 1990) are also controlled partly by DNA methylation. Like the *Spm* element, their activity state, which is associated with DNA methylation, can be altered during development, and that altered state can be transmitted to the next generation (Dennis and Brettell 1990). Regulation of transposable element activity by methylation changes that may become heritable has also been reported in *Antirrhinum* (Martin *et al.* 1989).
18. For a full account of these experiments and a discussion of their significance, see M.A. Matzke *et al.* (1989) and M.A. Matzke and A.J.M. Matzke (1990). Matzke and Matzke (1990) suggested that the phenotypic variability observed in plants with identical genotypes may be caused by epigenetic, potentially heritable variations such as those found in their experiments. They also suggested that somatic selection of epigenetic variants in the meristem is a strategy that enables plants to adapt rapidly to changing unpredictable environmental conditions, without waiting for genotypic change. This kind of adaptive strategy may have been one of the factors enabling the evolution of obligatory self-fertilization in plants: epigenetic variability may compensate for the reduction in genetic variability resulting from inbreeding. If so, obligatorily self-fertilizing plants may be found to have particularly high levels of heritable epigenetic variations.
19. It is noticeable that in almost all cases that have been studied at the molecular level, the heritable epigenetic mark has been the pattern or extent of DNA methylation. It is not clear whether methylation is really a major mechanism of epigenetic marking, or simply that heritable variations in DNA methylation have been detected because this type of mark is the only one that can be studied easily with the molecular techniques presently available. Unfortunately, at present there are no comparable techniques for the detailed study of marks involving DNA binding proteins.
20. As discussed in Chapter 2, in theory horizontal gene transfer from the soma to the germ line could lead to the inheritance of acquired variations.
21. In an interesting discussion of the role of cytoplasmic genes in the inheritance of acquired characters, Crosby (1956) reached exactly the same conclusion. He, too, argued that acquired adaptations are more likely to be found in plants than animals, and that they would be found in 'general characters of growth and vigour', although the reasons for his conclusions were rather different.
22. Maynard Smith (1990) has argued that marks are reset to the original state during gametogenesis, and because of this it is unlikely that inherited epigenetic variations are important in evolution.

7

Interactions between genetic and epigenetic inheritance

A goal for the future would be to determine the extent of knowledge the cell has of itself and how it utilizes this knowledge in a 'thoughtful' manner when challenged.
McClintock 1984, p. 798

In the first half of this book we have been at pains to demonstrate two things: first, that some non-genetic variations, epigenetic variations, are inherited both within cell lineages and from one generation of organisms to the next; second, that respectable biochemical mechanisms are known that show, or at least suggest, how these variations can be inherited. If it is accepted that epigenetic variations are part of what can be inherited, then it follows that they must also be part of evolution. The remainder of this book will be devoted to exploring the part epigenetic variations and their inheritance have played in evolution. Some of the topics we discuss are relatively uncontroversial. Those considered at the beginning of this chapter are an example, since no one denies that gene activity influences genetic change, even though little notice is taken of it in evolutionary theory. Some of the other topics we explore lead to conclusions and speculations that may seem a little more threatening to a conservative interpretation of Darwinism. However, all of our arguments are based firmly on Darwinian evolution. We are merely taking a somewhat less restrictive view of the nature and origin of inherited variations than that found in most present day evolutionary thinking.

In this chapter we explore the interrelationships between the genetic and epigenetic systems. Until now we have treated epigenetic variability almost as if it is independent of the underlying DNA sequence information, emphasizing how different heritable marks can be imposed on identical DNA sequences. Similarly, when discussing the way in which DNA sequences can sometimes change in direct response to environmental conditions, thus functioning as a response system as well as an inheritance system, the role of the epigenetic state of the locus in mediating the DNA sequence changes was largely ignored. Yet it is clear that, although the DNA inheritance system and EISs can behave autonomously, they usually interact. When looking at short-term evolutionary changes these interactions can be ignored, and each inheritance system can be considered in isolation, but when considering long-term changes, the interactions between inheritance systems become important.

We first review data showing how one type of epigenetic change, change in chromatin structure, affects the probability of changes in DNA sequences. The impact of chromatin variations on the rate of DNA evolution may be substantial. We then look at the effect of three aspects of DNA organization on epigenetic inheritance: first, the density of CpG dinucleotides in the promoter regions of genes; second, the presence of tandemly repeated sequences; and third, the regional division of chromosomes into euchromatic and heterochromatic blocks, and into light and dark G bands. We argue that these elements of DNA organization are vehicles of cellular memory through the chromatin-marking EIS; they are important in determining the 'memory spans' of loci and domains. The origin and distribution of these elements is related to the evolution of the length of cell memory. The interplay between the DNA inheritance system and the chromatin-marking system has been important in determining the evolution of chromosome organization.

The effects of chromatin structure on DNA sequence variation

How does the epigenetic state of a locus affect the probability that it will undergo a change in DNA sequence? Both old and more recent studies suggest that there is a correlation between the physiological state of a cell and the rate of mutation and recombination in its genome. Environmental factors such as temperature or diet, behavioural stress, and internal factors such as sex or age, have all been found to affect rates of mutation and recombination.[1] Often the effects have been rather general, with the rate of mutation and recombination in the genome at large being enhanced or reduced. But sometimes a particular chromosome or chromosome region has been found to be preferentially affected.

Recent studies have shown that altering chromatin conformation from condensed to extended makes the region more accessible to chemical mutagens, to repair enzymes, and to recombination enzymes. Conversely, a change to an inactive, condensed conformation may decrease the probability of DNA sequence changes. The processes that produce genetic variation—mutation, transposition and recombination—are all influenced by the gene's phenotype (Table 7.1). We shall look at each in turn.

1. *Mutation*

Some of the first studies showing a correlation between gene activity and mutability were made with inducible bacterial operons. For example, Herman and Dworkin (1971) found that in some strains of *E. coli* the rate of reversion of *lac*$^-$ mutations after treatment with the mutagen acridine orange was doubled when the lactose operon was induced. In another

Table 7.1 Environmental effects on genetic changes

Type of environmental stress or change	Change in chromatin structure	Activation of special enzymatic mutational machinery	Effects on mutation and/or repair	Effects on transposition	Effects on recombination
Any change leading to altered gene activity	Extended conformation; DNase-I sensitive if activated	None	Yes: Hanawalt (1987), Bohr and Wassermann (1988)	Yes: Jaenisch (1988), Bownes (1990)	Not known, but suggested by Thomas and Rothstein (1991)
Change in temperature	Change in gene expression[a]	Not known	Yes: Plough and Ives (1935), Evenson and Prescott (1970)	Yes: Ratner *et al.* (1992)	Yes: Plough (1917), Grell (1978*a,b*), Parsons (1988), Loidl (1989)
Change in nutritional conditions	Change in gene expression[a]	Not known	Yes: Durrant (1971), Cullis (1984)	Not known	Yes: Parsons (1988)
Starvation	Change in gene expression[a]	Yes (in bacteria)	Yes (in bacteria): Cairns *et al.* (1988), Hall (1988, 1990, 1991)	Yes (in bacteria): Shapiro (1984), Hall (1990)	Yes: Parsons (1988)
Demethylation	Extended conformation, DNase-I sensitive	Not known	Yes: Ho *et al.* (1989)	Yes: Fedoroff (1989)	Yes: Hsieh and Lieber (1992)
Infection by parasitic DNA	Not known	Yes: ripping	Yes: Selker (1990*a*)	Not known	Yes: Selker (1990*a*)
Behavioural stress	Not known	Not known	Yes: Belyaev and Borodin (1982)	Not known	Yes: Belyaev and Borodin (1982)

[a] Changes in gene expression are associated with changes in chromatin structure (Chapter 4).

study, Savić and Kanazir (1972) found that a constitutive state of the histidine operon in *Salmonella typhimurium* increased the rate of UV-induced frameshift mutations within the operon five- to eight-fold. As discussed in Chapter 3, transcription itself may be mutagenic, since, in both bacteria and yeast, transcriptional activity can enhance the spontaneous mutation rate.

The mutability of a region is determined by how accessible it is to both damaging agents and repair enzymes.[2] Not all mutagenic agents cause damage in the same way, so the features that make chromatin and DNA more sensitive or less sensitive to damage vary. In general, DNA in nucleosomes is more protected from damaging agents than linker DNA. The regulatory, DNase-I sensitive stretches which characterize the control regions of genes are often nucleosome-free and are particularly vulnerable to mutagenic damage (Bohr and Wassermann 1988).

The chromatin structure of active genes not only makes the DNA more likely to be damaged by mutagens, it also makes it more likely to be repaired.[3] Following UV irradiation, the distribution of genetic damage in the form of pyrimidine dimers is more or less random, but repair of this damage is not. In general, coding regions are repaired faster than non-coding regions, and active chromatin regions are repaired faster than inactive regions. For example, in human and Chinese hamster ovary cells, the active *dihydrofolate reductase* (*DHFR*) gene is repaired more efficiently than non-transcribed regions of the genome. Similarly, the inducible *metallothionein I* (*MTI*) gene is repaired more efficiently when induced (and expressed) than when repressed. For both the *DHFR* and *MTI* genes, a large chromatin region around the gene is also preferentially repaired. The size of this region, about 60–80 kb for *DHFR*, suggests that the unit of repair may be a chromatin loop, rather than a gene.

In both prokaryotic and eukaryotic cells, excision of the pyrimidine dimers formed after irradiation is more efficient in the transcribed strand of DNA than in the non-transcribed strand.[4] Taken together with the observation that repair is more efficient in active than in inactive chromatin, it suggests that transcription and repair are coupled. Mellon and Hanawalt (1989) suggested that a complex of repair enzymes detects when an RNA polymerase becomes blocked at a lesion, and this stimulates repair. They speculated that one of the reasons for the induction of transcription after UV irradiation is not so much that the gene products are needed, but rather that transcription ensures that damage is rapidly removed from important sequences.

Summarizing the data on the correlation between repair and gene expression in mammals, Hanawalt wrote:

> It is now clear that the repairability of damage in mammalian chromatin depends upon the type of lesion, its precise location in the genome, and the functional state of the DNA at that particular site. (Hanawalt 1987, p. 13)

Hanawalt's conclusion was based on DNA repair in cells in culture. If the same is true in the germ line, how does it affect the spectrum and frequency of mutations? If regions containing active genes are more mutable than those containing inactive genes, then regions that are active in the germ line or early embryo might show more inherited sequence variation than those that are not. Yeom and co-workers (1992) suggested that this is why the H-2K region of the major histocompatibility complex (MHC) in the mouse is so highly polymorphic. It contains a high density of genes, some of which are transcribed in the germ line or very early embryonic stages. The polymorphism could therefore result from an enhanced mutation rate caused by the open chromatin conformation in germ-line cells. The difficulty with this argument is, of course, that repair, as well as mutability, is greater in active chromatin. A case can therefore be made for the rate of sequence divergence being *less* in genes that are active in the germ line. Boulikas (1992) makes such a case.

Unfortunately, the spectrum of damaging agents and repair systems that have been compared in active and inactive genes is rather limited, so it is probably premature to try to reach any conclusions about the likely size and nature of differences in the divergence of DNA sequences that are and are not transcribed in the germ line. A further complication is that it may be wrong to assume that repair capabilities are the same in the germ line and soma. There are good evolutionary reasons for thinking that repair is probably more efficient in the germ line.[5]

2. *Transposition*

The activity of a locus is likely to alter its accessibility to viruses and transposons, as well as to chemical mutagens. Jaenisch (1988) showed that retrovirus integration occurs preferentially in the DNase-I hypersensitive sites associated with potentially active chromatin. Similarly, Bownes (1990) found that transposable P elements in *Drosophila* insert mainly into active genes, and therefore affect a somewhat different spectrum of genes in male and female germ lines. Either the more extended conformation of active chromatin, or some components that are bound to it, may facilitate the insertion of mobile elements. When Ratner and his colleagues (1992) studied transposition of a copia-like mobile element in *Drosophila* following a heat shock, which is known to change gene activity dramatically, they found transposition frequency in the germ line of treated males was increased by up to two orders of magnitude. The sites into which the transposons preferred to insert were not random; there were five hot spots for insertion, but the basis of this preference is unknown.

Transposition is affected not only by the activity at potential sites of insertion, it is also affected by the chromatin structure of the transposable element itself. For example, transposition of the *Spm* element in maize is affected by the extent of methylation of the element's control region (Chapter 6, p. 149)).

3. *Recombination*

Genetic variation does not result solely from changes brought about by mutation and transposition. It also stems from recombination. There are good reasons for thinking that the epigenetic state of chromatin affects the frequency of meiotic recombination, just as it does the frequency of mutation and transposition. For example, Grell (1971, 1978*a,b*) showed that heat stress affects recombination frequency in *Drosophila*. Exposing larvae to a high temperature during the premeiotic period increases crossing over in most regions. It can even induce recombination in the small fourth chromosome in which recombination normally never occurs. Remarkably, it can cause a more than 30-fold increase in recombination in centromeric heterochromatin.

Heterochromatin is highly condensed and transcriptionally inactive, and long ago cytogenetic studies suggested that recombination is suppressed in heterochromatic regions. In general, chiasmata do not form between segments of heterochromatin.[6] On the basis of an extensive review of the literature, Chandley (1986) concluded that genetic exchange and chiasma formation are restricted to early replicating chromosome regions. Heterochromatin (C bands) and the dark G bands of mammals, which are late replicating, are not sites of recombination. Direct microscopic observation by Ashley (1988, 1990) has shown that chiasmata are preferentially localized in regions corresponding to light G bands, and genetic analysis of recombination between rearranged chromosomes has confirmed that light G bands are preferred sites of crossing over. Light G bands contain active housekeeping genes (Table 4.2, p. 99), so the chromatin conformation associated with gene activity is also associated with recombination.

Recombination and chiasma frequencies are also influenced by the type of gametocyte in which meiosis takes place. For many species chiasma frequency in oocytes and spermatocytes is the same, but others show marked differences. In the most extreme cases there is no recombination in the heterogametic sex, usually the male, as in *Drosophila*. In other species, such as mice, both sexes have chiasmate meiosis, but there are differences in the frequency and distribution of chiasmata. Many alternative evolutionary explanations have been offered for these sex differences, but an explanation that has received attention recently is that they are associated with differences in gene activity in male and female gametocytes.[7] Oocytes and spermatocytes have different sets of active genes, and sex-specific differences in recombination could reflect this, with recombination being more frequent in females because they have a large number of active genes. Transcriptionally active chromosome regions may be more prone to recombination, either because they are more accessible to the enzymes that initiate recombination events, or because they suffer more mutagenic damage, and recombination is stimulated by repair processes. Thomas and Rothstein (1991) attributed the fact that the genetic map of human females is 90% longer than that of males to the greater gene activity in oocytes.

Comparison of the lengths of the genetic maps of different organisms also suggests that recombination may be restricted to transcriptionally active regions. The amount of crossing over that is characteristic for a species depends on several factors, including the number of chromosomes and amount of heterochromatin. No doubt it is fine-tuned by the species' evolutionary history. However, organisms with vastly different amounts of DNA and different chromosome numbers often have genetic maps of more or less the same size. Since the total lengths of the genetic maps are not very different, recombination per unit length of DNA must be much lower in organisms with large genomes. Because most of the differences in genome size reflect differences in the amount of non-coding DNA, Thuriaux (1977) suggested that the similarity in the size of the genetic maps may mean that recombination is confined to structural genes. Possibly it occurs mainly near the constitutively expressed housekeeping genes, the number of which is similar in all organisms.

The influence of the gene's phenotype on DNA sequence changes is undoubtedly complex. The way cytosine methylation affects DNA changes reveals the difficulty of making unqualified generalizations about the effect of chromatin structure on DNA sequence variation. Methylation at CpG sites has contrasting effects on mutability. On the one hand, a methylated cytosine is readily deaminated to thymine, so methylated CpG sites are hot spots for point mutations caused by C to T transitions; RIGS processes (Chapter 5, p. 124) may greatly enhance the frequency of these point mutations. On the other hand, CpG methylation participates in the organization of chromatin into an inactive 'closed' conformation, which is less accessible to mutagens than active chromatin, and is relatively protected from rearrangements and transpositions. Hence, in organisms that methylate their DNA, the quantity and distribution of methylated CpGs may be the result of a compromise between the mutability of methylated cytosines, and the protection afforded by methylated DNA from change in sequence organization through recombination and transposition.

The way that chromatin structure can influence the probability of changes in DNA sequence highlights the complexity of the role of the environment in evolutionary change. The environment is not just the agent of selection. Through its effects on the gene's phenotype, it also biases the direction, rate, and type of DNA changes at the locus. Consequently, very often the frequency of mutation and the frequency of recombination are *not* independent of selection.

Genetic assimilation mediated by chromatin marks

If a stimulus alters chromatin marks and they persist even when it disappears, and if altered marks affect the mutability or the reparability of the region, the genetic and evolutionary effects may be quite far reaching. For example, in some physiological conditions, a particular chromatin region may behave like a mutational or recombinational 'hot spot'.

Epigenetically-determined mutational hot spots may have results very similar to those observed in classical genetic assimilation. In Chapter 2 (p. 32) we discussed Waddington's genetic assimilation experiments and their interpretation in terms of selection of alternative alleles in an inducible polygenic genetic system. Through selection, such systems can be shifted from being stimulus-dependent to being stimulus-independent in a few generations. A similar transition to stimulus-independence can occur if the stimulus affects epigenetic marks and thereby alters the probability of mutational changes. Consider assimilation of the bithorax phenocopy produced by ether treatment of *Drosophila* embryos. Assume that, in embryonic cells that are going to contribute to germ cells and adult somatic structures, ether treatment affects the chromatin of loci influencing the bithorax complex. In the germ line these loci may become 'hot spots' for mutation and recombination. The ether-induced DNA changes may affect regulation of the bithorax complex so that the system becomes constitutive and produces the bithorax phenotype without the ether stimulus. In other words, a 'genocopy' of the phenotype would be seen; the originally stimulus-dependent response would be genetically assimilated. The difference between this type of assimilation, 'mutational assimilation', and the classical, recombination-dependent type is that recombination-dependent assimilation depends on the presence of genetic variability within the treated population, and on the shuffling of genes at meiosis. It could not occur in an asexual species, or in highly inbred lines. Mutational assimilation, on the other hand, can occur in inbred lines, and should occur almost as easily in an asexual species as in a random bred sexual population. 'Almost' because both the chromatin structure and DNA sequence of an allele influence its mutability, and genetic variation within a population may mean that some alleles mutate more readily than others because of their DNA sequence.

In Chapter 6 (p. 146) we referred to experiments of Ho and her colleagues (1983) that showed that genetic assimilation of bithorax can occur in inbred lines of *D. melanogaster*, and that it occurs in the absence of selection. We suggested that this could be caused by induced epialleles at the *bithorax* locus. However, mutational assimilation mediated by changed chromatin marks is another possible explanation of the results. It is not without interest that in his original experiments with bithorax phenocopies,

Waddington reported that new 'bithorax-like' mutations cropped up during the course of the experiment.[8] Were these purely 'random' mutations? If they were not, and appeared as a result of mutational assimilation, the ether treatment would have had to have been given very early in development. Mutational assimilation can occur only if the affected cells can produce germ cells, so it is most likely to occur before germ-cell determination. In the case of the bithorax assimilation experiments, ether treatment was given very early, two to three hours after the eggs were laid, which was probably just after germ-cell determination. However, the speed of events during the early embryonic stages of *Drosophila* development means that very precise timing of the treatment would be required to be confident that mutational assimilation had or had not taken place. Mutational assimilation is in fact far more likely to occur in organisms with late germ-line segregation or somatically derived germ cells, rather than in organisms like *Drosophila*.

Heritable marks and DNA sequence changes

If induced changes in chromatin marks are heritable, the mutational effect of a 'hot spot' will be enhanced because it will remain 'hot' even when the environment reverts to normal. Unfortunately, as far as we know, there are no studies exploring the relationship between *heritable* epigenetic marks and the probability of DNA sequence changes in that region. However, the type of effects that changed marks could have on allele frequencies can be seen by considering some simple computer simulations.

Consider a gene G in a haploid asexual population. It has two alleles, G^1 and G^2. When the gene carries mark *m1* it is inactive, and mutation from one allele to the other is infrequent. When it carries mark *m2* it is active, and the mutation rate from G^1 to G^2 and vice versa is much higher. Assume that initially all individuals in an infinitely large population carry allele G^1 and it is inactive. The population is then subjected to a different environmental condition that activates the gene, changing all marks from *m1* to *m2*. The new environment persists for a number of generations, but eventually conditions change so that the gene becomes inactive again. What effect does the period in which the gene was active have on the frequency of alleles G^1 and G^2? The answer to this question depends on the length of time the population spends in the changed conditions, the relative mutation frequencies, whether or not chromatin marks are inherited, whether there is selection for or against gene activity, and whether there is selection for one or other of the alleles. Fig. 7.1 gives the basic parameters of a model used to simulate changes in gene frequencies under certain conditions, and Fig. 7.2. gives the results of some of the simulations.

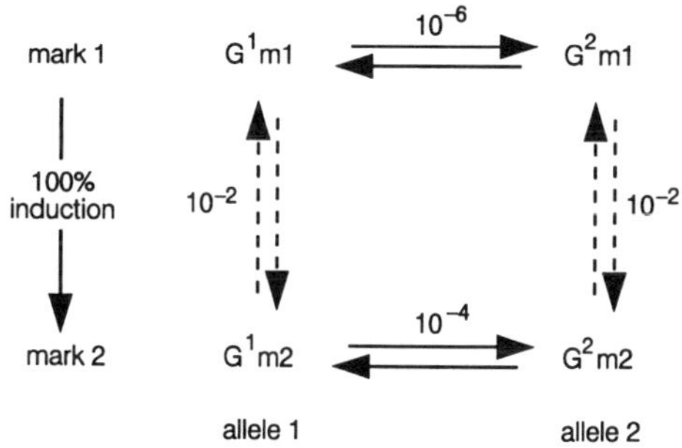

Fig. 7.1 Model used to simulate the effects of epigenetic marks that influence mutation rate on allele frequencies in a population. The mutation rate between G^1 and G^2 is 10^{-6} per generation when the gene carries mark *m1* and is inactive, and 10^{-4} when it carries *m2* and is active. When chromatin marks are not inherited, genes immediately adopt the marks and state of activity determined by the environment. When marks are inherited, the induced state *m2* is retained in the absence of the stimulus, and spontaneous changes between *m2* and *m1* occur at a rate of 10^{-2} per generation.

Figure 7.2a shows that if a gene can spontaneously change from one heritable state to another, it affects the rate of allele substitution. Even without an environmental change and without selection, there is a more than 40-fold increase in the frequency of G^2 when mark *m1* can spontaneously change to the more mutable state *m2* and that state is inherited. Figure 7.2b shows that if epigenetic marks are inherited, when activity is induced in every gene (*m1* is induced to *m2*), it causes only a small further increase in the frequency of G^2. Even when the inducing environment lasts for 50 generations, the frequency of G^2 is only slightly above the level reached through spontaneous transitions to *m2*. In other words, with the fidelity of mark transmission used in the model (10^{-2}), spontaneous changes result in a sufficient accumulation of genes in the *m2* state to have a larger effect on mutation accumulation than even substantial periods of induced activity. If the epigenetic memory had been better, i.e. the spontaneous rate of transition was lower, then the effect of induced changes in marks would be more pronounced. The effect of induced gene activity is greater when selection is applied. This can be seen in Fig. 7.2c where it is clear, once again, that inheritance of marks has a substantial effect on the rate at which the frequency of G^2 increases. After 300 generations of mild (1%) selection for G^2, its frequency is more than ten times greater than without epigenetic inheritance, both when activity is induced and when only spontaneous changes occur. Figure 7.2d shows the obvious: selection for the state of activity (mark *m2*) has no effect on the frequency of G^2 unless marks are inherited. Even when marks are inherited, the effect on G^2 of selection for *m2* is small, because most genes will be G^1m^2.

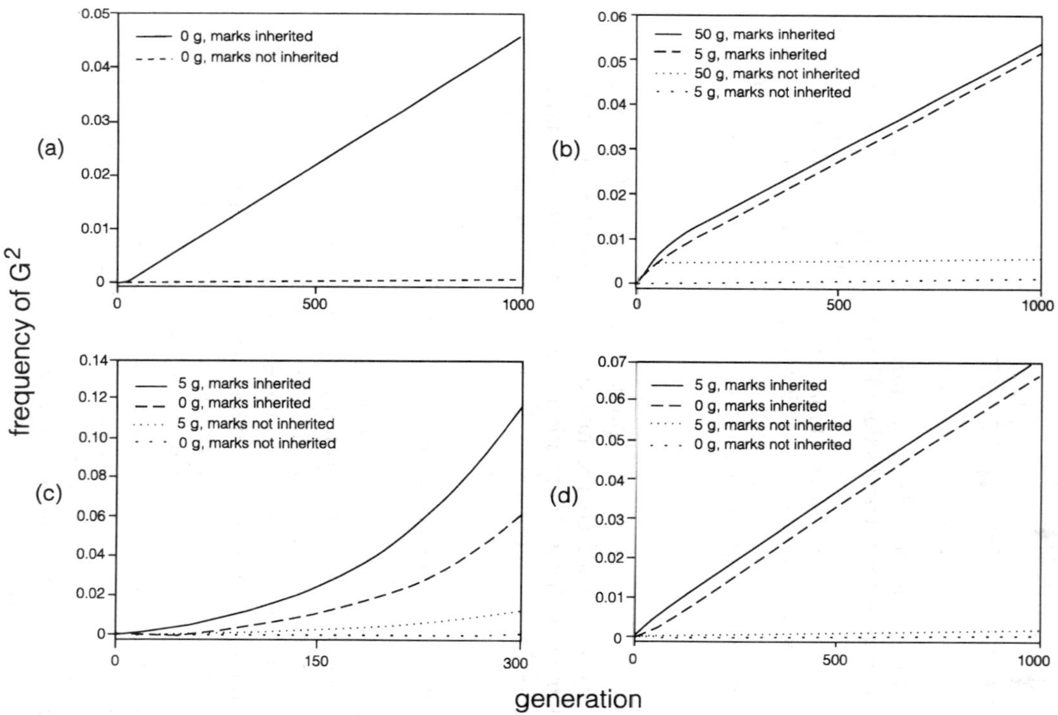

Fig. 7.2 Simulations of the effect of inherited marks on the frequency of allele G^2 in an infinitely large population. Initially all individuals had allele G^1 in an inactive state (with mark *m1*). In changed environmental conditions, all genes become active and acquire mark *m2* and, unless there is epigenetic memory (i.e. marks are inherited), they revert to *m1* when the environment no longer induces activity. **g** is the number of generations in the environment inducing gene activity. (a) The effects of spontaneous changes in marks. (b) The effects of exposure to an environment that induces activity (changes marks to *m2*). (c) The effects of inheriting marks when there is continuous selection (1%) for gene G^2. (Note the change in scale.) (d) The effects of continuous selection (1%) for gene activity (mark *m2*).

From the simulations shown in Fig. 7.2 it is clear that the inheritance of epigenetic states can have substantial effects on the rates of change of gene frequencies. Both the activity of a gene (its epigenetic state) and the stability of its epigenetic memory influence the rate and direction of genetic change.

The evolution of genomic responses

In normal environmental conditions, preserving, rather than changing, existing DNA sequences and organization makes evolutionary sense. Preferential repair of damage in active chromatin is probably a selective advantage, since active genes are engaged in physiologically relevant processes and their integrity is important. Provided that the metabolic cost is not too high, systems that ensure efficient error-free repair of active genes should be favoured. In a similar way, in unchanging conditions, forms of asexual reproduction, which preserve existing DNA sequences and genomic organization, should be preferred to sexual reproduction and extensive recombination. However, in conditions of stress, when by definition normal functions are inadequate, retaining existing DNA sequences and organization may not be advantageous. Indeed it can be argued that organisms are expected to evolve mechanisms that generate new DNA variation in conditions that threaten their survival. They should have what Koch (1993) has called 'catastrophe insurance'. One way of generating the diversity that might produce genomic variations appropriate for new conditions is through sexual reproduction. There is ample evidence suggesting that in organisms capable of both sexual and asexual reproduction, sexual reproduction is resorted to in adverse conditions such as crowding and starvation (Bell 1982, pp. 370–371). The switch to sexuality is an evolved response to recurring environmental stress, and is under epigenetic control. Bacteria have another type of evolved response to recurring stress: high mutability. Loci that have the potential to provide an adaptive solution to stress have a high mutation rate (Moxon *et al.* 1994).

As McClintock (1984) and Wills (1984) have emphasized, cells experience different types of stress, and their genomic responses are not the same for all types. Stresses can be recurrent, having happened many times in the evolutionary past, or they can be unique and unanticipated; they can be caused by accidents within the cell, or by external conditions. Cells have some way of sensing the stress and responding. Their genomic responses are the result of evolved mechanisms, so the way new genomic variations are generated, and the frequency and extent of the changes, will depend on the organism, the type of stress, its severity, and its familiarity.

In most cases, the molecular mechanisms that produce genomic changes are unknown. What is clear is that it is not just the local condition of a gene that determines whether or not it is changed. A locus is not isolated from the rest of the cell: the chances of a locus changing also depend on the type of enzymatic mutation- and repair-systems that are in operation in the cell. The cell's responses alter according to the nature of environmental conditions. They may vary from non-specific, 'desperate' responses, to very specific programmed responses. It is possible to imagine a set of 'rules' that determine the genomic response to internal and external challenges:

(1) If the change in environment is not stressful, but alters the state of activity of some genes, DNA sequences will change in a way that depends on the altered chromatin conformation of the loci. The other rules are all superimposed on this rule.
(2) If a stress is familiar and specific, evolved mutational systems with specific effects may be mobilized to deal with it (e.g. directed mutational changes such as those reported, but not understood, in bacteria, or mutational systems such as those involved in ripping; see Chapter 3).
(3) If a stress is not very acute but general, mutational systems that affect loci having general effects on cell physiology are activated (e.g. amplification of ribosomal DNA, such as is found in flax; see Chapter 3).
(4) If a stress is unfamiliar and acute, error-prone repair systems are activated (the 'desperate' measure), or the rate of variation in the genome is increased non-specifically by other means (e.g. by increasing the rate of transposition, as suggested by McClintock (1984)).

Environmental change and stress therefore affect the probability of a DNA change in two somewhat different ways. First, under conditions of stress, different types of DNA-damaging and repair systems may be activated. Second, local chromatin conformation may be changed, thus altering the probability of the locus being mutated. The two phenomena may or may not be causally related, but both will contribute to an altered mutation frequency at a locus.

The interactions between the epigenetic state of a locus and the generation of DNA variations in that locus mean that the feedback between the expression of a gene and its DNA sequence is much more direct than is usually assumed: gene expression affects not only the probability of fixation of a DNA variation through selection, but also the probability of its occurrence. The chances that the DNA sequences of a locus will change in a new environment depend on the effect of the new conditions on its epigenetic state, and on how long the environmental change lasts. Clearly, the longer it lasts, the stronger the effect. As the model showed, environmental effects on a DNA sequence are determined by the length of the period of environmental induction, the extent of epigenetic heritability, and the mutation rate, which depends on the epigenetic state.

The effect of DNA organization on epigenetic inheritance

It is clear that through their effects on mutation and recombination, chromatin structure and chromatin-marking EISs influence the way DNA sequences in chromosomes are organized. It is equally clear that DNA organization must affect chromatin structure and the chromatin-marking EISs. Inherited methylation patterns and protein marks are replicated and segregate with DNA during cell division, and the number and nature of the DNA sites that are methylated or bind proteins must influence the stability of clonal inheritance, or what we shall call clonal 'memory span'. By memory span we mean the average number of cell divisions through which a particular epigenetic state is faithfully transmitted to daughter cells.

Holliday (1990*a*) has pointed out that one aspect of development that has been neglected is the way in which cell lineages measure time. He suggested that there are molecular mechanisms that measure time by counting cell divisions. After a certain number of divisions, a lineage switches to a different developmental state. Holliday calls the mechanisms for counting cell divisions 'developmental clocks', and believes that they may be based on progressive changes in the epigenetic state of a chromatin region, with the number of methylation sites and number of repeated sequences being important. What we are calling a 'memory span' is very similar to Holliday's epigenetically-determined developmental clock. We prefer the term memory span partly because we want to avoid the flavour of precision and cyclicity that may be associated with a clock, but mainly because we believe that epigenetic memory is important in evolution as well as in development. Epigenetic memory may affect lineages of organisms, as well as lineages of cells. We shall argue that the evolution of some aspects of genome organization reflect past selection for loci and domains to have particular memory spans. The organization of the genome can be understood as an evolved memory apparatus.

As with all Darwinian evolution, the evolution of different memory spans requires variability on which selection can act. There is ample evidence showing that such variability exists. Populations are polymorphic for the amount of heterochromatin and the number of repeated sequences at particular sites, and there is variation in the patterns of methylation both within and between individuals. Like many other polymorphisms, such as allozyme variations, in most circumstances the polymorphisms are probably of no selective importance. Stabilizing selection weeds out extreme variants that have substantial effects on gene expression, but a range of variants is tolerated. It is only when the environment changes that the existence of variation becomes important.

1. *Clonal memory span and CpG clustering*

As discussed in Chapter 4 (p. 97), the activity of some genes is associated with methylation of cytosines in the CpG dinucleotides of their control regions. Usually when CpG sites are methylated, genes are inactive; when demethylated, they are potentially active. It is now becoming clear that for some genes the stability of inactivation is determined by the density of methylated sites in their promoters: the higher the density, the more stable the inactive state (Boyes and Bird 1992).

Methylated CpG sites can repress gene activity either directly by inhibiting the binding of transcription factors, or indirectly by binding with proteins that interfere with transcription. Many sequence-specific binding proteins are known to be influenced by the state of methylation of their target sites: usually, methylation inhibits binding. However, there are some proteins that bind specifically to methylated CpGs, and do so regardless of their sequence context. One of these, MeCP-1 (methyl-CpG binding protein), is particularly interesting. It binds to DNA containing several methylated CpGs. For tissue-specific genes, the number of methylated CpGs determines the stability of the inactivation mediated by MeCP-1. When there are few CpGs, inactivation by methylation is unstable and easily overcome by a strong enhancer, whereas when the density of CpGs in the promoter region is higher, inactivation cannot be relieved by a strong enhancer. Fig. 7.3 illustrates this effect of methylated CpG density on the stability of gene inactivation. The scheme suggests that the evolution of stable and heritable gene activity may be related to changes in the number of CpGs clustered in control regions.[9]

Tissue- and stage-specific genes have promoters in which the CpG sites are methylated in tissues in which they are not expressed. Each tissue-specific or stage-specific gene has to have a characteristic ‘memory span’. During development, the genes need to become active after a certain period in which they were inactive, or vice versa. They may have to respond to spatial gradients of activators or repressors. The density of CpG sites could influence the memory span of these genes. If a mutation occurs in a promoter such as one of those shown in Fig. 7.3, so that it has more CpGs, it will probably be more resistant to reactivation, or be inactive for a longer period. For example, if the probability of spontaneous demethylation is 10^{-2} per site, having three CpGs decreases the probability that all three have lost the methyl group (and the binding protein) to 10^{-6}. A mutation changing the number of CpGs in a promoter region may thus cause a shift in the time during development at which the gene controlled by this promoter is activated. Alternatively, it could cause a shift in the spatial pattern of activity, since the response to a gradient of activator may depend on the promoter's binding strength.

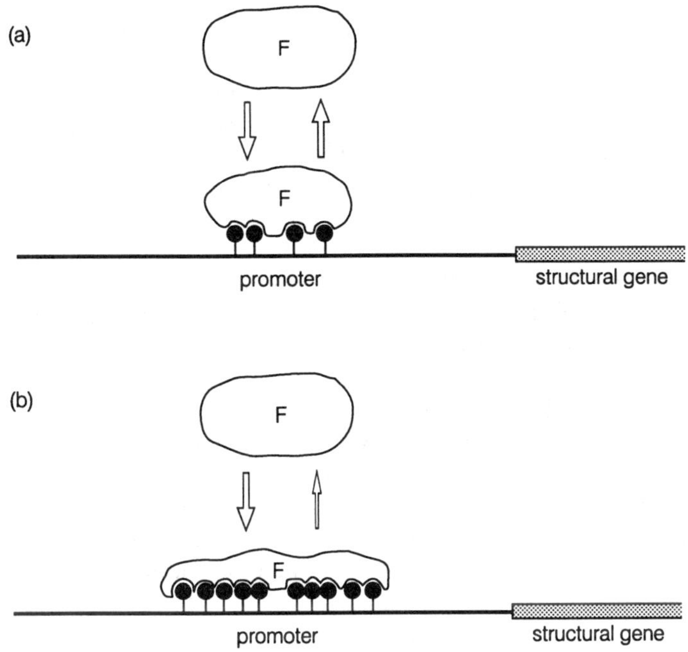

Fig. 7.3 The effects of CpG density on the binding of proteins to methylated DNA. The initial binding of protein **F** is assumed to be the same for both a low-density promoter (a), and a high-density promoter (b), but the low-density promoter releases the protein more readily. The likelihood of binding and release is indicated by the thickness of the arrows. (After Boyes and Bird 1992.)

2. *Tandem repeats as vehicles of clonal memory*

Short, tandemly repeated sequences are now playing an important role in forensic science and population biology. They form the basis of the 'genetic fingerprinting' technique. Since the number of repeats varies from individual to individual and is inherited, ancestry can be determined. But the number of repeats is of more fundamental importance than this: they play a role in the regulation of gene expression. They are found near most genes, and many control regions contain repetitive sequences that act as binding sites for regulatory proteins.[10] On the basis of an extensive review of the literature, Vogt (1990) suggested that repetitive sequences adopt local and specific folded conformations, which assemble proteins to form characteristic chromatin structures. He called the ability of a particular type of repeat to fold and organize chromatin in characteristic way, the 'chromatin folding code', and suggested that this folding code affects transcriptional activity, time of replication, and recombination. The folding code acts as an expression code because the proteins assembled in control regions determine the regulation of adjacent coding sequences.

It is not difficult to see how, if DNA binding sites are repeated in tandem, the stability of states of activity may be increased. Proteins that bind to the repeated sequences could form highly stable multimeric complexes. Fig. 7.4 illustrates how the number of tandem repeats in the control region could affect the stability of the functional state of a neighbouring gene.

The way in which the DNA-protein interactions that determine gene activity are transmitted to daughter cells is unknown, but mechanisms similar to those that transmit methylation patterns have been proposed (Chapter 4). Many binding sequences are symmetrical on the two DNA strands, and the old chromatin structure can be retained if protein subunits that are symmetrically bound to DNA remain bound to the parent strands when DNA replicates, and nucleate the formation of a new complex (Fig. 4.12b, p. 101). If tandemly repeated motifs are present, it may not be necessary for the binding site to be organized symmetrically. The protein complex of newly replicated DNA could be re-assembled gradually at the

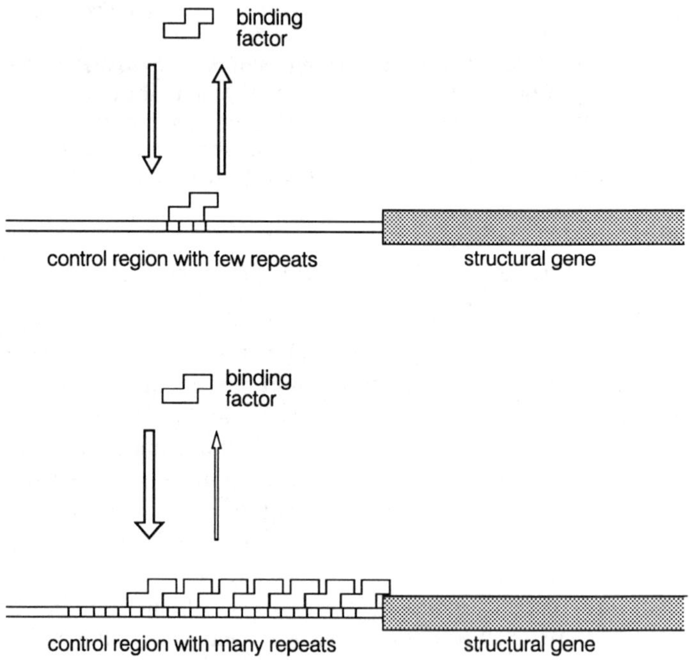

Fig. 7.4 The influence of tandem repeats on the binding of proteins repressing gene activity. The more tandem repeats, the more likely that additional factors will bind, and the less likely they are to be released once bound. The likelihood of binding and release is indicated by the thickness of the arrows.

replication fork, using the proteins bound to the repeats of the double-stranded, yet-to-be replicated DNA to nucleate assembly on the newly replicated regions. As with the simpler model described in Chapter 4, in this model a type of structural inheritance occurs, with a three dimensional, pre-existing structure directing the assembly of a new identical structure in the next generation. The fidelity of transmission—the memory span—would be increased with increasing number of tandem repeats. Tandem repeats may therefore play a role in ontogeny, and in the evolution of ontogeny. Repeat clustering, like CpG clustering, may reflect past selection for a particular cellular memory span during development.

Chromosomal organization and clonal memory

One of the more obvious features of chromosome organization, particularly in vertebrates, is the presence of visible chromosome bands. These large genetic units, some of which contain many genes, may be units of regulation with features that determine the memory spans of genes. In this section we describe two ways in which bands may affect memory span. First, there is a genome-wide effect, mediated by repeated sequences in constitutive heterochromatin; second, there is a regional control of memory span, which operates within domains and within euchromatin bands. We suggest that in all cases the mechanisms underlying the stability of chromatin structure, and hence the fidelity of epigenetic memory, depend on the binding of multimeric protein complexes.

1. *Constitutive heterochromatin and clonal memory*

Repeated sequences seem to play a role in the large-scale control of chromatin organization. As discussed in Chapter 4, chromosomes show various types of bands, the most easily detectable being the C bands which distinguish heterochromatin and euchromatin. Regions of constitutive heterochromatin are commonly found around the centromeres and at the ends of chromosomes. Frequently they show ectopic pairing, with aggregations of heterochromatic regions forming 'chromocentres' visible in interphase cells. With some minor variations, the amount and distribution of heterochromatin is characteristic for the chromosome, and for the species. The maintenance and transmission of these heterochromatic regions provides the most striking example of the long-term inheritance of states of chromatin organization. Very few genes are located in constitutive heterochromatin; most of the DNA is satellite DNA. It is made up of a variety of tandemly repeated sequences, with the repeat unit varying from a few to several thousand base pairs. The DNA is not transcribed, usually does not take part in meiotic recombination, is late replicating, and in many organisms it is highly methylated. Commonly it is regarded as 'junk' DNA, with scarcely any effect on fitness, although Zuckerkandl (1986, 1992) has argued that this may be an inappropriate way to look at it. Some heterochromatic regions certainly do have an effect on fitness. For example, the *Responder* locus of *D. melanogaster* is located in centromeric heterochromatin, and is made up of a tandem array of 120 bp repeats, yet it affects fitness both through being the target locus of *Segregation Distorter*, and through effects on viability in the pre-adult stage.[11]

Although most constitutive heterochromatin has no known phenotypic effect, it can dramatically change the expression of euchromatic genes (Chapter 4, p. 92). When a chromosomal rearrangement brings a gene next to constitutive heterochromatin, in some cells the inactive state of heterochromatin spreads into the euchromatic genes and inactivates them. The effects of this invasive spreading, position effect variegation (PEV), were first detected in *Drosophila*, where a normally euchromatic gene, such as the wild-type allele of *white* eyes, when moved near to centromeric heterochromatin, becomes heterochromatinized and inactive in some cells (Fig. 7.5). Once inactivation occurs, the inactive state is clonally inherited. The result is a mosaic, or variegated, phenotype.[12] If a further chromosome rearrangement moves the variegating gene back into euchromatin, the wild-type phenotype is restored, showing that inactivation is caused by sequences located in heterochromatin.

The extent of variegation depends on both environmental and genetic factors. Inactivation can spread over distances greater than 1000 kb, but how far it spreads depends on the site of the junction between euchromatin and heterochromatin. Genes near the junction are more subject to variegation than those further away. Variegation is suppressed by butyrate, an inhibitor of histone acetylation, and by deletions of histone genes. It is also influenced by the amount of heterochromatin in the genome, with additional heterochromatin suppressing variegation, and by numerous modifiers that can enhance or suppress it.

How is PEV explained, and what do the factors that influence variegation actually do? The formation of heterochromatin, and the spreading of the heterochromatic state into adjacent euchromatic genes, seems to be dictated by the tandemly repeated DNA sequences of heterochromatin. The interaction of these repetitive sequences with specific combinations of proteins leads to chromatin condensation. Some of the mutations that suppress or enhance PEV are in loci coding for DNA-binding proteins. With most of these loci, deletions suppress variegation and duplications enhance it, suggesting that their gene products play a role in the packaging and organization of heterochromatin. A few loci have the opposite effect, and these probably either organize euchromatin, or inhibit the formation and spreading of the heterochromatic state. The structure of the proteins encoded by some of the loci is known, and points to a DNA-binding

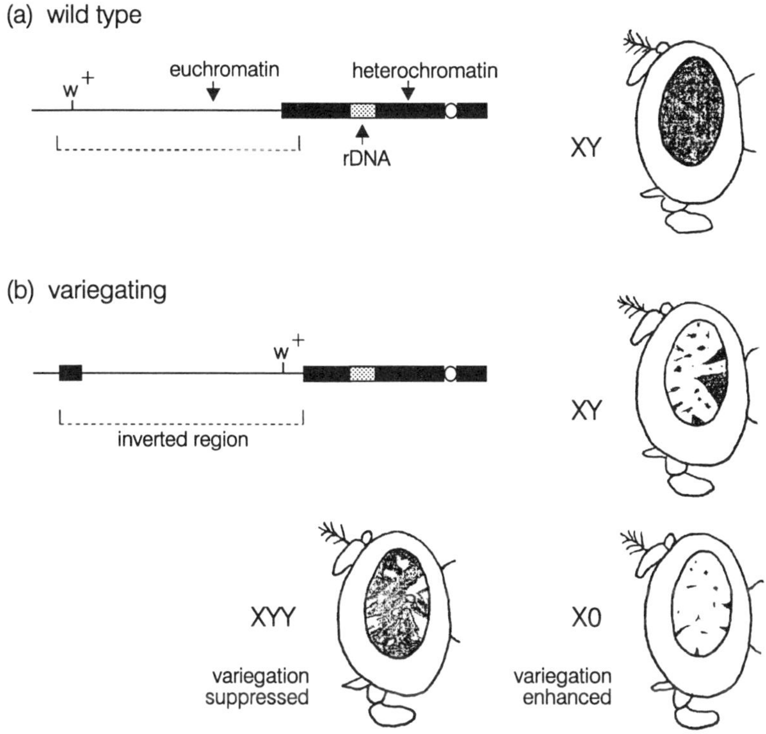

Fig. 7.5 Position effect variegation (PEV) in *Drosophila*. (a) The normal X chromosome and wild type, red-eyed phenotype. (b) An inverted chromosome with the eye-colour locus moved to near the centric heterochromatin; the eye is variegated, a mosaic of red and white clones of cells. The number of Y chromosomes modifies expression.

function; the proteins associate preferentially with heterochromatin and a few other chromosomal sites.

Locke and his colleagues (1988) suggested a simple and elegant model, based on the law of mass action, to explain how the products of modifier genes form heterochromatin. They believe that the modifiers code for components of a multimeric complex that spreads progressively by the addition of new units. It is a type of directed self-assembly process. The number of copies of each modifier gene affects the concentration of each component, and hence the assembly and spread of heterochromatin into neighbouring regions.[13] Given the stochastic nature of the processes that establish the 'invasive' chromatin conformation, it is remarkable that once a certain chromatin structure has been established, it is inherited in the cell lineage. The EIS is very efficient, ensuring the stability of transmission of the established state.

One of the particularly interesting features of PEV is that it shows parental origin effects. The extent of variegation depends on whether the rearranged chromosome was transmitted by the mother or by the father, and in males, on the source of the heterochromatic Y chromosome. Using various genetic tricks, it is possible to produce male *Drosophila* with Y chromosomes inherited from their mother, rather than from their father. They show a different degree of variegation. It is thought that Ys from males and females, and from different strains, use different amounts of the various chromatin components for their heterochromatinization. By competing for chromatin components, Y chromosomes, and any additional heterochromatic segments in the genome, influence the expression of variegating genes because they alter the availability of the chromatin proteins. A change in the chromatin structure of the Y can have effects on many generations of male descendants. Dorn and colleagues (1993) showed that a mutation in an autosomal locus coding for a DNA-binding protein that affects chromatin conformation at various loci, also affects the Y chromosome. The modified chromatin of the Y influences the extent of PEV. Remarkably, in progeny in which the autosomal gene was not present, the Y retained its altered chromatin structure through several generations, and its phenotypic effect on PEV persisted.

The expression of a eukaryotic gene is therefore affected not only by the nearby linked repeated DNA sequences that constitute heterochromatin, but also by the amount and conformation of heterochromatin elsewhere in the genome. All contribute to determining the heritable state of activity or inactivity of the gene.

The role of additional heterochromatin in determining the extent of PEV may provide a clue to its function. Heterochromatin has had many different functions attributed to it, all based on the general effects it has on regulation, recombination, chromosome segregation, and so on. The additional role proposed here does not exclude these other roles. The proposal is that heterochromatin has a general effect on memory spans in the genome. Repeats in heterochromatin act as sinks for the various DNA-binding proteins that determine the chromatin structure and stability of many loci throughout the genome. Large heterochromatin blocks may have evolved because large blocks of repeats are more efficient as sinks than short sequences; they allow nucleation of DNA-protein complexes and their spread along the chromosome. These properties may make blocks of heterochromatin general determinants of cell memory.

The concentration of repeated sequences into large blocks, rather than being dispersed throughout the genome, may avoid some of the perils presented by multiple small blocks of repeats. The greatest potential hazards of having dispersed multiple repeats are interchromosomal recombination, and interference with the regular chromosome pairing and consequent segregation during meiosis. If repeats are clustered near the centromeres or telomeres, however, the potential hazards are reduced: recombination between non-homologous chromosomes often will not disrupt chromosome structure in a way that seriously interferes with chromosome behaviour. Moreover, clustering of repeats may reduce recombination: methylation of multiple sites, or cooperative binding of proteins to clustered repeats, allows efficient condensation of the region and makes the repeats inaccessible to recombination enzymes. Large blocks of repeats may therefore be less likely to undergo recombination than scattered repeats.

2. *Inactivation of chromosome domains*

There is evidence from *Drosophila* suggesting that repeated sequences other than those in heterochromatin may influence the formation of repressed chromosome regions. Many years ago, some regions of *Drosophila* polytene chromosomes were dubbed 'intercalary heterochromatin', because they shared certain properties with centric heterochromatin. They had a tendency to pair ectopically, to be under-replicated, and to form weak points susceptible to breakage. Subsequently, it was shown that intercalary heterochromatin is also late replicating, and that its under-replication is enhanced by the same environmental and genetic factors that enhance PEV (Zhimulev *et al.* 1989). Unlike heterochromatin, however, sites of intercalary heterochromatin are not sites of satellite DNA. Nevertheless, several of them are known to be the sites of repeated euchromatic genes, such as the multiple copies of the histone genes, the 5s RNA genes, and some tRNA genes. It is possible that their 'heterochromatic' properties are associated with the presence of repeats that cause their chromatin to be packaged in a 'heterochromatic' way.

Some of the gene products that are important in packaging heterochromatin also bind to normal euchromatic regions, suggesting that they may play a role in the structure of euchromatin. They may be partly responsible for organizing euchromatic regions into heritably silent states. For example, Paro (1990) suggested that they could be involved in the maintenance and inheritance of the functional states of homeotic genes. The homeotic genes of *Drosophila* are concerned with the characteristic structures associated with each body segment. They are organized into two large clusters: the Antennapedia complex, ANT-C, is responsible for the head, the first thoracic segment, and the anterior part of the second thoracic segment; the Bithorax complex, BX-C, is responsible for the posterior part of the second thoracic segment, the third thoracic segment, and the abdominal segments.[14] One of the intriguing features of ANT-C and BX-C is that the order of the genes on the chromosome corresponds to their order of expression in development. Furthermore, a mutation in one gene tends to inactivate genes immediately distal to it.

Analysis of mutants has revealed that there is a gene, *Polycomb* (*Pc*), that is necessary for perpetuating, but not establishing, the repressed states of genes in the homeotic complexes. Mutations in *Polycomb* and similar genes lead to instability in the transmission of repressed determined states. Genes are expressed in the wrong places; anterior segments tend to be transformed into posterior segments. The protein encoded by *Polycomb* has a region of homology with one of the proteins that organizes heterochromatin. It is possible that this region, called a 'chromodomain', is important in assembling and transmitting inactive chromatin. The products of *Polycomb* and similar genes are thought to preserve the pattern of gene expression established in the early embryo by forming multimeric complexes that spread along the homeotic gene clusters until they encounter genes that are being transcribed.[15] Genes covered by the protein complex are not expressed, and the extent of the region covered by the complex is inherited by daughter cells. Whether or not repeated sequences play a role in the assembly of the heritable repressed state of *Drosophila* BX-C and ANT-C genes in the same way as they do in heterochromatin is not clear; however, the relationship between the sequence of the genes in the chromosomes and the sequence of their expression domains suggests very strongly that the organization of DNA sequences is influencing epigenetic states and their inheritance. This DNA organization may be associated with DNA sequences that determine features such as the bendability of DNA, which enable the enucleation and spread of protein complexes to organize a heterochromatin-like structure.[16] The principle is similar to the spread of heterochromatin in PEV, although the nature of the DNA-binding sequence is different.

Regional repression is not a peculiarity of *Drosophila*. Yeast also shows position effects.[17] The position effects exerted by yeast telomeric sequences are similar to those exerted by heterochromatin in *Drosophila*. Gottschling and his colleagues (1990) found that when an active gene is moved to a site near a telomere, it is often repressed, and the repression can spread for distances of 3–4 kb.[18] Another position effect is seen with the mating type loci of yeast. The heritable alternative states of these loci (discussed in Chapter 4, p. 107) also depend on where they are located. When silent, they are in regions of general repression. Many of the molecules associated with this state are the same as those that mediate repression in telomeric regions. The repressed state is not very stable, but can be clonally inherited for 10–20 cell divisions.

Spread of heterochromatin following a change in chromosome position is also seen in mammals. Autosomal regions present in the inactive X chromosome are often inactivated and heterochromatinized. As described in Chapter 4, clonal transmission of the inactive state of X chromosomes in female mammals is the best understood example of the clonal inheritance of repressed states. Such facultative heterochromatinization also occurs in all the paternally-derived chromosomes in male coccids, and in the paternally-derived X chromosomes of *Sciara*. DNA methylation may participate in the packaging process associated with this heterochromatinization, but methylation is not essential, since species such as *Drosophila* and *Sciara* do not methylate their DNA.

Is there any reason for thinking that repetitive sequences are important in facultative heterochromatinization? The precise role of repetitive sequences is not clear, but there are hints that they may be important. For example, at the mammalian X-inactivation centre, from which inactivation spreads, there is a gene containing tandem repeats that could serve a role in the initiation of heterochromatinization.[19] In the X chromosome of *Drosophila*, Lowenhaupt and colleagues (1989) found that the concentration of simple repeats of the mononucleotide C/G and dinucleotides CA/GT and CT/GA is at least twice as high as that in other chromosomes. They suggested that these sequences may be involved in dosage compensation, which in *Drosophila* is achieved through an increase in the activity of the single X in males, but the repeats could also facilitate preferential X-inactivation in the male germ line. The coccid genome has been found to have long runs of adenines, and it has been suggested that this may play a role in heterochromatinization of the paternal chromosome set (Epstein *et al.* 1992). In *Sciara*, Crouse (1960) found that the elimination of the paternal Xs is determined by regions of heterochromatin in the X chromosome. In all these cases, it is at least plausible that neighbouring repeated motifs bind proteins that initiate the assembly of multimeric complexes of DNA and proteins, and these determine and maintain the state of activity of the whole chromosome in cell lineages.

3. *Epigenetic memory and chromosome banding in vertebrates*

Vertebrate chromosomes show another type of banding, G banding, which is not usually considered to be related in any way to the constitutive heterochromatin typical of the centromeric and telomeric regions (C bands), or to the facultative heterochromatin of the sex chromosomes. Mammals have about 2000 light and dark G bands, each of which can include over 1000 kb of DNA and cover several looped domains (Chapter 4, p. 93). Light and dark G bands have distinct structural and functional properties. They differ in their base composition, the type of gene within them, the types of repetitive sequences they contain, and the time at which their DNA is replicated (Table 4.2, p. 99). Like C bands, G-banding patterns are stable, heritable, species-specific features of chromosomes.

What are these bands? How did the banded chromosome structure evolve? In this section we suggest that the G-banded chromosome structure reflects the evolution of regional regulation of vertebrate genomes.

The sequence composition of the band, the nature of its repeated motifs, and its overall organization determine the memory span of the genes located within it. With the increase in the number of tissue-specific and stage-specific genes in vertebrates, selection favoured a clustered organization that enables the simultaneous and long-term suppression of many genes.

Holmquist (1989) believes that dark G bands should be regarded as facultative heterochromatin, and that their distinctive properties are associated with the time at which they replicate DNA. He suggested that mechanisms similar to those that enable insects and other non-chordates to heterochromatinize whole chromosomes facultatively have been adapted by vertebrates for repressing small groups of genes. He imagined an evolutionary scenario that begins with vertebrate ancestors in which all DNA is early replicating. As the complexity of development and number of cell types increased, so did the number of tissue- and stage-specific genes. The number of cell types in higher vertebrates is at least twice and probably more than three times greater than that in higher invertebrates (see Table 8.1, p. 202), so the demands of regulation and coordination are much greater. Late replication evolved as a mechanism for repressing the activity of stage- and tissue-specific genes at times and in places where they were not needed. Tissue-specific genes that had to be subject to long-term inactivation became clustered in dark G bands. Housekeeping genes, which are active in all cell types, remained early replicating and were clustered in light G bands. Once bands with different times of replication had been established, other changes occurred within them.

One of the changes was in base composition. The genome of vertebrates is a mosaic of isochores (Bernardi 1989, 1993). Isochores are stretches of DNA, more than 300 kb long, of homogeneous base composition, which are interspersed with other regions of a different base composition. In mammals, isochores that are rich in G and C are found in light G bands. The GC-richness is not just the result of the CpG islands associated with housekeeping genes: both coding sequences and non-coding sequences, such as introns, also have a GC bias. In contrast, dark G bands have GC-poor isochores.

Why are some regions richer in G and C than others? How did isochores evolve? There are several possible explanations. Filipski (1988) and Sueoka (1992) suggested that they are the result of mutational bias. As we discussed earlier in this chapter, the rate and nature of the damage and repair of genes that are active in the germ line are different from those of inactive genes. Consequently, because they contain housekeeping genes that are active in germ-line cells, light G bands are subject to a different type of damage and repair, and this causes them to acquire a more GC-rich spectrum of base changes than dark G bands. An alternative explanation for the existence of isochores was suggested by Wolfe and co-workers

(1989); this explanation is based on differences in the time at which light and dark bands replicate, and changes in the chemical environment during the cell cycle. The availability of the bases needed for DNA replication is not always the same, so the type of replication error likely to occur will change during the period of DNA synthesis. Genes that replicate early accumulate more changes to C and G, whereas late-replicating genes accumulate more changes to A and T. A third explanation of the evolution of isochores, suggested by Bernardi (1989, 1993), is that they have evolved through their effect on thermal stability. GC-rich DNA is more stable than AT-rich DNA at the body temperature of warm-blooded vertebrates. The same changes that make the DNA more stable also increase the stability of the proteins the DNA encodes, because many of the codons containing G and C code for amino acids that contribute to the thermostability of proteins. Selection has therefore favoured AT to GC changes in the genes of functionally important proteins.

One of the elements that contribute to the GC richness or poorness of isochores in mammals is the type of repeated sequences that they contain. There are two types of interspersed repetitive sequences: long interspersed repeated elements (LINEs) and short interspersed repeated elements (SINEs).[20] Both are types of retrotransposons, capable of amplification. SINEs are derived from small RNA molecules that have been reverse transcribed and integrated into the genome. The predominant SINE family in humans is the *Alu* family, with nearly a million copies of a sequence of about 300 bp dispersed in the genome. LINEs result from the movement of transposable elements. They are much larger than SINEs and encode some or all of the enzymes required for their transposition. The major human LINE, L1, can be up to 7 kb in length, although many of the 50 000–100 000 copies in the genome have deletions. The distribution of LINEs and SINEs in the mammalian genome is not random. Most SINE sequences reside in light G bands and have a similar base composition, whereas most LINEs are found in, and have a similar base composition to dark G bands. LINEs and SINEs therefore match the sequence environment in which they are found. Zuckerkandl (1986) suggested that the non-random distribution of LINEs and SINEs is the result of the need for retrotransposons to be 'polite'. 'Polite' DNA leaves the sequence composition into which it inserts undisturbed. For example, an AT-rich transposon will be found in AT-rich regions, because it 'prefers' this region, and the region is more 'hospitable' towards such a transposon.

There is another possibility, however. It may be that rather than merely being tolerated by the regions into which they insert, LINEs have been important in bringing about the stable repression of genes in dark G bands. LINEs could be sequences that facilitate the binding of proteins that repress gene activity. This could happen in two ways. First, LINEs could bind repressor proteins and act as nucleation centres for the spread of protein complexes along the band. Second, they could affect chromatin structure through pairing ectopically and thereby initiating the formation of multimeric complexes of the factors that stabilize repression and increase memory span. Whether LINEs are important or not, the existence of G bands containing clusters of inactive genes and having a different, characteristic base sequence and time of replication, suggests that long-term repression may be easier when the unit is a block of genes rather than an individual gene. Both coding and non-coding DNA is organized in ways that are shaped by the need for stability and long-term epigenetic memory. The complexity of development in vertebrates is reflected in the organization of their chromosomes.

Evolution and epigenetic memory span

If, as we have suggested, clustered CpG sites, heterochromatin blocks, and arrays of tandem repeats are vehicles of cellular memory, these elements will be important in the evolution of ontogeny. For example, tandem repeats may play a role in genomic imprinting. We suggested in Chapter 6 that imprinted genes have particularly 'stubborn' marks—marks that survive the restructuring of chromatin in the early embryo and have a particularly long memory span. The length of the memory may be determined by repetitive motifs in the neighbourhood of the genes. The *Igf2r* gene has a CpG-rich region containing several repeats that are methylated in a parent-specific manner (Stöger *et al*. 1993). Barlow (1993) pointed out that GC-rich regions are also found in other imprinted genes, and may serve as carriers of parent-specific marks. If the number of repeats in such regions is indeed important, strain differences in the presence or extent of imprinting may occur not only because of differences in the modifiers that code for chromatin proteins, but also because of differences in the number and sequence of the repeated motifs to which these proteins bind.

The mutability of methylated CpG sequences and the extreme instability of tandem repeats, especially of simple repeat arrays, allows for rapid variation in repeat number and in the memory span of the adjacent genes. Unequal crossing over, gene conversion, and slippage during replication, generate variations in the number of copies in clusters of repeats, and act to homogenize the repeat sequence (see Chapter 9, p. 264). Such variations may often be neutral or have small effects, affecting the patterns of determination and differentiation in quite subtle ways, but a large variation in repeat number, or the introduction of new sequences that cause alterations in the organization of chromatin, may have more dramatic phenotypic effects. They could affect the time at which gene activity changes during development, the ease with which it is changed, and the relative stability of the alternative states of activity.

We know from some pathological conditions in man that changes in the number of repeats can have profound phenotypic effects. Several genetic diseases are associated with the expansion of trinucleotide repeat sequences. For example, in the fragile-X syndrome discussed in Chapter 5 (p. 125), normal non-carriers have 6–60 copies of a CCG triplet in a particular gene, but affected people, who have severe mental retardation, have over 600 copies. Similar findings have been reported for Huntington's disease, myotonic dystrophy, spinobulbar muscular atrophy, and spinocerebellar ataxia type 1.[21] With the fragile-X syndrome, amplification of the repeat sequences is associated with changes in methylation and a decrease in the amount of gene product. One of the hallmarks of all the diseases in which there is expansion of triplet repeats is that the number of repeats tends to increase, and the illness becomes progressively more severe, in successive generations. Once a certain number of repeats is present, the DNA seems to be very unstable.

There must be constant interactions between the genetic and epigenetic inheritance systems. Chromatin structure determines the likelihood of genetic events such as slippage during replication, unequal crossing over, and gene conversion, all of which alter the number of repeated sequences. Since these sequences are vehicles for epigenetic memory, they can alter the epigenetic state of a locus. Similarly, the chromatin regions rich in methylated CpGs are hot spots for C to T transitions, so the methylation marks are changed. There are perpetual feedback loops between marks and the DNA sequences that carry them.

Changes in the DNA sequences that influence cellular memory may not always be spontaneous and random. Since control sequences respond to intracellular conditions and modulate gene expression through changes in chromatin structure, and since chromatin structure affects the probability of changes in DNA sequences, new environmental conditions may alter the sequences in control regions. In other words, an environmental change can lead to a DNA change that alters the memory span of genes whose activity is affected by the new environment. As shown in the general model described earlier in this chapter (p. 169), differences in memory span can affect the rate at which genes are fixed.

Another important aspect of environmentally-driven changes in DNA is the targeted nature of the changes—their locus specificity. Even if locus-specific changes are not adaptive, if only loci that are responsive to the environment are affected, the 'search space' for adaptive changes is greatly reduced. When an environmental change persists, and there is selection for a new memory span, evolution can proceed very rapidly and very efficiently. The targeted formation of variations is much more efficient than a general increase in the overall rate of variation, because it does not impose such a high genetic load on the population.

Summary

The interplay between the DNA inheritance system and EISs is very complex. There is little doubt that environmental factors can directly influence DNA sequence divergence through their effects on chromatin structure: active and inactive genes differ in their rates of damage, repair and recombination. In addition, cells seem to have a repertoire of responses to stressful conditions that influence overall genomic behaviour. Therefore, the heritable epigenetic marks that genes carry can influence the occurrence and spread of allelic variants, even in the absence of selection. Since the environment influences epigenetic marks, and epigenetic marks influence changes in DNA sequence, the role of the environment in evolution is not solely that of an agent of selection. It is also an agent of variation.

DNA variations obviously affect chromosome marks. The stability with which marks are transmitted to daughter cells probably resides in part in features such as the density of CpG dinucleotides and the number of repeats of various sequence motifs. These affect both the local chromatin conformation and regional chromatin structure. The fidelity with which an epigenetic state of gene activity is transmitted—its memory span—may depend on the size of blocks of repeats both in its neighbourhood and elsewhere in the genome. The banded organization of the vertebrate genome may reflect regional control stemming from selection for long-term suppression of tissue-specific genes.

The marks that genes carry, and the fidelity with which they are transmitted, are important in determining the stability of cellular inheritance and the ease of induction during development. The evolution of the DNA sequences that carry chromatin marks is therefore intimately linked with the evolution of ontogeny.

Notes

1. The effects of environmental factors on mutation and recombination are reviewed by Hoffmann and Parsons (1991).
2. Boulikas (1992) gives a wide-ranging review of factors affecting DNA damage and repair, and the evolutionary consequences of differences in repair and damage rates.
3. Concise reviews of the effect of transcriptional activity on DNA repair are given by Hanawalt (1989), Bohr and Wassermann (1988), and Bohr (1988).
4. Mellon and Hanawalt (1989) showed this for the *lac* operon of *E. coli*, and Mellon *et al.* (1987) for the *DHFR* gene in cultured human and hamster cells.
5. Kirkwood and Holliday (1979) have argued that repair in somatic cells is likely to be less efficient than in germ cells because natural selection will favour less

investment in the repair of the mortal 'disposable' soma than in potentially immortal germ cells.

6. Reviewed by Stack (1984) and John (1988).
7. Burt *et al.* (1991) have assembled much of the data on differences in chiasma frequency between the sexes, and discussed the theories put forward to explain the differences. The marsupials are an exception to the general trend for the heterogametic sex to have less recombination than the homogametic sex (Hayman *et al.* 1988). Jablonka and Lamb (1990*a*) suggested that this is associated with the non-random X-inactivation found in marsupials. Bernstein *et al.* (1987) and Thomas and Rothstein (1991) have discussed the relationship between gene activity in gametocytes and recombination frequency.
8. See Waddington (1961). Similar new 'spontaneous' dumpy mutations were found in experiments on the assimilation of an induced dumpy phenotype, but unless parallel induction occurred, these were unlikely to be the result of mutational assimilation, because the treatment was given to pupae, long after germ-line determination.
9. The relationship between CpG density in promoters and the potential stability of gene inactivation is not entirely straightforward. It holds only for some genes. The reason for this is that there are two types of promoters: those associated mainly with housekeeping genes, and those associated with most tissue-specific genes. Housekeeping genes, which are constitutively active, have promoters that are very CpG rich but, for reasons at present unknown, they are not methylated, except when in the inactive X chromosome. Such non-methylated, GC-rich regions are known as CpG islands. They are found in organisms with large genomes, such as plants and vertebrates. There seems to be some special mechanism by which cells recognize the islands and either refrain from methylating them, or actively de-methylate them (Hergersberg 1991). It has been suggested that the mechanism that prevents housekeeping genes from becoming methylated also protects them from being imprinted in the germ line.
10. See Vogt (1990) for examples. Bodnar and Ward (1987) analysed the base sequence of a number of genes and found that they contained multiple copies of 7–11 bp sequences. Although not arranged in tandem, the closeness of these short repeated sequences suggests that proteins binding to them could cooperate and form multimeric complexes which influence the stability of gene expression.
11. See Lyttle (1991) and Chapter 9.
12. Position effects in general are reviewed by Wilson *et al.* (1990). Reviews of PEV in *Drosophila* are given by Henikoff (1992), and Reuter and Spierer (1992).
13. For details of this model and the supporting evidence, see Locke *et al.* (1988), Tartof *et al.* (1989), and Tartof and Bremer (1990).
14. Lewis (1985) reviews some of his own and other classical work on the Bithorax complex.
15. See Gaunt and Singh (1990), but also Paro (1990) for a contrary opinion.
16. For a discussion of 'bendability' and the binding to DNA of multimeric complexes with low sequence specificity, see Serrano *et al.* (1993).
17. Pillus (1992) reviews position effects in yeast.
18. It has been suggested that the effect of telomeric heterochromatin on the activity of neighbouring genes may be a cause of ageing in mammals. The progressive shortening of telomeres with age allow heterochromatinization to

spread to the adjacent chromatin, and eventually cause serious damage by inactivating essential genes (Wright and Shay 1992).

19. See Brockdorff *et al.* (1992) and Brown *et al.* (1992).
20. For a brief review of SINEs, see Okada (1991); LINEs are reviewed by Martin (1991).
21. Recent work is summarized in Davies (1993) and Kuhl and Caskey (1993).

8

The role of epigenetic inheritance systems in adaptive evolution

What is that which grows?
Aristotle: *On generation and corruption*
Chapter 5, 321^{a}, line 30

Our aim in this chapter and the next is to show how, by broadening Darwinian theory to give a role to the biochemically well-recognized EISs discussed in Chapter 4, it is possible to give an explanation of some aspects of evolutionary history that is more satisfying than that based on natural selection of random DNA variations. Both the direct and indirect effects of epigenetic inheritance have been important in evolution. We first discuss the way in which high mutation and epimutation rates may have been moulded by natural selection, and how the frequency of epialleles changes in populations. We then argue that the evolution of complex multicellular organisms was made possible by EISs that already existed in unicellular organisms, and that EISs have shaped some of the developmental strategies seen in higher animals and plants. We also discuss the role of epigenetic variations in macroevolutionary changes.

The evolution of cellular memory and memory span

The study of the role of EISs in evolution is in its infancy, so there is very little that can be said at present about the evolutionary origin of the different types of EIS. It seems likely that self-sustaining systems—what we have called steady-state EISs—are very ancient. Dyson (1985) suggested that steady-state systems preceded the emergence of self-replicating molecules. He proposed that a complex of self-sustaining biochemical reactions was the first 'living' system, thus identifying life with self-regulating and self-sustaining metabolism. According to this idea, a self-maintaining system that is not based on nucleic acids is older than the nucleic-acid based inheritance system which characterizes life as we know it today.[1] Even if they did not evolve before the nucleic-acid based inheritance system, steady-state EISs were almost certainly part and parcel of the first living cells.

Structural EISs probably also had a very early origin. A three-dimensional template, capable of directing the assembly of complex organic molecules to create a copy of itself, may also have preceded the emergence of an inheritance system based on nucleic acids. Cairns-Smith (1985) based his theory of the origin of life on the self-assembly and growth of clay crystals, which then directed the self-assembly and reactions of complex organic molecules. Whether or not this idea is correct and structural EISs preceded inheritance systems based on nucleic acids, the fundamental role of structural EISs in preserving the internal structure of modern cells suggests that this type of EIS had a very early origin.

By definition, the chromatin-marking EIS depends on the existence of the DNA inheritance system, so the emergence of this EIS must have followed the evolution of DNA-based inheritance. The presence in bacteria of a defence system against parasites, which is based on sequence-specific methylation of cellular DNA and enzymatic degradation of inappropriately methylated foreign DNA, suggests that methylation-marking originated fairly early in evolutionary history. The type of methylation-dependent restriction-modification system found in bacteria may have been an evolutionary precursor of the methylation system adopted for the regulation of gene expression (Bestor 1990). The inheritance of epigenetic information in unicellular eukaryotes (described in Chapters 4 and 6) makes it likely that EISs evolved in single-celled organisms. Their complexity in present-day protists suggests that they have a selective advantage for single-celled organisms, and underwent evolutionary elaboration at the unicellular stage.

Unicellular organisms can transmit both phenotypes and genotypes to daughter cells. It is therefore worth asking under what circumstances the inheritance of phenotypes rather than genotypes is advantageous. What has determined the rate of epigenetic switching between phenotypes—the length of cell memory? Nanney (1960) suggested that because heritable epigenetic variations occur at a high rate, they will be an advantage to organisms living in traumatic and unpredictable environments. A population that can produce and inherit epigenetic variations, and thereby maintain heterogeneity, may be able to adjust and survive erratic environmental shifts, whereas a population lacking this facility becomes extinct. If such epigenetic variations are essentially random, they differ from classical mutations only in the rate at which they are produced. Many will be maladaptive, so the cost to the population of a high rate of random epigenetic variation will be substantial.

Although randomly generated epialleles (epimutations) may have been an important source of variability in the unpredictable environment of early unicellular organisms, the refinement and exploitation of EISs was probably an evolutionary response to more predictable aspects of the environment. Primitive organisms lived in environments that fluctuated. They experienced cycles of change in light intensity and temperature associated with night and day, cycles associated with tides, and the longer cycles associated with months and years. In such fluctuating environments it takes time to adapt to conditions in different phases of the cycle, so it may have been an advantage to inherit a phenotype as well as a genotype. When the length of the environmental cycle is greater than the generation time of the organism, but shorter than the average time taken to adapt through fixation of classical mutations, epigenetic inheritance may be beneficial. Lachmann-Tarkhanov and Jablonka (personal communication) have suggested that this type of environment be called an Intermediate Length Cycle (ILC) environment. It is intermediate in length between a short cycle, which is repeated many times during the life of an individual and to which organisms adjust rapidly through physiological changes, and a long cycle, occurring over many generations, to which populations adapt through selection of classical mutations. A unicellular organism that transmits its adaptive phenotype to progeny in this type of predictable, fluctuating environment will have an advantage because, for the progeny, the cost of being transiently in a non-adaptive state is usually avoided. Selection will favour a transition rate from one epigenetic state to another that reflects the periodicity of the environmental fluctuation.

The optimal transition rate can be modelled quite easily. Consider a rapidly reproducing unicellular organism in an environment with low night temperatures and high day temperatures. At night it is best to have gene *A* on, and by day to have it off. The number of generations in the night and day periods are roughly the same. Since the diurnal cycle is much longer than the generation time of the organism, it is an advantage to transmit the current state of activity to the progeny for a certain period. According to Lachmann-Tarkhanov and Jablonka, in general, in this type of fluctuating ILC environment, the best spontaneous rate of change from one state to another will be close to $1/n$, when the total length of the cycle in generations is $2n$. For example, in an environment that changes every 25 generations ($n = 25$), simulations show that the best transition rate is 0.05 per generation, which is close to the predicted value of $1/n$. Each locus will evolve a transition rate that reflects the environmental periodicity to which it responds. The relationship between the cycle length and the best transition rate for a locus is shown for a very simple case in Fig. 8.1. In this example, there are only two types of environment, an equal number of generations is spent in each, and the environment does not induce changes in the activity of the locus.

The simple situation depicted in the figure, where the environment is merely the selective agent, is essentially one of classical neo-Darwinian selection. Shifts from one epiallele to another occur spontaneously approximately every *n* generations, and are based on EISs rather than DNA mutations, because EISs are more appropriate for rapid and middle-term rates of change. Selection should also favour *directed* epigenetic changes, in which the heritable change in phenotype is induced by the environment as conditions fluctuate.

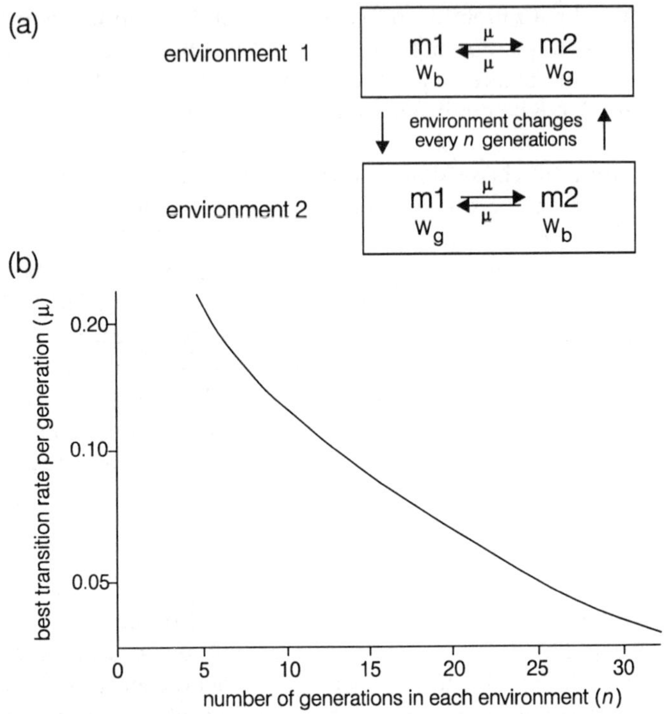

Fig. 8.1 The best transition rate for a locus in a regularly fluctuating environment. (a) The general model, in which periodic symmetrical shifts occur between two environments. Epiallele *m1* confers greater fitness in environment 2, whereas *m2* is superior in environment 1; epialleles *m1* and *m2* have fitnesses W_g in their good conditions, and W_b in their bad conditions. (b) The results of a simulation based on the model shown in (a) when $W_g = 1.1$ and $W_b = 0.9$.

Optimal, locus-specific, rates of transition between epialleles will also evolve in more complex situations, where there are several types of environment and the time spent in each is different. The type of gene expression found in these ILC environments is neither a short-term, stimulus-dependent response, nor a constitutive, long-term, stimulus-independent response, but a response that is intermediate between the two. It is a medium-term response which is 'remembered' by progeny cells, even in the absence of the stimulus that originally induced it, but it is not remembered very well. The memory lasts longer than the generation time, but not indefinitely.

If the evolution of the memory span of a locus or a phenotype in early eukaryotes was determined by the periodicity of an ILC environment, what exactly was selected? Were the selected variations differences in the epigenetic marks themselves, or in the DNA sequences that carried these marks? As we argued in the last chapter, at the DNA level the memory span of a locus is determined by features such as the CpG density in promoter regions, the number and character of tandem repeats near coding sequences, and the distribution of sequences in heterochromatin blocks that can act as 'sinks' which bind regulatory factors. Selection of DNA sequences would certainly influence the memory span of epigenetic states in unicellular organisms.

The elaboration and exploitation of EISs that occurred in primitive single-celled organisms was probably a response to selection pressures very similar to those experienced by parasites today. Parasites also commonly experience ILC conditions, for example by regularly going through several generations on one host before moving to another. Often the phenotypic changes seen during parasite life cycles are induced by transient stimuli, but some seem to be the result of spontaneous transitions that occur at characteristic rates. This is the case for the antigen switching found in many pathogens: antigenic variants are generated at a high rate and enable the pathogens to evade the host's immune system. Reversible DNA rearrangements underlie some of these changes.[2] In other cases EISs affecting transcription seem to be important. For example, in some uropathogenic *E. coli* the expression of the operon that controls pilus type is regulated by adenine methylation patterns in the control region.[3] The state of activity of the operon is inherited and switches between active and inactive states.

A comparable type of switching is seen in the infectious yeast *Candida albicans*, which can switch between several alternative phenotypes and produce colonies with different characteristic forms.[4] One of the best-studied transitions is the switch between 'white' and 'opaque'. It involves a dramatic change in the cell phenotype, which is reflected in changes in the morphology and colour of colonies. Switching from white to opaque occurs less frequently than in the opposite direction, and the rate of switching is affected by environmental factors. The precise mechanism underlying switching is not known, but it seems to involve heritable changes in chromatin structure affecting gene expression. The switch between some different forms of *C. albicans* is known to be associated with different levels of methylation (Russell *et al.* 1987). Probably, as with other pathogens, switching of phenotypes in *C. albicans* has evolved as a response to changing environments within the host, so the frequency of switching and the rate of change of the environment should be correlated.[5]

The fate of changed epigenetic marks in populations

Epigenetic variants were probably common in early eukaryotes. They could arise spontaneously, or be induced by environmental stimuli. Some would be transmitted to daughter cells, others would be temporary. Some would be adaptive, others maladaptive. The frequency of a particular epiallele in a population would depend on the spontaneous transition rate to and from the variant, on its selective value (if any), on the persistence of the inducing environment if the variant was induced, and on population size.

It is possible to construct simple models showing how epialleles should behave in asexual populations.[6] The two models described here are based on the chromatin-marking EIS, which is the easiest to model. For both the *two-state model* and the *multi-state model* we assume that:

(1) there is a finite number of gene phenotypes;
(2) there is a characteristic spontaneous rate of change from one gene phenotype to another;
(3) in the presence of an inducing stimulus, the rates of change are altered, with some phenotypes being produced much more frequently.

What we look at is how the epigenetic marks, induced by exposing a population to an environmental stimulus for one or a few generations, fade away after the population is removed from the inducing conditions.

The *two-state model* describes the simplest case of inherited epigenetic marks. There are only two possible phenotypes for the gene: it carries either mark *m1* or mark *m2*. The spontaneous rate of change from *m1* to *m2* is u, and from *m2* to *m1* is v. If the population starts with a frequency p_0 of mark *m1*, then after t generations, the frequency p_t of *m1* is given by:

$$p_t = \frac{v}{u+v}\left(p_o - \frac{v}{u+v}\right)(1-u-v)^t$$

This equation is, of course, exactly the same as that for changes in allele frequencies brought about by mutation. If the changes in marks are random, the interpretation is also the same, since in the absence of selection or other factors that change gene frequencies, the population will slowly move towards an equilibrium in which the frequencies of the two epialleles depend on the rates of change from one mark to another. However, if changes in marks are induced by changes in the environment, the situation is rather different. Whereas mutation pressure is usually low, the pressure of epimutation may be high. Figure 8.2 shows what would happen to a population in which initially the rate of change u is 5×10^{-3} per generation and v is 10^{-2}, when it is exposed to an environment that induces *m2*. It is assumed that the inducing conditions increase the rate of change from *m1* to *m2* to 0.6 per generation, but the change lasts for only two generations. After two generations, conditions return to normal. The figure shows that induced marks linger for many generations after the environmental stimulus is applied, and the population only gradually returns to its original condition, even though the marks are selectively neutral.

In the second model, the *multistate model*, we assume that a cell can exist

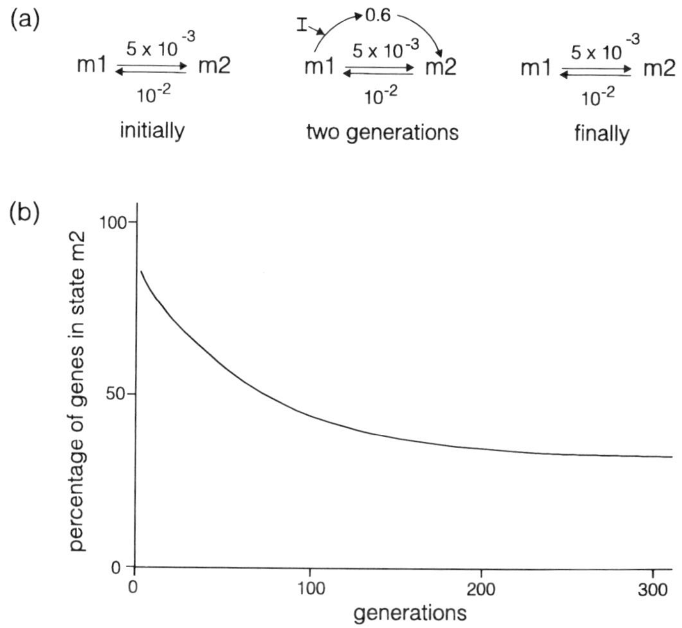

Fig. 8.2 Change in the frequency of a mark after two generations of exposure to an inducing stimulus. (a) The spontaneous and induced (**I**) transition rates between marks *m1* and *m2*. (b) Results of a simulation based on (a).

in several different epigenetic states and there are progressive changes from one to another. The different epigenetic states may reflect the state of one particular locus, with four epialleles, or they may represent different functional states of the cell, with several genes involved. In order to simplify the discussion, we shall talk of epialleles at a single locus. An example of the model is shown in Fig. 8.3 where it is assumed that the gene has four epigenetic states, *m1*, *m2*, *m3*, and *m4*, which in normal environmental conditions change from one to another at a rate of $u = v = 10^{-2}$ per generation. *m1* and *m2* are inactive states of the gene, *m3* and *m4* are active. In the presence of an inducing stimulus, genes in epigenetic states *m1* and *m2* are induced to become *m3* at a rate of 0.6 per generation, but all other spontaneous rates of change remain the same.

We have looked at the behaviour of this gene in populations experiencing different conditions. First we asked how, starting from a population in which all genes had epiallele *m1*, the proportion of active genes would increase if there was no inducing stimulus. The result of a computer simulation is shown in Fig. 8.4a. Gradually, over many generations, the population moves to an equilibrium, with 50% of the genes active and 50% inactive. Of more interest are the situations shown in Fig. 8.4b, c and d, in

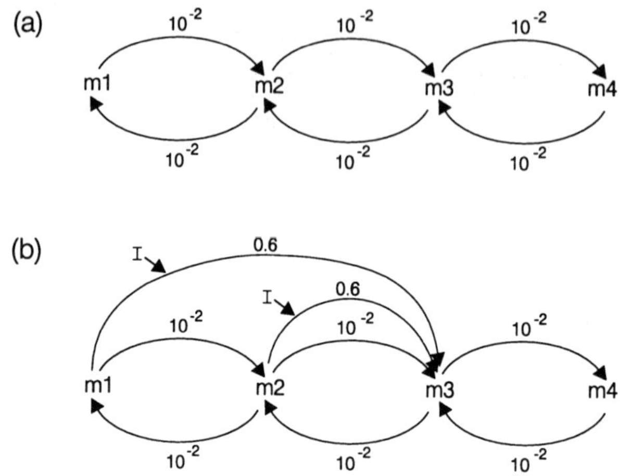

Fig. 8.3 An example of the multi-state model with four different epialleles. (a) Changes that occur in the absence of an environmental stimulus. (b) The effect of a stimulus **I**.

which the behaviour of populations exposed to an inducing stimulus for one, two, and five generations has been simulated. It can be seen that after the environmental stimulus is applied, a high proportion of the genes become active, and this state persists for many generations after the removal of the stimulus. The length of exposure to the inducing stimulus affects the length of time the induced state lingers. The similarity of these simulations to the results Jollos called 'Dauermodifikationen' (Chapter 6, p. 137) is obvious. It should be noted that in none of the examples illustrated in Fig. 8.4 is there any selection for or against the induced epialleles. Yet it is clear that even without selection, a change in the environment can have a profound and long-lasting effect on the population.

The simulations show the behaviour of newly-induced marks in unicellular organisms, or in organisms that reproduce asexually by budding or fission. With small modifications, the models can also be applied to sexually reproducing organisms, if it is assumed that some marks are stubborn, and are not erased during development or gametogenesis. The models can also be applied to developmental processes in multicellular organisms, in which once induced, determined and differentiated states are maintained through successive generations.

EISs and the transition to multicellularity

The evolution of the machinery responsible for the transmission of epigenetic variations, and the development of locus-specific transition rates,

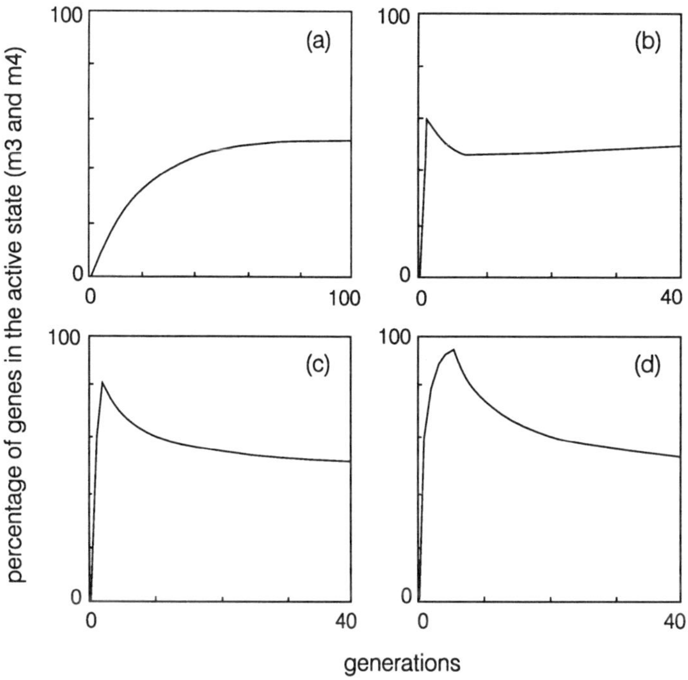

Fig. 8.4 Examples of simulations based on the multi-state model when there are four epialleles and the rates of transition between them are those shown in Fig. 8.3. Changes in the number of active genes are shown (a) when the is no environmental stimulus; (b) when the environmental stimulus lasts for one generation; (c) when the stimulus lasts for two generations; (d) when the stimulus lasts for five generations.

were probably pre-requisites for the evolution of complex multicellular organisms. EISs are essential because EISs maintain the determined state of cell lineages and allow different lineages to retain their specificity in spite of cell turnover. Transition rates are important because they determine the ease with which lineages change their phenotypes. But how did multicellularity itself evolve? Did EISs play a role in the evolution of the first multicellular forms?

Prokaryotes are believed to have evolved about 3500 million years ago, and for the next 2000 million years they were the only form of life. The oldest eukaryotes are found in geological formations dating from 1400 million years ago, but it is not until the late Precambrian, 700 million years later, that the first multicellular organisms are found. It seems that it took a very long time for multicellular life to evolve, but once it had done so, there was a rapid and dramatic radiation of multicellular forms. Many types of complex multicellular animals, some of which are clearly related to present-day groups but many of which are not, form the Cambrian explosion of 570 million years ago.[7]

According to Bonner (1974, 1988), multicellularity has evolved many times. There are at least seventeen groups with some multicellular taxa, and multicellularity probably had an independent origin in each group (Fig. 8.5). Only three of the present-day groups, plants, animals and fungi, have a substantial number of highly differentiated cell types, but sponges, cellular and acellular slime moulds, some green algae (*Volvox* and its relatives), and members of some other groups all show multicellularity involving a degree of cellular differentiation and cellular interdependence.

What selection pressures led to the frequent origin of multicellularity? Bonner has argued that there are probably several different reasons why multicellularity is an advantage. One is that a group of cells can feed more effectively than isolated single cells: by cooperating and producing a large quantity of digestive enzymes, larger particles of food can be broken down. A second reason is that multicellularity may lead to more effective dispersal: the fact that so many of the fungi and slime moulds have small stalked fruiting bodies suggests that these multicellular structures have promoted greater reproductive success, probably by making dispersal easier. A third reason is protection: large organisms can probably evade predators more effectively simply because they are too big to be eaten. Bell and Koufopanou (1991) suggested that large size may also be an advantage

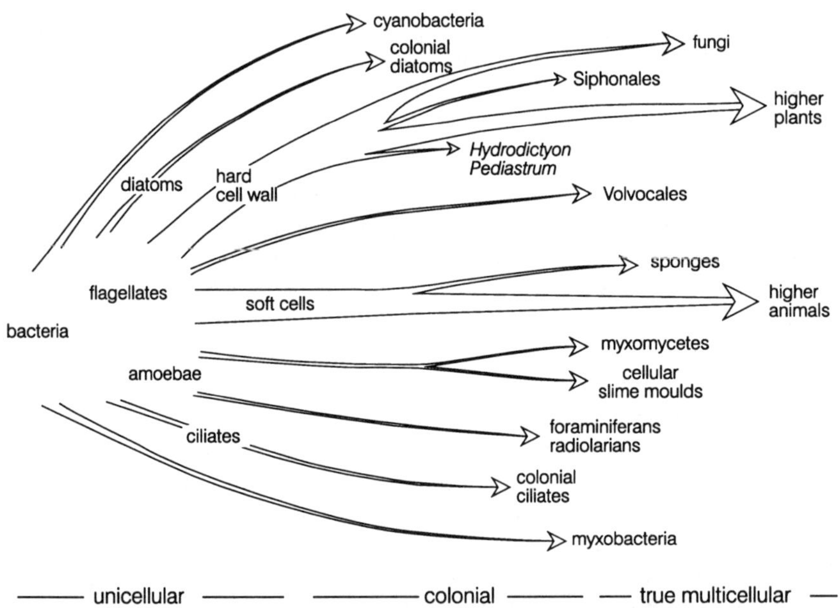

Fig. 8.5 The many origins of multicellularity. (Based on Bonner 1958 and 1974.)

in environments that vary in space or time. When resources are sometimes plentiful and sometimes scarce, large organisms do better because in times of scarcity they take longer to starve to death. If resources are plentiful in some areas but not in others, there may be benefits in having a period of prolonged growth followed by multiple fission and dispersal.

Although it is possible to imagine a number of circumstances in which an increase in size is beneficial, it is clear that there is also a price to pay for becoming large. Bell and others have collated data showing that as size increases, reproductive rate decreases (Fig. 8.6).[8] In an 'ideal' environment, the most successful organism would be small and reproduce by binary fission, just like prokaryotes and some small unicellular eukaryotes. In non-ideal environments, large size has some advantages, but when single-celled organisms are large, they become less efficient. The decrease in the ratio between their surface area and volume causes transport problems, and there are also problems in producing proteins at a sufficiently high rate to maintain the cytoplasm. Large unicellular organisms have overcome these problems in various ways,[9] but a different, and apparently more successful, way of solving the problems of being large is to be multi-

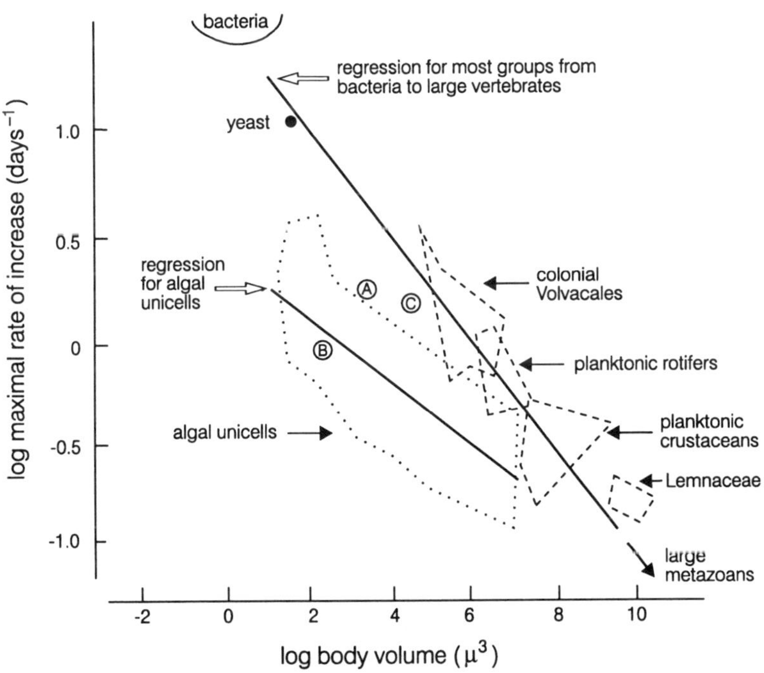

Fig. 8.6 The effect of increase in size on reproductive rate. The dotted outlines surround data points for large numbers of species of the different groups indicated. **A** marks the average for amoebas, **B** for *Chlamydomonas*, **C** for ciliates. (Redrawn from Bell 1985.)

cellular. Multicellularity can be achieved either by independent cells aggregating and cooperating, as happens in the cellular slime moulds, or, more commonly, by daughter cells failing to separate following cell division.

As Bonner (1988) has shown, the increase in size brought about by multicellularity has usually been accompanied by an increase in complexity. Larger organisms have a division of labour between different cell types; the larger the organism, the more types of cell there are (Table 8.1). But what is the advantage of this division of labour? What is the advantage of going beyond a loose association of equal cells? Bell (1985) suggested that in colonies the division into somatic and reproductive cells increases metabolic efficiency because somatic cells act as a source, and germ cells as a sink for metabolites.[10] By moving materials from somatic cells to germ cells, concentration gradients between the external medium and the cytoplasm of the somatic cells remain steep, and a high rate of synthesis can be maintained.

In addition to the physiological advantages, division of labour may also solve another problem for eukaryotic multicellular organisms. Mitotic spindles, cilia, and flagella are homologous structures, each dependent on a microtubule organizing centre (MTOC). Margulis (1981) argued that since the unicellular ancestors of multicellular organisms probably had only a single MTOC per cell, these cells could either possess cilia and move, or they could divide; they could not do both at the same time.[11] Consequently, selection in early multicellular organisms favoured a division of labour because without it they, like their unicellular ancestors, would have to give up motility during reproduction.

Table 8.1 The number of cell types in different organisms (Based on data in Bonner 1988.)

Number of cell types	Organisms
1	Some bacteria and cyanobacteria (prokaryotes); some asexual green algae (probably descended from more complex ancestors)
2–3	Spore forming bacteria; some cyanobacteria; myxobacteria
4	Green algae such as *Volvox* and *Ulva*; filamentous fungi such as *Mucor*
about 7	Some fungi (mushrooms) and brown algae (seaweeds)
9–12	Sponges; cnidarians
about 30	Higher plants
about 55	Higher invertebrates: annelids, molluscs, echinoderms, arthropods
over 120	Vertebrates

With the exception of some groups such as the cellular slime moulds, where multicellularity is the result of cell aggregation, the genomes of the functionally different cell types that make up a simple multicellular organism are identical. How did the regulated differentiation and division of labour between cells evolve? According to Wolpert (1990), all of the basic processes needed for development—differentiation, spatial patterning, and change in form—are already present in single-celled organisms.[12] Even cell adhesion and cell-to-cell signalling would require little modification of the membrane systems existing in unicellular eukaryotes. Wolpert suggested that there is only one new property that needs to be evolved in metazoans–EISs or, as they are usually called when talking about development, cell memory systems:

> There is, perhaps, one cellular process that may require a novel evolution in metazoans and that is cellular memory. Liver cells breed true. The inheritance of the differentiated state through a cell cycle might require new methods for controlling gene action, as the inactivation of the X chromosome suggests. (Wolpert 1990, p. 111)

We agree that cellular memory was essential for the evolution of complex multicellular organisms, but believe that it did not require novel evolution. The evidence in Table 6.1 (p. 134) suggests that it was already present in their single-celled ancestors.

Multicellularity could have evolved in the following way. In situations in which large size was favoured, mutations that made daughter cells remain together were advantageous. Once cells remained together, a primitive division of labour may have been an inevitable result of the spatial arrangement within the group: cells on the surface were different from cells inside the cell mass, simply because their environment was different. For example, cells within the cell mass may have benefited from the synthetic activities of their neighbours, and therefore divided earlier. Similarly, cells in one region of the group might, in response to environmental cues, have become specialized for feeding, while those in another region became specialized for locomotion; cells at the surface may have produced flagella and engaged in locomotion, while those within the cell mass retained the ability to divide, thus overcoming the conflict between motility and reproduction. The efficiency of the cell group would be increased, because activities that originally had to be sequential could now occur in parallel. Specialization of cells in different positions could lead to inductive interactions between different cell types. For example, cells that are exposed to the external environment and specialize in motility might, as they differentiate, produce substances that act as inducers of genes in internal cells.

Up to this stage in the evolution of multicellularity, sophisticated EISs may not have been necessary. Each cell, regardless of lineage, would differentiate in the way dictated by its neighbours and by its position in relation to the external environment. The cellular specializations and behaviour of early multicellular eukaryotes probably resembled those seen today in bacterial colonies that have different cell types and quite complex multicellular interactions.[13] For example, in swarming bacteria such as *Proteus mirabilis*, new colonies are at first made up of short vegetative cells, but as the colonies grow, cells at the edge differentiate into long, multinucleate, hyperflagellated, swarm cells, which rapidly migrate away from the colony. The migratory behaviour is a coordinated action by a group of cells, and depends on cell-to-cell contact. When a certain distance from the main colony, swarm cells revert to the vegetative form. Myxobacteria and *Caulobacter* species show even more elaborate differentiation and multicellular behaviour. Even bacteria such as *E. coli* and *Bacillus subtilis* show quite complex patterns of colony organization: in *E. coli* colonies there are well-delineated zones of differentiation associated with differences in cell size, cell shape and patterns of alignment; in *B. subtilis*, colony morphology can vary, and some variations have been found to be heritable, and potentially reversible, responses to the environment.

Even when heritable, cell specialization of the type seen in bacteria probably requires only simple steady-state and structural EISs. The same is true for primitive multicellular eukaryotes: the early stages of evolution could have occurred without EISs, or with just very simple ones. However, a division of labour that depends on the continuous presence of the inducing stimulus is unlikely to provide the structural and functional stability of cell lineages in more complex organisms. For these, sophisticated cell memory systems are essential.

Although the first stages in the evolution of multicellularity could have occurred without EISs, we think it more likely that EISs played an important role right from the start. Since EISs already existed in unicellular ancestors, the transition to multicellularity could have happened by the following, somewhat different, route. First, selection for large size resulted in a tendency for cells to remain together in groups after division. Spatial or temporal patchiness in the environment then caused functional phenotypic differences between the loose groupings of cells. Such environmentally induced phenotypes would tend to be preserved by the EISs during cell division. Because of the EISs, phenotypic similarity within groups would be greater than that between groups, even if the environment became more homogeneous. A group of cells, characterized and unified by its heritable epigenetic state, then functioned as a unit of selection. Selection between the epigenetically distinct cell groups led to the evolution of group properties, with variations that promoted group advantage being selected. For example, genetic or epigenetic variations affecting growth patterns, the presence of cytoplasmic bridges between cells, the rate of fragmentation, and group motility, were all subject to selection. Selection for cell-to-cell communication was strong, and elaboration of structural EISs to organize an extracellular matrix was favoured, since it enhanced the integration of cell groups.

Selection of and for such group properties led to a further elaboration of adaptations that maintain the cohesiveness of the community of cells, including division of labour. Cell activities came to rely less on the external environment. Interactions between different lineages within the group became important, with products of one cell type altering the heritable epigenetic state of a sister lineage. The evolution of such interdependence and integration between the cell lineages within a group of cells resulted in a new entity—the complex multicellular integrated individual.

An important property of an integrated multicellular organism is that it maintains its coherence despite turnover of the component cells. Cells divide and die, yet the organism retains its identity and functional integrity. In other words, the system is more enduring than its component parts. The maintenance of this coherence in the face of constant cell turnover means that newly produced cells must be similar to the cells they replace. Some kind of transmission of old states must occur. It is the EISs that make it possible for organisms to survive longer than the life span of their component cells. If the original unit was an individual cell, with EISs it becomes the cell lineage; the functional state it is not limited by cell division, but by the epigenetic memory span.[14] EISs thus enabled the evolution of a new unit of function, the cell lineage, which is also a unit of hereditary variation. The memory span, or transition rate, of gene phenotypes evolved in unicellular organisms as a response to an ILC environment; the 'failure' of epigenetic memory after a number of generations was a passive way of resetting the epigenetic state of the gene. Multicellular organisms retained this system and use locus-specific epigenetic memory to maintain determined states, and 'failures' of memory for changing direction in development.

The role of EISs in the transition to complex multicellular organisms was therefore twofold. First, they enabled the emergence of a new unit of structure and function, the phenotypically distinct cell lineage. Second, they allowed the formation of the stable interdependences between epigenetically distinct cell lineages, which resulted in the evolution of integrated organisms from loose groups of cells.

Competition and cooperation in cell lineages

Multicellular organisms exist because their component cells cooperate rather than compete. Yet within the individual organism, genetic and epigenetic cellular variants that are potentially capable of destroying the integrity of the whole are inevitable. In particular, a mutation in a cell with somatic functions could lead to its lineage dominating the developing individual and invading the germ line. Buss (1987) has interpreted early metazoan development in terms of past competitions between mutant cell populations, in which successful variants became germ-line cells and by chance produced superior individuals. This concept is important, but as discussed in Chapter 2, there are problems with it. In particular, it is difficult to see why a mutation that makes a lineage successful in competition with genetically different lineages in the parent should also be advantageous in progeny: all cells in the progeny are genetically the same, so the new variant has no selective advantage. This difficulty is to some extent removed if instead of thinking of genetic variants, one thinks about epigenetic variants. A lineage in which a chromatin mark or steady-state system has been changed is usually not permanently modified: the change could be wholly or partially reversed if the lineage comes to contribute to the germ line, and reappear in the same situation and circumstances in the progeny. Successful epigenetic variants can therefore compete for germ-line status, and through the EISs transmit the information that made them successful to the next generation.

Thinking about the evolution of development in terms of variants transmitted through EISs, rather than in terms of mutations, circumvents some of the difficulties in Buss's model of the evolution of early ontogeny through competition between cell lineages. However, rather than thinking in terms of competition, we prefer to think about the evolution of development in terms of cooperation between cell lineages that are genetically identical, even if epigenetically different. Changes that cause some cells to forgo reproduction and take on somatic functions are favoured if the behaviour of those cells confers additional reproductive advantages on sister cells, and therefore on the organism as a whole.

The changes that led to a division of labour between cells in early multicellular organisms were probably initially epigenetic changes that arose within the cell group either by chance, or as a response to conditions in the external environment, or as a consequence of interactions with neighbouring cells. Whatever their origin, if new epigenetic variants contributed to the successful survival and reproduction of the cell group (the individual organism), selection between groups would favour genetic changes that facilitated the maintenance and perpetuation of the new variants. For example, a new mark might be more easily acquired in a somatic lineage and lost in germ cells if a particular DNA base-change occurred. Although what is ultimately selected is a DNA sequence, it is selected for its epigenetic potential. As discussed in Chapter 7, often the important DNA change is not in a coding sequence: it can be a change in the number of repetitive sequences, a change in the number of CpG sites, changes in the size and distribution of heterochromatin blocks, or changes in the amount or nature of binding factors. What is selected are the genetic elements responsible for carrying and generating marks. This sequence of events contrasts with the scenario Buss envisaged: Buss suggested that the evolution of metazoan development involved chance mutations affecting the behaviour of cell lineages, the effect of which is then reflected in the epigenetic behaviour of those lineages in development. We see the primary change as the epigenetic event, which is then stabilized by genetic changes.

In some ways the evolutionary processes we believe shaped early development are comparable to those described in Chapter 2 when considering Waddington's ideas about canalization and genetic assimilation. At first the cells of early multicellular organisms were able to fulfil many functions, but in response to environmental cues, or to their position in relation to other cells, a particular set of genes in some lineages was more likely to be expressed. If this was beneficial to the individual, changes in DNA sequences that influence marks and made the response more predictable were selected. Initially, therefore, the changes were phenotypic, and were transmitted by an EIS. Later, the responses became more canalized through genetic changes that made them appear in the appropriate place and at the appropriate time in a more regulated way. The feedback between the epigenetic state of a locus and the probability and nature of genetic changes in that locus, which were described in Chapter 7, contribute to the assimilation process.

Sister cells carrying identical hereditary information are expected to behave altruistically towards each other, but cells carrying newly arisen 'selfish' mutations or epimutations must constantly threaten to jeopardize the harmony within organisms. As Buss (1987) argued, it is therefore not surprising that some taxa have strategies for resisting or suppressing selfish cell lineages. In the next sections we look at four strategies that are found in some, but not all, highly differentiated multicellular organisms: first, beginning development from a single cell; second, maternal control of early development; third, a sequestered germ line; finally, meiotic sex. Buss (1987) and Maynard Smith (1988*a*) have argued that these all help to maintain the integrity of multicellular organisms by minimizing the opportunities for competition between genetically different cells. However, these strategies have an additional role: they secure the integrity of multicellular organisms by enabling each generation to have a fresh epigenetic beginning. They can be thought of as evolutionary responses to the potential hereditary persistence of epigenetic variations.

Beginning from a single cell

Most multicellular organisms begin their life cycle as a single cell. In a way this is puzzling: if the early stages of the evolution of multicellularity involved selection for large size, why should development be initiated from just one cell? Ancestral cell groups could probably reproduce by fragmentation, a form of reproduction that maintains relatively large size, so why was it abandoned? There are several possible reasons. Perhaps the simplest is that suggested by Bonner (1988): that multicellular organisms inherited from ancient eukaryotes a genetic system involving mitosis, meiosis and fertilization, and it is a requirement of that system that there is a unicellular stage. However, the evolution of obligate clonal reproduction in some groups argues against this position. An alternative suggestion is that starting from a single cell is a way of reducing competition between cell lineages (Maynard Smith 1988*a*). If new individuals are initiated from a group of cells, there is the danger that the cells are genetically different. This danger exists even if the cells come from the same parent, because there may have been mutations in some parental lineages. Initiating development from a single cell reduces the likelihood of genetic differences between cells. The descendants of a single cell are genetically similar, so are likely to cooperate, rather than compete. Buss (1982) showed the importance of this in his study of one of the slime moulds that becomes multicellular by cell aggregation.[15] In this species, *Dictyostelium mucoroides*, mutant cells that are incapable of forming reproductive fruiting bodies on their own can combine with normal cells and become part of a communal spore mass, thereby achieving their own propagation. They are selfish somatic-cell parasites. When a life cycle begins from a single cell, cellular parasites cannot originate in this way.

We believe that there may be another reason why beginning development with a single cell is an advantage. The division of labour among the cells of primitive multicellular organisms probably depended largely on their spatial positions with respect to each other and to the external environment. Although EISs operated, inherited epigenetic states could be reversed quite readily. All cells remained totipotent. If cell groups reproduced by fragmentation, the group could reorganize to produce a coherent whole in much the same way as some early embryos can regulate when parts are removed. However, as the division of labour became more complex, involving secondary differentiation and cell–cell interactions, isolated groups of differentiated cells could no longer re-adjust to form an integrated entity. A single-celled beginning allows a fresh epigenetic start and the opportunity to repeat the basic inductive interactions of early development.

Maternal control of development

The sophisticated division of labour seen in complex multicellular organisms can develop in different ways. At one extreme is mosaic development, the type of development seen in animals such as molluscs and nematodes. In these animals, the spatial distribution of morphogenetic determinants in the egg cytoplasm decides the developmental potential of different regions of the embryo. By furnishing the egg with cytoplasmically localized information, the mother decides the fate of the embryonic cells in her offspring. Commonly, development at first runs on the products of maternal genes, with transcription from zygotic genes not beginning until quite late in embryogenesis. This maternal control of early development is in some ways comparable to the control seen in some social insects, particularly some ant species, in which the caste of daughters, whether they become workers or reproductives, is manipulated by the queen. She controls caste either directly through the materials she puts into the egg, or indirectly through pheromones. With both the individual animals of social insects and the individual cells of multicellular organisms, maternal control makes daughters assist their sisters to reproduce. The opportunities for genetic or epigenetic variants to rebel and themselves contribute to the next generation are greatly reduced because they are not allowed to develop autonomously.

Initially, maternal control of development limits the potential of individual cells. But what of later development? Stable and reliable epigenetic inheritance systems enable cell specialization and efficient division of labour, but they also pose a threat. The more stable and reliable the inheritance systems are, the greater the chances of transmitting a new variant to the next generation if cells carrying the variant become part of the germ line. One of the reasons for the effectively irreversible somatic differentiation found in many animal species may be that there has been selection against the ability of somatic cells to invade the germ line and form germ cells. Selection against de-differentiation and a change in role may have been quite strong. In plants, where there is no functionally segregated germ line, there is far less restriction on the potency of differentiated cells. A single cell can often generate a whole plant. In metazoan development, on the other hand, the potential to form germ cells is often very restricted. The extreme can be seen in copepods such as *Cyclops strenuus*, nematodes of the genus *Ascaris*, and the midge *Wachtiella persicariae* where, early in embryogenesis, the somatic lineages shed much of their genetic material.[16] This chromatin diminution depends on cytoplasmic factors and involves mainly heterochromatic segments. Although its functional significance is unknown, removal of heterochromatin and rearrangement of chromosome segments are likely to affect gene expression. It is a little surprising that modulating gene expression through chromatin elimination is not more common,[17] since so long as cells retain a complete genome, there is always a possibility that rogue cells will invade the germ line and contribute to gametes. Chromatin elimination in somatic cells is a very good way of subjugating potentially deviant daughter cells.

Not all multicellular organisms have mosaic development. At the other end of the spectrum are groups such as echinoderms, vertebrates and plants, in which there is far less maternal control of development. Developmental decisions are based more on cell–cell interactions. Cells are responsible not only for inducing new inherited functional states in their neighbours, they also seem to restrict each other's potency, thereby ensuring that they cooperate to produce an integrated individual. The chances of epigenetic or genetic variants disturbing the cohesiveness of the developing embryo and taking on germ-line status are reduced by the control exerted by sister cells.

The difference between the regulative and mosaic modes of development is certainly not absolute. Most zygotes have some localized cytoplasmic information, and all embryos are able to regulate to some extent. There seems to be an association between the mode of development and use of different EISs. Species with more regulative development tend to have higher rates of cell turnover and higher levels of methylation than those with more mosaic development (Table 8.2). For example, plants have very high levels of cytosine methylation, and among the metazoa, vertebrates and echinoderms also show relatively high methylation levels. In these groups, development is very regulative, and there is substantial cell turnover in determined and differentiated lineages during the predominant stage of the life cycle, usually the adult stage. In contrast, small animals with short life spans, such as nematodes and *Drosophila*, have strongly mosaic development, with little or no cell turnover as adults, and virtually no methylation. Their cell lineages have very few divisions. In the Insecta, where cell turnover and methylation levels are both generally low, those species in which cell division occurs during the adult stage seem to have higher levels of methylation. For example, in the locust, cells of the caecum and the intestine divide during adult life, and the level of DNA methylation is significantly higher than that found in *Drosophila*, in which there is no cell turnover in the adult. The coccids and aphids are exceptions to the low cell-turnover–low methylation correlation. In spite of the lack of cell division in adults of these two groups, their methylation levels are relatively high. Nevertheless, as we show in what follows, there are good reasons why coccids and aphids should not conform with the general pattern.

Table 8.2 Adult cell turnover and cytosine methylation

Group or species	Cell turnover	% Cytosine methylation[a]	Source of methylation data[b]
Mollusc (snail)	Yes	2.9	Adams and Burdon (1985)
Nematode	None	Not detectable	Simpson *et al.* (1986)
Crustacean (brine shrimp)	None	Not detectable	Warner and Bagshaw (1984)

Table 8.2 (*continued*)

Group or species	Cell turnover	% Cytosine methylation[a]	Source of methylation data[b]
Insects			
Drosophila	None	Not detectable	Patel and Gopinathan (1987)
Sciara	Probably none	Not detectable	Gerbi (1986)
Mosquito	None	Not detectable	Proffitt *et al.* (1984)
Locust	In some cell types	0.95	Adams and Burdon (1985)
Coccid	None	0.4–1.4	Scarbrough *et al.* (1984)
Silkworm moth	Probably none	0.18	Patel and Gopinathan (1987)
Aphid	Probably none	Yes (unquantified)	Field *et al.* (1989)
Sea urchin	Yes	2.9–6.5	Adams and Burdon (1985)
Fish			
Salmon	Yes	9.1	Adams and Burdon (1985)
Herring	Yes	8.9	Adams and Burdon (1985)
Amphibian (*Xenopus*)	Yes	6.9	Hergersberg (1991)
Birds			
Pigeon	Yes	4.5	Adams and Burdon (1985)
Chicken	Yes	3.7	Hergersberg (1991)
Mammals			
Mouse	Yes	4.6	Hergersberg (1991)
Rabbit	Yes	4.2	Hergersberg (1991)
Man	Yes	4.2	Hergersberg (1991)
Plants			
Bracken	Yes	33	Adams and Burdon (1985)
Wheat	Yes	25.5	Hergersberg (1991)
Barley	Yes	22.6	Hergersberg (1991)
Corn	Yes	29.3	Adams and Burdon (1985)
Bluebell	Yes	~40	Adams and Burdon (1985)
Pine	Yes	19.7	Adams and Burdon (1985)

[a] For some species different authors give different values; the most recent is used in the table.
[b] Most data for cell turnover is taken from Finch (1990); that for the brine shrimp is from Tate and Marshall (1991).

What is the significance of the correlation between mode of development, cell turnover in the adult, and the level of DNA methylation? The extent of methylation is roughly correlated with genome size, which is influenced by the amount of repetitive sequences (Hergersberg 1991). Since repetitive sequences are commonly methylated, it is usually assumed that the low level of methylation in organisms such as nematodes and *Drosophila* is directly related to their small genomes. However, as we see it, having a small genome with low amounts of repetitive DNA, and having little or no DNA methylation, are probably both associated with the mode of development, in particular with cell turnover. Repeated sequences are important carriers of both protein and methylation marks. In groups with relatively low amounts of repeated sequences and little methylation, the role of the chromatin-marking EIS in development may be small, because development is mosaic, with little cell turnover. Where there is little cell turnover, steady-state EISs can be very efficient, since there is little chance that the regulatory balance will be disrupted by repeated cell division. Such organisms therefore may not need the methylation-dependent marking system, and we speculate that in some groups it was lost. The methylation system has potentially damaging mutagenic effects, both because methylated cytosines are mutational hot spots, and because methylating enzymes can damage DNA bases. This disadvantage is a particular handicap in organisms with low cell turnover, where there can be little somatic selection to eliminate mutant cells. Therefore, the advantage of methylation as a reliable EIS is outweighed by its disadvantage when organisms have little cell turnover in the adult.

The apparent exception seen in the coccids, which have no cell division in the adult, but do have methylated DNA, makes sense because coccids have a sex-determining system based on inactivation of the paternal set of chromosomes in males (see Chapter 5, p. 112). The inheritance of the parental imprints that underlies this requires a chromatin marking system, so methylation has been retained. Aphids are also an exception to the general rule that organisms that lack cell turnover in the adult also lack methylation. Here again there may be an evolutionary reason for retaining methylation. Aphids have ameiotic parthenogenesis: the clones of individuals resulting from parthenogenesis form an extended soma. Many hundreds of cell generations may separate two meiotic events. During the aphid life cycle, phenotypically different forms are produced, and the phenotypes are inherited for several asexual generations. Efficient EISs, such as methylation marking, are probably essential for this type of reproduction.[18]

In contrast to insects and organisms with mosaic development, organisms in which cell–cell interactions are important in early development, and where cell turnover is high, must have developmental systems that are able to respond to stimuli flexibly, and to transmit the response stably through many cell divisions. The chromatin-marking EISs, particularly DNA methylation, which exploit the precision of chromosome replication, are particularly well suited for transmitting epigenetic information through many cell divisions. For organisms with regulative development, the advantages of the methylation EIS outweigh its potential dangers.

The segregated germ line

Buss (1987) emphasized that early determination and sequestration of a distinct population of cells to form the germ line is one of the main mechanisms preventing newly arisen variant somatic cells from becoming germ cells. The segregated germ line is one of the principal props of the argument against the inheritance of acquired characters (Chapter 2, p. 39). However, a segregated and sequestered germ line is not universal. In a multicellular organism such as that represented in Fig. 8.7a, a

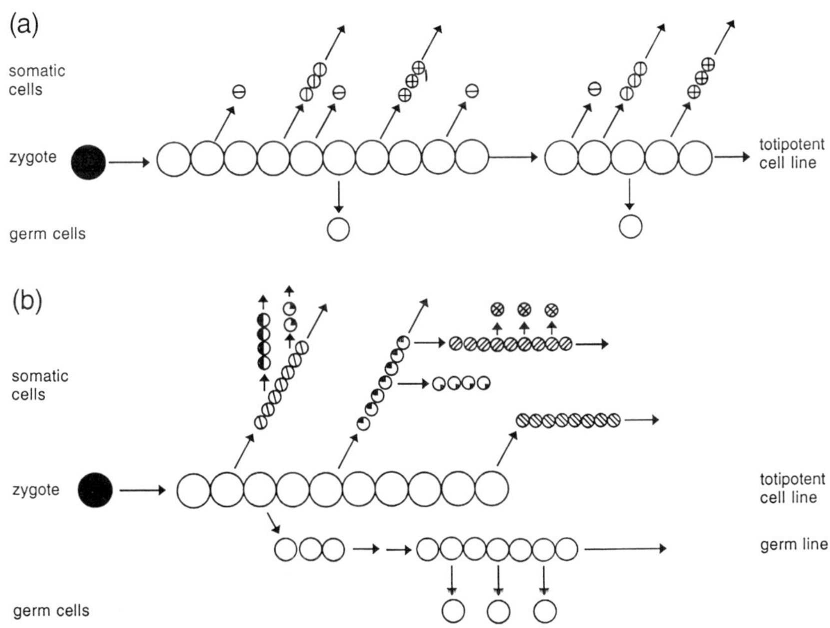

Fig. 8.7 Development in simple and complex multicellular organisms. (a) A simple multicellular organism which maintains a totipotent cell line capable of giving rise to all types of somatic cells and to gametes. Some differentiated cells divide further, others do not. (b) A more complex organism with progressive somatic differentiation. The totipotent cell line gives rise to multipotent stem cells, which give rise to more specialized cells, some of which are capable of further division. Once determined the germ line is sequestered and divides very little. (Loosely based on Buss 1987.)

totipotent cell lineage is retained, and supplies somatic cells throughout ontogeny. In contrast, in multicellular organisms such as that represented in Figure 8.7b, differentiation is progressive: totipotent lineages give rise to multipotent lineages with a more restricted capacity for producing new cell types. In organisms of this type, it is possible for a group of primordial germ cells to be sequestered and remain relatively inactive throughout development. Even if a new variant somatic cell lineage competes successfully with other cell lineages, it will not contribute to gametes if the cells cannot physically invade the germ line. A sequestered germ line therefore reduces the transmission of both genetic and epigenetic variants. The possibility of a new variant arising in the germ line itself is reduced if there is little cell division in the lineage during development. A quiescent germ line has little epigenetic information imposed on it, and little memory to erase before the next generation.

Segregation of soma and germ line may therefore have evolved as a response to the selection pressure imposed by cells with persistent epigenetic marks, which were able to contribute to the germ line and disrupt development of the next generation.

Meiosis and gamete production

Even in organisms with a segregated germ line, genetic and epigenetic changes in the germ-cell lineage are inevitable. Variants can arise either before the germ line is determined, or during the development of the germ cells themselves. However, probably only a small fraction of these variants is transmitted to the next generation. Most variants are eliminated either through selection between germ-line cells, or during meiosis and gametogenesis.

Since according to Buss (1988) and Ruvinsky (1990) there are no groups of differentiated multicellular organisms that are not primitively sexual, it is not surprising that several people have suggested that meiotic sex may have been a prerequisite or preadaptation for the evolution of multicellularity. As Szathmáry (1994) pointed out, sexuality certainly has properties that are excellent preadaptations for development. These include a sophisticated genetic programme producing different cellular states, and the use of cell-specific cell-surface molecules to establish contacts between cells. Szathmáry suggested that multicellularity evolved by using and modifying these features of sexuality in single-celled organisms.

An old, but still controversial idea is that the adaptive advantage of sex is associated with the role meiosis plays in rejuvenation and repair. Bell (1988*a*) has gone so far as to suggest that the reason why the first two-thirds of the history of life is occupied exclusively by unicellular organisms is that until sex had evolved as a means of repair, large multicellular organisms were impossible. The type of repair Bell considered important was what he termed exogenous repair. In this type of repair, the shuffling of genes during meiosis and production of haploid gametes results in parents being able to produce progeny that carry fewer harmful mutations than they do themselves. Without sex, the load of deleterious mutations transmitted to progeny cannot be reduced, so the evolutionary potential of asexual organisms is limited.

A rather different view of the role of sex is that its importance lies in endogenous repair processes. The origin and maintenance of sex are thought to be associated with the repair of molecular damage to DNA.[19] Homologous pairing of chromosomes during meiosis enables a damaged chromosome region to be repaired using the undamaged region of the homologue as a template. The role of meiosis in repair may go even beyond this, however. Ettinger (1986) suggested that meiosis has a role in maintaining the overall pattern of DNA organization in chromosomes, preventing selfish DNA from disrupting gene functions. Holliday (1984, 1988) argued that not only does meiosis help keep germ cells free from genetic damage and selfish DNA elements, it also helps free them from epigenetic defects, that is, from epimutations. During homologous pairing, the recombination repair pathway enables errors in methylation patterns to be corrected. According to Holliday, some of the detrimental effects of inbreeding may be consequences of the accumulation of epigenetic defects that cannot be removed. When inbred, organisms become homozygous, and error correction is impossible because there is no longer a non-defective chromosome to act as a template.

The adverse effects of continuous asexual reproduction can be seen in the senescence of clones of protozoans.[20] In *Paramecium aurelia* and some other ciliates, the rate of fission declines as cultures get older, until eventually the line dies out. The finite life span of such clones resembles that of normal diploid mammalian cells in culture. When the protozoan cultures are allowed to reproduce sexually, they are rejuvenated. Although other explanations are possible, most of the data are consistent with the idea that in the absence of sex, genetic and epigenetic damage and defects accumulate, but the accumulated faults can be repaired or eliminated through meiosis.

A lot of attention has been focused on the role of sex in the removal of genetic errors and damage, and this may indeed have been its original function. Its role in generating genetic diversity has also been considered in detail, and many people think this is the most significant aspect of sexual reproduction.[21] Its possible importance in epigenetic inheritance is usually ignored, yet meiosis could play a crucial role in reprogramming epigenetic systems. It may be that it is not, as Bell suggested, that the evolution of differentiated multicellular organisms was impossible without sex, but that multicellularity was impossible without efficient control of epigenetic inheritance, and efficient control of epigenetic inheritance was impossible without sex. Meiosis may have been essential for the efficient restoration of the epigenetic marks required to return cells to the ground state.[22] We do not want to argue that the *origin* of meiosis was associated with its role as an epigenetic renovation system, but we do believe that this role has been important in the evolution of multicellular organisms and in the shaping of the sexual processes seen today. During the formation of gametes, chromatin is restructured. This is seen most dramatically during spermatogenesis when in most species existing histone and non-histone proteins are extensively reorganized. DNA methylation patterns are also altered in both male and female gametogenesis. Therefore, whatever the source of the cells undergoing meiosis, whether from a sequestered germ line or from a somatic lineage, most existing chromatin marks will be changed. Meiosis and gametogenesis re-mark DNA. The chances of somatic invaders of the germ line transmitting soma-specific marks are therefore greatly reduced.

In addition to reorganizing chromatin marks, the sexual system of multicellular organisms has features that are important for structural and steady-state inheritance systems. Higher plants and animals are anisogamous, the male gamete being small relative to the female gamete. Selection for the small size of sperm may have occurred for energetic or mechanical reasons. Alternatively, selection for small size and the shedding of cytoplasm before entering the oocyte may have occurred because it prevents the transmission of intracellular parasites and avoids the possibility of conflict between competing maternal and paternal cytoplasmic organelles.[23] Whatever the evolutionary origin of the inequality in size of male and female gametes, one consequence of it is that there is an asymmetry in the information transmitted to the next generation. Cytoplasmic organelles are largely maternal in origin, as are steady-state and structural inheritance systems.

Although steady-state and structural inheritance systems have to be taken into account when considering the evolution of sex and meiosis, chromatin marks are probably of more evolutionary significance. We see meiosis as one of the pre-conditions for the evolution of complex multicellular organisms. Altering marks during gametogenesis enables a fresh epigenetic start to each generation, and suppresses newly arisen heritable epigenetic variants that compete for the status of gametes (Jablonka and Lamb 1989).

Exploiting marks and imprints for adaptive evolution

The studies of parental imprinting discussed in Chapter 5 show that new sex-specific marks are acquired during gametogenesis; genes and chromosomes inherited from the mother differ from those inherited from the father in ways that affect their activity. The possible function or functions of such imprinted genes and chromosomes has been the subject of much speculation. How did imprinting evolve? What is the selective advantage of imprinting? Answers to these questions illustrate some of the ways in which epigenetic marks are thought to have been exploited in adaptive evolution.

Initially parental imprinting probably had no function. It was an accident, the pleiotropic effect of changes in chromatin structure that were selected for other purposes. The differences between homologous chromosomes in the zygote originated as an indirect consequence of selection in parents for the chromatin changes necessary for meiotic pairing, compact sperm, and large oocytes. The modifications required for spermatogenesis and oogenesis were simply carried over into the zygote where, in many cases, they were probably of no selective importance whatsoever. Small differences in the marks on paternally and maternally derived chromosomes leading, for example, to slight differences in the time at which maternal and paternal alleles are activated, are probably normally completely unimportant. Only in the unusual circumstances resulting from experimental procedures in which zygotes are produced with two maternal or two paternal chromosomes, do most imprints become apparent. However, if an initial difference between maternal and paternal marks had a more substantial functional effect, it would have been subject to further selection. If an imprint was detrimental, then selection would favour changes in the parents which modified the imprints, or changes in post-fertilization events which ensured the imprint was rapidly erased. However, if the initial differences between the chromosomes benefited the offspring, selection would ensure that the initial imprints were preserved, perpetuated, and perhaps enhanced.

In what ways could parental imprinting benefit the offspring? How have different organisms capitalized on the chromatin differences established in gametogenesis? There is no shortage of explanations for the selective advantage of imprinting, and no reason to believe that all groups of organisms exploit the differential marking of their chromosomes in the same way. Even within a species, imprinting may be used in several different ways. The explanations offered for its existence and maintenance fall into two broad, non-exclusive, categories. The first includes those explanations that suggest that imprinting is important because homologous chromosomes are labelled as coming either from the father or from the mother; the second includes explanations suggesting that the significance of imprinting is that it generates differences between the homologues, rather than that it causes chromosomes to carry marks identifying their parental origin.

The first category of explanations obviously includes those in which imprints determine the preferential inactivation or elimination of paternal chromosomes associated with sex-determination and dosage compensation.

Since imprints can ensure that the appropriate chromosomes are eliminated or inactivated in a controlled way, initial differences between maternally and paternally derived chromosomes may have been enhanced by natural selection. Differences in the marks of sex chromosomes are particularly likely to be influenced by selection. Males always give their X chromosomes to daughters, whereas females give theirs to both sons and daughters; therefore, selection should favour changes in spermatogenesis that produce imprints on the paternal X that are specifically favourable to daughters (Jablonka and Lamb 1990*a*).

Haig and his colleagues suggested that many cases of imprinting are associated with conflicts of interest between the parents over the size of their offspring.[24] The offspring of mammals and flowering plants are nourished directly from maternal tissues, so genes expressed in the embryo can influence the amount of nourishment embryos receive from their mothers. Where the mother has offspring from more than one father, paternal genes that are programmed to obtain as much nourishment as possible for the embryo will be favoured. A father's genes have more chance of being passed to future generations if his offspring are large; if they become large at the expense of their half-sibs or mother, it is unimportant to the father, since the mother and half-sibs do not carry his genes. However, it is in the interests of the mother that her genes are programmed to counter this effect, so that her own survival, and that of present and future offspring, are ensured. Therefore, paternal imprints that promote the acquisition of nutrients by the embryo have a selective advantage, whereas selection favours imprints on maternal genes that counter the effects of paternal imprints. This war of attrition between maternal and paternal imprints can lead to the establishment of imprints with opposite effects on rates of growth and resource exploitation.[25]

Like the first, the second category of explanations for the importance of imprinting contains a mixed bag of ideas. What most have in common is the notion that imprinting is important because the differences between alleles enable better control of gene activity. One suggestion is that if two alleles are differently imprinted and therefore have different affinities for binding proteins, only one allele need be used. The second copy of the gene is held as a reserve, available if the first is damaged, or for use during periods of rapid growth. Genes are more controllable when imprinted, because instead of both alleles being on or both off, there is an intermediate state with one on and the other off. Another possible advantage of imprinting is that imprinted alleles may replicate their DNA at different times, so the shut-down of transcription which accompanies DNA replication does not happen in both alleles at the same time. An intriguing variation of the idea that imprinting is selected because it enables only one allele to be active is Holliday's (1990*b*) suggestion that *all* developmentally important genes are subject to allelic exclusion and become functionally haploid. He argued that this is necessary for the precise control of differentiation, because in diploid cells, when genes are switched from inactivity to activity during development, there is a possibility that daughter cells will inherit one switched and one non-switched allele, rather than both switched or both non-switched. Inactivating one allele eliminates this possibility. Monk (1990) suggested another role for imprinting, based on the possibility that there are mechanisms that allow homologous regions of chromosomes to communicate with each other and, if epigenetic differences are detected, modify alleles in ways that affect their functioning. A rather different proposal is that imprints may have been important in the early evolution of anisogamy: where zygotes formed from gametes of the same sex had lowered fitness, the difference between maternal and paternal imprints made selection against the union of like-sex gametes possible (Chandra and Nanjundiah 1990).

As can be seen from the examples outlined above, there is no lack of ideas about the role of imprinting in the evolution and functioning of organisms.[26] Exactly what was selected when imprints and imprinting processes were shaped by natural selection has received less attention. The obvious answer is that what was selected was DNA sequences that carried the epigenetic marks, or sequences coding for DNA-binding factors that influence the stability of marks. Such DNA sequences were selected for their potential to produce a stubborn mark: a mark that during embryogenesis in the offspring does not lose all traces of its past. However, there is another possibility: the initiating event in the evolution of a persistent sex-specific mark might have been purely epigenetic: the mark was changed, and the change was transmitted to the offspring. No change in DNA sequence occurred initially, although subsequently DNA changes may have been selected to stabilize marks.

Induced epiallelic variations and adaptive evolution

If epigenetic variations occur and can be selected and inherited without any initial changes in DNA sequences, in what way is the contribution of purely epigenetic variations to adaptive evolution different from that of genetic variations? In asexually reproducing unicellular organisms, transmission of epigenetic information is relatively straightforward. Induced marks or steady-states that represent an adaptive response to environmental or developmental challenges can be inherited. In contrast to mutations that are random and rare, the response to the environment will typically involve a coordinated change in the marks of many genes. Some of these are likely to be transmitted to the next generation and facilitate the response of the descendants to the same stimulus. The result is a progressively more efficient response to the environment.

In clonal organisms that are reproducing asexually, the same is true, but since they are multicellular, somatic selection of the induced epialleles may also play a role in adaptation (see Chapter 6). Many plants show clonal variation, but usually the causes of the variation are poorly understood: the variations do not behave like classical genetic mutations, yet some of them are heritable.[27] Breese and colleagues studied somatic variations in various characters in populations of the perennial rye-grass *Lolium perenne* (Breese *et al.* 1965; Hayward and Breese 1968). They compared sexual and asexual populations, and found that somatic variation occurs in clones of this species, and that the amount is related to genotype, to the age of the clone, and to environmental conditions. The variation is selectable, with different variations having different heritabilities and transmissibilities. One of the most intriguing results of the studies was the marked differences found within asexual populations. Epigenetic variation followed by somatic selection could be the explanation of variability in these asexual populations.

Selectable, seemingly non-genetic, somatic variability has been found in other clonal populations. It is likely that epigenetic variation is very important in clonal plants and organisms that are obligately asexual. It may enhance the speed of adaptive evolution, especially of specialization. The evolutionary potential of heritable variations which can be selected somatically is enormous. As Silander pointed out:

> Somatic variation in clonal plants may provide an opportunity for rapid evolution. Somatic effects that are deleterious may impose little or no genetic load on the population or on clonal ramets (cf. Whitham and Slobodchikoff 1981). The defective organs (or ramets) could simply be shed. Adaptive changes, whether small or saltational, could be perpetuated clonally even if they are not immediately incorporated into the germ line. Many of the somatic effects described for agricultural species represent dramatic changes in form or structure, literally saltational. . . . Analogous changes in natural populations that are advantageous may spread rapidly and be transmitted in the germ line as well as clonally. The general phenomenon of somatic variation represents an area that has been seriously neglected. . . (Silander 1985, pp. 126–127)

It is not clear whether the inheritance of induced epialleles is widespread and important in organisms that have a sexual phase. However, as we have stressed repeatedly, many organisms do not have a segregated germ line, and in those that do, it is commonly formed late in development. There are therefore many opportunities for epialleles to be included in the lineages that will produce gametes. The potential to pass on induced epigenetic information is particularly strong in plants, where apical meristems retain the ability to produce germ cells as well as other cell types. Work with plant cuttings has shown that individuals raised from cuttings often differ from each other more than do those raised from seeds (Schmid 1992). The reason for this is thought to be that when raised from cuttings, plants have a longer post-zygotic period during which epigenetic divergence can occur.

Development is not reset in cuttings, so they retain epigenetic memories of their origin. Significantly, it has been found that some of the non-genetic phenotypic differences between cuttings are transmitted to their sexually produced progeny: somatically acquired epigenetic variation is inherited. How common this is, is unknown. Somatically derived germ cells are also found in fungi and throughout metazoan groups such as the Cnidaria and Porifera (see Table 2.2, p. 45). Any epigenetic changes that affect these cells can be passed on to the next generation unless they are erased during gamete formation.

The problem with accepting the importance of the inheritance of induced epigenetic marks in the evolution of organisms that have a segregated germ line is twofold. First the marks have to be preserved throughout meiosis, even though the reprogramming processes in meiosis are likely to be very efficient, having been selected to ensure that each developmental cycle has a fresh epigenetic beginning and the zygote is totipotent. Nevertheless, as the model in Chapter 6 (p. 155) shows, a new epigenetic mark could become heritable in a sexually reproducing multicellular organism, in spite of efficient reprogramming processes, without compromising the zygote's totipotency. The second problem is that to have evolutionary effects, a variation must occur before soma and germ line become segregated. When segregation is early, the potential for somatic cell variants to contribute to germ cells is almost eliminated. However, selection within the germ line itself still occurs, so induced epigenetic variations are of evolutionary importance, even in organisms with a segregated germ line. Because the type of cell that can become a gamete is restricted, the only induced epigenetic variations with potentially adaptive effects are those that affect germ-line functions. The difference between organisms with a segregated germ line and those without is that in the former the evolutionary effects of heritable variations and selection between cells are restricted to fewer cell lineages. Marks affecting somatic adaptations cannot be transmitted to progeny when there is a segregated germ line; only marks affecting germ-line adaptations can be passed to future generations.

For example, imagine an environment that changes hormone production. The change affects germ-line cells by altering chromatin marks on some of the genes expressed in the germ line, thereby accelerating germ-cell differentiation in females. The result is smaller and more numerous eggs, and a smaller and morphologically simpler larval form. This effect of the hormone could be adaptive in the environment that induced it, and also in some other, non-inducing, environments. Assume that the marks are not reset to the normal state during oogenesis, either because they are 'stubborn' to start with, or because the inducing environment persists for several generations and there is a progressive build up of marks, which eventually leads to the construction of stubborn marks. If the new marks persist in both sexes, they will be transmitted to the next generation without any further problems, even in the absence of hormonal induction. (If the marks are erased in male gametogenesis, the second generation of females will be heterozygous for the marks and additional assumptions have to be made for the marks to persist.[28]) The stimulus-dependent phenotype could thus become stimulus-independent, purely by epigenetic means, without a change in DNA sequence. The phenotype could then be further stabilized by selection for DNA sequences that ensure a more reliable transmission of the marks in both sexes. If the marks enhance the mutability of the loci, the chances that genetic assimilation of the epigenetic state occurs may be increased further. This hypothetical example is illustrated in Fig. 8.8.

The main kinds of adaptive changes stemming from induced epigenetic variations in organisms with early and rigid germ-line segregation are therefore those involving germ-line cells or the early stages of embryogenesis. This may sound as if induced epigenetic variants can make only a very modest contribution to evolution, but of course such adaptations could have tremendous effects on embryogenesis and on adult characters. Even such seemingly 'trivial' characters as the size of the egg can have far-reaching effects. In sea urchins, experimental manipulation of the egg size of one species has resulted in concerted changes in larval morphology and physiology that make it resemble the larval form and development of another species. It was concluded that:

> Substantial differences in larval form and development rate, sufficient to characterize different species, need not require allelic substitutions into the larval genome. (Sinervo and McEdward 1988, p. 897)

Hope for the hopeful monster? [29]

Currently, one of the most fashionable explanations of macroevolutionary changes is that they result from heterochronic changes—from alterations in developmental timing.[30] Simple shifts in the relative time at which events occur early in development can affect several features simultaneously. For example, Thomson (1988, 1992) suggested that a small change affecting the timing of blastema maturation at the pre-chondrogenic stage of limb development could underlie the evolution and adaptive radiation of reptilian ankles. Because of the integrated nature of development, several phenotypic features are affected. As Goldschmidt recognized many years ago, mutations occurring early in development may produce a fairly coordinated novel organism—a hopeful monster—that could be the basis of macroevolutionary change.

Goldschmidt believed that what he called 'systemic' mutations, that is mutations resulting from chromosome 'repatterning', were responsible for the changes in timing of developmental events (Goldschmidt 1940).

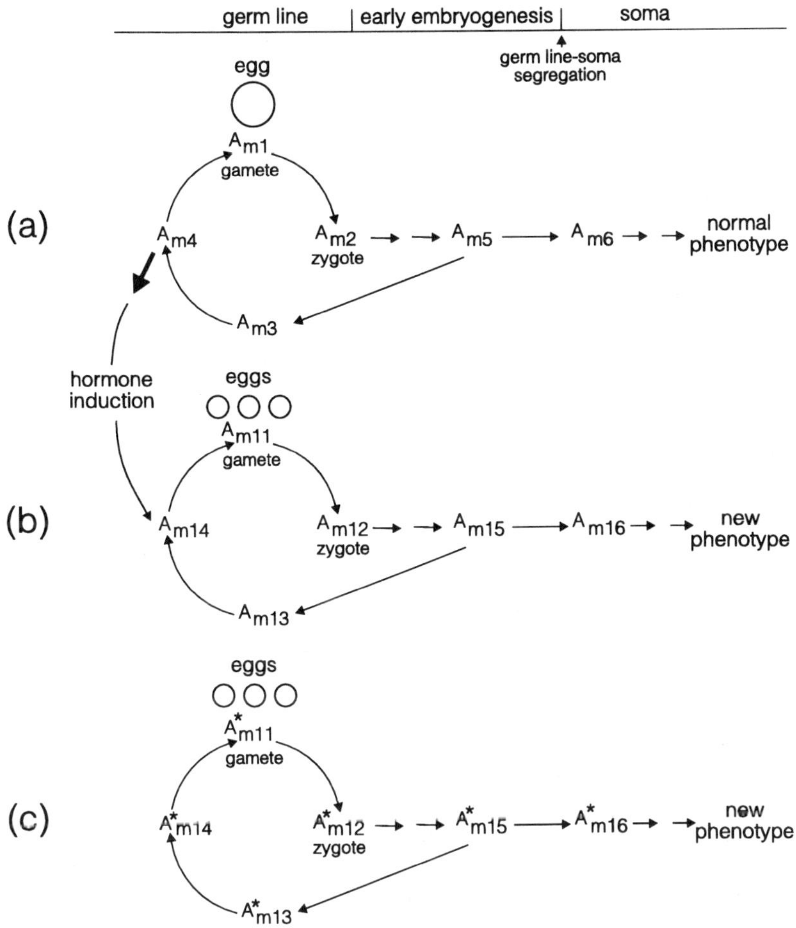

Fig. 8.8 Hormone-induced changes in marks leading to a phenotypic change in egg size and larval form (cf. Fig. 6.5, p. 155). (a) The normal developmental cycle of changing marks on gene *A*. (b) Following hormone induction, mark *m4* is changed to *m14*, and this affects subsequent marks and egg size. (c) A genetic change from *A* to A^* stabilizes the new marks.

Nowadays, the mutations that are thought to underlie the heterochronic shifts causing macroevolutionary changes are generally believed to be changes at regulatory loci. The types of change that may be involved are those discussed in Chapter 7. They are changes that affect the stability and fidelity of cellular memory—the ease with which gene activity can be altered. They could be increases or decreases in the number of methylation sites, or in the number of repetitive sequences. They could also be changes in the amount of neighbouring heterochromatin, in which case the mutations might be the result of a chromosomal repatterning, as Goldschmidt suggested.[31] It is possible, however, that sometimes the initiating events in heterochronic shifts do not involve mutations at all. Changes in temperature or salinity could easily affect a gene's phenotype in ways that make it more or less easy to switch on, and therefore affect the time in development at which it is expressed. Even if such changes are not heritable, if the changed environment persists, and if the organism with the changed phenotype is viable, genetic changes moulding and canalizing the new phenotype would follow.

The attraction of heterochrony as the basis of evolutionary change is that it offers an explanation for novelties arising in many aspects of the phenotype as a result of a single genetic change. However, the same effect could be brought about by epigenetic changes that affect many loci. One of the important differences between mutations and induced epialleles is that whereas random mutations occur infrequently, several epialleles can be induced independently and simultaneously in the same individual. Changes in temperature, salinity, or nutrients might affect epigenetic marks in different systems, even though the systems are not closely coordinated genetically or developmentally. If the new marks are inherited, as is particularly likely in lower organisms, the genes may respond more or less readily in the next generation. If they are not inherited, but the environmental change persists, the phenotypes could be genetically assimilated.

Epigenetically produced hopeful monsters differ from genetic monsters in one important respect: epigenetic changes in response to environmental challenges can result in the novel phenotype appearing in *many* individuals in the population at the same time. The problem of the 'lonely' hopeful monster struggling to survive and find a mate, does not exist. 'Monster' may be the predominant phenotype in a population in a new environment, so it does not take time for the new phenotype to become common.

Summary

Epigenetic inheritance systems were present in the unicellular ancestors of multicellular organisms. They had a selective advantage for cells living in environments that fluctuated regularly with cycles that were longer than the generation time of the cells. The fidelity of transmission of epigenetic information, that is, the length of clonal memory, was moulded by natural selection to fit the frequency of cycles of environmental fluctuations.

The presence of EISs in unicellular organisms enabled the transition to multicellularity, and ultimately the evolution of complex multicellular organisms. Sex was probably a precondition for such evolution, because during meiosis epigenetic marks can be reset, enabling each new generation to have a fresh epigenetic start.

Many fundamental features of the ontogenies of multicellular organisms, such as beginning development from a single cell, maternal control of early development, irreversible differentiation, and the segregation of the germ line from the soma, can be seen as the indirect effects of EISs: the benefits of EISs have been exploited, and the potential threats of epigenetic and genetic inheritance have been countered in various ways. Imprints, the different marks resulting from differences in chromatin restructuring in male and female gametogenesis, have also been exploited in some organisms, sometimes playing a role in sex determination or dosage compensation.

The spectrum of EISs used by different organisms is not always the same. It has been influenced by the type of development; in particular, adult size and cell turnover affect the use of the methylation marking system. DNA methylation is a particularly reliable system for stabilizing suppressed or active states of gene activity, and is abundant in large organisms with high cell turnover. Organisms that have little or no cell turnover as adults have no need for the methylation EIS, and may have dispensed with it because of its mutational hazards.

Since new epialleles can be induced by the environment, they may occur at several different loci in a genome, and be directly responsible for a change in phenotype. Moreover, similar changes are expected to occur in many, or even most, individuals in a population. Such changes would be most apparent in asexual organisms, but would also occur in those with sexual reproduction. In the latter only changes induced in cells that are capable of producing germ cells are of long-term evolutionary importance.

Epigenetic inheritance has both indirect and direct effects on adaptive evolution. Inevitably, therefore, it has also had effects on speciation and on the evolution of higher taxa.

Notes

1. The definition of life as depending on self-maintenance is very ancient, and its beginning can be traced to Aristotle's *De Anima*. The biological philosophy of Claude Bernard, with his great emphasis on homeostasis as the essence of biological systems and of life, also follows this line.
2. Reviewed by Borst and Greaves (1987).
3. Reviewed by Van de Woude *et al.* (1992).
4. Reviewed by Soll *et al.* (1993).
5. Similar mechanisms of epigenetic switching presumably underlie the life-cycle changes seen in multicellular parasites, such as trematodes, which exploit several hosts sequentially, have different phenotypes in each host, and often reproduce asexually in the different hosts.
6. The models described are taken from Jablonka *et al.* (1992).

7. Gould (1989) provides a fascinating account of early metazoan evolution.
8. See Bell (1985) and references therein.
9. Some unicellular organisms have overcome the problems of large size by adopting shapes that increase their surface area; others have polyploid nuclei or are multinucleate, thus having more DNA templates for transcription, and a greater potential for meeting the demand for proteins needed to maintain their large bulk. The large ciliated protozoa have a 'somatic' macronucleus containing many copies of the genetic information and a germinal micronucleus that functions only in reproduction.
10. Bell (1985) made a detailed study of the Volvocales, flagellated green algae showing all degrees of size and complexity from small unicellular forms such as *Chlamydomonas* to the large and truly multicellular forms such as *Volvox*. This study has shown convincingly that the unicellular algae have a very much lower capacity for increasing their numbers than do the related colonial algae and other planktonic plants and animals of a similar size.
11. Margulis (1981, pp. 267–272) pointed out that no plant or animal cell retains cilia or flagella while undergoing mitotic cell division. Although some protozoan groups can divide while remaining ciliated, she believes that the ancestors of multicellular groups could not. This argument has been criticized by Bell (1989) who has questioned both the basic generalization that commitment to a cilium or flagellum precludes cell division, and the assumption that the ancestral metazoan had only a single MTOC.
12. The same view has been expressed by Raff and Kaufman (1983) who suggested that the intracellular mechanisms that coordinate the activities of the nuclear and organelle genomes within single-celled organisms are a pre-adaptation for multicellular life, where coordination and cooperation of different cell genomes is essential.
13. Details of the work on *Proteus mirabilis* are given by Allison and Hughes (1991) and Shapiro (1988), who also discusses work on *E. coli*, *Pseudomonas*, and other species. *B. subtilis* colonies are discussed by Ben-Jacob *et al.* (1992).
14. Jollos suggested long ago (1921) that the term 'individual' need not be used only for a single paramecium, but was also appropriate for a lineage of paramecia integrated by its response to a stimulus, and delimited by the duration of the Dauermodifikation, i.e. by the length of the epigenetic memory of the response, which usually persists from one meiosis to the next.
15. Buss (1982) also collated data showing how organisms that are able to form chimeras usually have somatic compatibility systems that reduce the opportunities for competition between genetically different cell lineages.
16. See Gilbert (1991*a*, pp. 273–275) for a brief discussion and references on chromatin elimination. Chromatin elimination in *Ascaris* was known to Weismann and was used by him as evidence for his hypothesis that differentiation is the result of quantitative changes in the hereditary material in the nucleus.
17. Spradling *et al.* (1993) have found evidence that segments of heterochromatin are eliminated during polytenization in *Drosophila*. (They were previously thought to be under-replicated.) They suggested that DNA elimination may be a quite common way of modulating gene expression during development.
18. The brine shrimp *Artemia* has no cell division in the adult and no methylation, and some populations reproduce parthenogenetically. However, in this case parthenogenesis is meiotic, so there is no extended soma, and the number of

divisions between meioses is limited. There is therefore no advantage in having a methylation-dependent memory system.

19. The first person to consider sex as a cellular repair process seems to have been Dougherty (1955), but the idea has been made more well-known and popular by Bernstein and his colleagues in a series of papers from 1977 onwards. See Bernstein *et al.* (1988) and Maynard Smith (1988*b*) for a discussion of the pros and cons of the idea that meiotic sex is primarily an error-correcting mechanism for DNA.
20. See Bell (1988*b*) for a thorough description and discussion of the complex and contradictory evidence of clonal senescence in Protozoa.
21. The adaptive value of sex has been the subject of debate and controversy for a long time, and many competing explanations have been offered (see Bell 1982). After reviewing many of the proposed advantages of sex, Gould and Gould concluded: 'We have, it seems, an embarrassment of plausible hypotheses to account for the evolution and maintenance of sexual recombination. Which are the most likely to be correct? It seems reasonable to suppose that sex evolved originally to deal with one problem, and has since been exploited to solve quite another'. (Gould and Gould 1989, p. 65)
22. The existence of asexual multicellular taxa shows that meiosis is not essential to organisms with complex ontogeny. However, the fact that most such taxa are evolutionarily short-lived may be in part due to the accumulation of defects resulting from the failure to reset epigenetic programmes.
23. See Hurst (1992) for a discussion of the evolution of anisogamy.
24. See, for example, Haig and Westoby (1989), Haig and Graham (1991), and Moore and Haig (1991).
25. Haig and Graham (1991) show how their hypothesis offers a very satisfying explanation of the pattern of imprinting on the mouse *Igf2* and *Igf2r* loci discussed in Chapter 5. An obvious way in which to test the hypothesis would be to study differential imprinting of growth-affecting genes in monogamous and polygamous species of mammal. According to Haig and Graham's hypothesis, fewer cases of differential imprinting are expected in monogamous species because there is no asymmetry in the interests of the parents.
26. It is not difficult to think of possible selective advantages of genes being imprinted. For example, one which we believe has not been suggested before is that differently imprinted alleles may allow the production of qualitatively different protein products. It is known that some genes are differently expressed in different tissues, or at different stages of development, depending on which promoters are used, or where the polyA sites are. Is it possible that there are genes of a similar type for which parental imprints affecting the binding of regulatory proteins determine which promoter is used? If so, imprints could result in the use of alternative alleles to produce different products in different tissues, or at different times. A single allelic difference involving imprinted genes might in this way generate greater functional diversity than is expected. The difficulty with this idea, like so many of the others, is that there is almost no evidence for it, and it is not easy to see an evolutionary route leading to it.
27. Silander (1985) reviewed some of the data on variability in clonal plants.
28. In this case, if the new marks are to persist in future generations in the absence of hormonal induction, it has to be assumed either that selection for the new phenotype is very strong, or that some kind of mark conversion to the new state takes place during female gametogenesis (a type of imprinting occurs), so

that all female gametes carry the new marks, and all progeny carry them on at least one homologue.

29. The term 'hopeful monster' was coined by Goldschmidt who said 'In a former paper (Goldschmidt, 1933) I used the term "hopeful monster" to express the idea that mutants producing monstrosities may have played a considerable role in macroevolution.' (Goldschmidt 1940, p. 390).
30. Gould's book *Ontogeny and phylogeny* (1977) was influential in re-awakening interest in heterochrony. Hall (1992*b*) describes and discusses many examples of what may be heterochronic shifts.
31. A large part of Goldschmidt's argument for the importance of 'chromosomal repatterning' was centred on position effect variegation in *Drosophila*, which results from structural changes in chromosomes (see Chapter 7).

9

Heredity and the origin of species

> . . . the origin of species—that mystery of mysteries, as it has been called by one of our greatest philosophers.
>
> Darwin, Introduction to *The origin of species*

Ever since Darwin wrote *On the origin of species by means of natural selection*, there has been a more or less continuous debate about the nature of species, the circumstances that lead to speciation, the nature of the speciation events, and the evolutionary significance of species and speciation. Many different definitions of species have been used, and this has complicated discussion of the origin of species, but since the advent of the Modern Synthesis, the species concept that has dominated evolutionary thinking has been that known as the 'biological' species concept, associated with the names of Dobzhansky, Muller, and Mayr. According to Mayr, species are:

> groups of actually or potentially interbreeding natural populations, which are reproductively isolated from other such groups. (Mayr 1942, p. 120)

Species are thus defined in terms of reproductive isolation. Although there are many difficulties associated with this species concept, not least of which is that it cannot be applied to asexual organisms, it is at present the most widely used, and the one that will be used here.[1]

Gene exchange between species is prevented by pre-zygotic and post-zygotic isolating mechanisms. Pre-zygotic mechanisms, such as breeding at different times of the year or in different habitats, or having different courtship behaviour or incompatible genitalia, prevent zygotes from being formed. If zygotes are produced, post-zygotic mechanisms, such as sterility or inviability, reduce the success of the hybrids. Once present, reproductive isolation allows two species living in the same area to diverge even further: genetically their future evolutionary paths are independent of each other.

It is generally thought that most speciation occurs by accident: while populations are isolated from each other by a geographic barrier of some kind, they diverge genetically, and reproductive isolation arises, or at least is initiated, as a by-product of this divergence. Initially, therefore, there is no selection for reproductive isolation, and the intrinsic isolating mechanisms are irrelevant. They originate as side-effects of adaptation for other purposes, or as a result of non-adaptive genetic changes.

In this chapter we look at the origin of species in the light of epigenetic inheritance and directed variations. We consider the role of EISs in the origin of the isolating mechanisms that develop while populations are geographically separated, and discuss the possibility that non-genetic mechanisms could initiate sympatric speciation, i.e. speciation without geographic isolation. The effects of EISs help to explain the differences in the number of species in different taxa, the important role the sex chromosomes play in speciation, Haldane's rule, and meiotic drive. We follow Darwin and begin by looking at domestication.

Domestication as a model for speciation: back to Darwin

Darwin began *The origin of species* with a chapter entitled 'Variation under domestication'. In it he used domestic animals and cultivated plants as a model for what he believed happens in nature. The changes brought about by animal breeders and horticulturists using artificial selection were used as an analogy for the changes wrought by natural selection. In addition, although it is often forgotten, Darwin used the variability seen in domestic species as the basis for his discussion of the origin and inheritance of variation. In the first chapter of *The origin* and throughout the book Darwin returns to the problem of why domestic animals are so variable. He did not believe that the variations arose by chance. He wrote:

> I have hitherto sometimes spoken as if the variations—so common and multiform with organic beings under domestication, and in a lesser degree with those under nature—were due to chance. This, of course, is a wholly incorrect expression, but it serves to acknowledge plainly our ignorance of the cause of each particular variation. . . . variability is generally related to the conditions of life to which each species has been exposed during several successive generations. In the first chapter I attempted to show that changed conditions act in two ways, directly on the whole organisation or on certain parts alone, and indirectly through the reproductive system. (Darwin 1872, *The origin of species*, 6th edn, p. 100)

Darwin certainly believed that changed conditions caused variability. This belief is repeated and reinforced by the wealth of empirical evidence assembled in his massive treatise *The variation of animals and plants under domestication*, first published in 1868. Darwin suggested that excess of nutrients is one of the main causes of variability in domestic species, but other environmental factors are also important. He commented on the fact that sometimes several generations of exposure to changed conditions are necessary before variability becomes apparent. He claimed that use and disuse played a role in modifying parts of organisms, and that such modifications can be inherited. The pangenesis hypothesis, developed in Darwin's book on variation, was an attempt to explain the origin and maintenance of the explosive variation found when animals and plants are domesticated. Darwin firmly believed that the conditions of life during domestication produced specific and non-specific heritable variations which were then selected by man. The environment can induce new variations, as well as select them.

Possible explanations for the many observations Darwin described became clear after the Mendelian basis of heredity was established. It was realized that apparently new phenotypic variation could arise through rare recessive alleles becoming homozygous through inbreeding in small domesticated populations, or through epistatic interactions or pleiotropic effects of the new gene combinations resulting from drift and the relaxation of natural selection. Although it continued to be acknowledged that artificial selection in animals and plants is a valuable analogy for natural selection, interest in domestication as a model for the origin of variation declined. When the Modern Synthesis made acceptance of hard inheritance almost universal, Darwin's argument that domestication also had relevance to the generation of new heritable variation was generally forgotten, ignored, or assumed to be wrong. However, there were exceptions to this. In the USSR attitudes to animal and plant breeding were different, and when genetics emerged from the Lysenko era, interesting facts and ideas about the process of domestication were put forward.

One of the leading lectures at the International Congress of Genetics held in Moscow in 1978 was given by Belyaev, director of the Institute of Cytology and Genetics in the Siberian branch of the USSR Academy of Sciences.[2] He devoted his talk to domestication, which he described as one of the greatest biological experiments, pointing out that in the course of 15 000 years there has been an increase in the rate and range of variation in domestic animals far greater than that seen at any other time in evolutionary history. He outlined the results of many years' work on the domestication of silver foxes. Foxes had been selected for tameness—for a calmer, more dog-like behaviour. The selection was successful, but it was not just behaviour that changed. Many other aspects of the phenotype were affected: the reproductive season of females was prolonged, and some mated twice a year; the time of moulting changed; the levels of corticosteroids, oestrogens, and progesterone were different; there were karyotypic changes involving additional heterochromatic microchromosomes; the way in which the tail and ears were carried was modified; piebald animals, known as 'Star', often appeared. The frequency of aberrant phenotypes, 10^{-2}–10^{-3}, was very high. These changes were not the result of inbreeding, since care was taken to avoid this, and many characters were in any case semi-dominant; they were unlikely to be the result of new mutations in structural loci, since the rate of occurrence was high, and several changed phenotypes were often found in the same animal.

The explanation Belyaev proposed for the novel phenotypes was that selection for tameness caused changes in the neurohormonal system, and that under the influence of an altered hormonal equilibrium, dormant genes were activated. The idea of dormant genes, and the experiments that describe their properties, were discussed in Chapter 6 (p. 140). Essentially, studies of *Star* in foxes and *Fused* in the mouse suggest that genes can be transmitted from generation to generation in an inactive, dormant state. Altered conditions, such as those leading to changes in the hormonal levels of the parents, can change the heritable state, making the genes heritably active. Belyaev and his co-workers suggested that since the domesticated foxes often had additional heterochromatic microchromosomes, modification of the genes might be associated with the degree of heterochromatinization in the region in which they were located. It should be remembered that it was not just a single character that was affected: the stressful conditions of domestication released a spectrum of dormant genes. As suggested in Chapter 7, additional heterochromatin could act as a sink for chromatin binding factors, and thereby activate several previously inactive genes. Alternatively, any changes in environmental conditions that produce metabolic disturbances could affect the marks on many genes, making them more easy or less easy to activate.[3]

Changes in phenotype that are difficult to explain in conventional terms have been reported in other newly domesticated species. For example, there are several species of ducks in which one of the first visible changes following breeding in captivity is a change in plumage.[4] This can occur after only two or three generations, so is unlikely to be the result of new mutations. Although it could be the result of inbreeding making existing recessives homozygous, there is evidence from one species that makes this doubtful. The Laysan duck is found on a small island where its numbers have probably never exceeded about 600. It is known to have passed through a bottleneck of 10 individuals in 1909. A few individuals were taken into captivity in 1963, and after about 15 years, new colour variants appeared in their descendants. The history of inbreeding prior to captivity makes it unlikely that the plumage variants arose through alleles for recessive characters becoming homozygous, and the small numbers make it unlikely that new mutations occurred. Were dormant genes reactivated in the stressful conditions of domestication?

According to the Russian workers, domestication, like other stresses, is not simply a selection process in which animals with poor adaptations and adaptability to the domestic environment are eliminated. It also induces extensive and explosive variability. Consequently, changed conditions accelerate evolution not just because selection pressures change, but also because they widen the range of hereditary variation on which selection can act. As we described in Chapter 7, under conditions of stress the frequencies of mutation, recombination and transposition increase. The genome acts as a response system that produces variation in adverse conditions. Domestication may generate new variation because the imposed conditions are a severe ecological and behavioural stress. In addition, new variation could result from genomic stresses caused by changes in the normal breeding patterns. As McClintock and others have emphasized, genomic stresses can mobilize transposable elements and cause other changes that increase gene and chromosome mutation, and cause a profound reorganization of the genome.[5]

If new environmental challenges produce new genetic and epigenetic variation, Darwin was right in thinking that domestication is a useful analogy for the origin of variation as well as for the effects of selection in transforming characters. It is usually thought that domestication does not produce new species, because new strains happily mate with each other. But, since breeders select for non-discriminating mating behaviour by encouraging mating with the chosen form, however bizarre it may be, it is hardly surprising that reproductive isolation is rarely seen. Nevertheless, as Hemmer (1990) has emphasized, some domestic animals probably are quite effectively reproductively isolated from their wild ancestors, and by most species' criteria should be regarded as incipient species:

> Domestication is accompanied by incipient speciation characterized by the occurrence of pre-mating isolation mechanisms in the form of temporal and ecological isolation, social isolation and structural isolation. In particular, the mechanisms caused by the special behavioural syndrome of domestic animals restrict free interbreeding with the original wild species, even where feral animals occur alongside them. (Hemmer 1990, p. 191)

Epigenetic inheritance and modes of speciation

In theory, in natural populations, reproductive isolation between different parts of an initially interbreeding population could evolve in several different ways, and there is fierce argument about which of the possible models of speciation is the most likely.[6] It would be inappropriate to discuss the fine details of these arguments here, so we shall simply outline some of the main ways in which speciation could occur and consider how EISs may be involved in each. We then discuss in more detail the role that EISs play in speciation.

1. *Allopatric modes of speciation*

It is widely accepted that most speciation occurs allopatrically, that is, while parts of the ancestral population are separated by geographical barriers. Gene flow between the parts is impossible or greatly restricted, so they diverge genetically, and often also phenotypically. Sometimes genetic divergence is sufficient to make the formerly separated populations incapable of interbreeding if they come into contact again.

Two somewhat different ways in which allopatric speciation can happen are recognized. The first involves division of the ancestral species into fairly large populations by a geographical barrier such as a river or glacier. While separate, each part is subject to different selective forces, different mutations, different population fluctuations, and is likely to show slow adaptive divergence. As a by-product of this, partial or complete reproductive isolation may occur. The isolation could be the result of divergence in pre-zygotic factors, such as habitat preferences, that prevent mates meeting, or of accumulated genetic differences that lead to the inviability or infertility of any hybrids that are produced.

An additional factor that could be responsible for hybrid inviability or sterility in this, as in other modes of speciation, is accumulated differences in chromatin marks. These may (but need not) be associated with DNA changes. Three ways in which differences in the chromatin structure of a gene could arise are illustrated in Fig. 9.1. First, there could be random or directed changes in the marks themselves, producing new heritable epialleles. Second, changes in DNA bases that affect the marks the gene carries could occur. Finally, DNA changes elsewhere in the genome, such as increases or decreases in heterochromatin, may affect the chromatin structure of the gene. In each case the chromatin differences could affect the time or place of gene activity and have phenotypic effects, although they need not do so. More importantly, as we shall discuss later, differences in chromatin structure could underlie hybrid sterility.

A second type of allopatric speciation involves what Mayr originally called speciation through 'the founder effect', but later referred to as peripatric speciation.[7] He suggested that a small number of individuals, perhaps even a single gravid female, colonizing a new area at the edge of a species' range, could initiate quite rapid and dramatic restructuring of the genome. Mayr argued that a number of factors contributed to what he called a 'genetic revolution': the selection pressures on an isolate are different, because a newly occupied habitat is likely to be both different from, and more uniform than, the old habitat; the isolate may be genetically less variable than the original population, and have different allele frequencies that change epistatic interactions; there is likely to be more homozygosity. Old co-adapted genomes are broken up, and variability is released so that new ones can be formed.

Mayr's ideas on founder effects have been influential, but have been criticized on theoretical grounds.[8] Nevertheless, the wealth of evidence showing that island populations are often highly divergent suggests that colonizations have been important in speciation. In many ways, colonizing a new island is comparable to domestication: there is a bottleneck in numbers and selection pressures change and are more uniform. Undoubtedly, there is also an increase in stress, and it seems likely that the stress-induced changes described earlier will also occur in new colonizers.

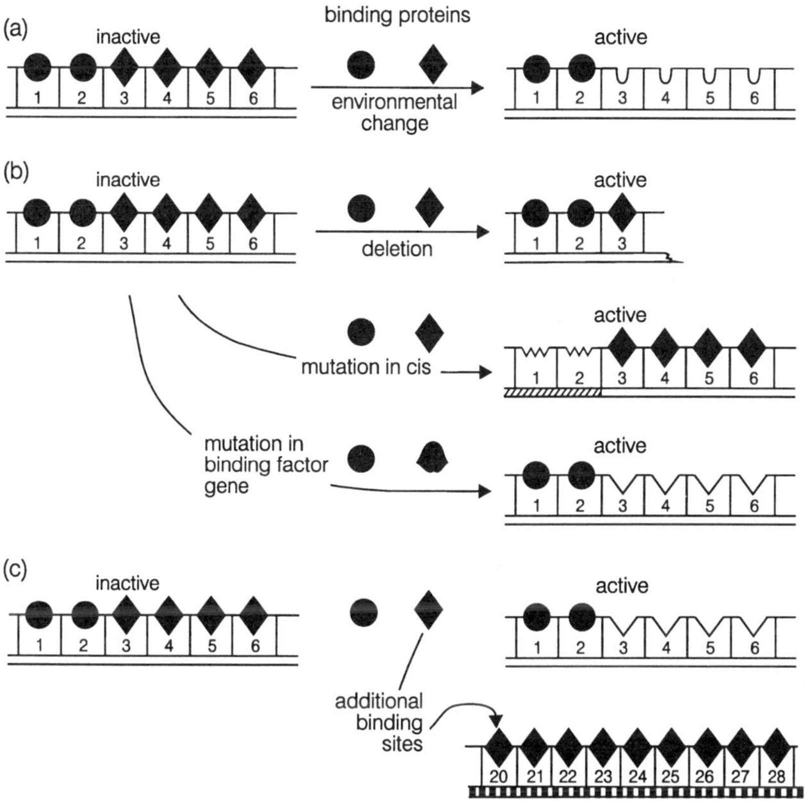

Fig. 9.1 Three ways in which chromatin marks can diverge. The numbered blocks represent different marks in a control region. Solid shapes are DNA-binding proteins. The gene is active unless all sites are bound. (a) Heritable, environmentally-induced changes that cause epiallelic changes; a heritably inactive gene becomes heritably active without a change in DNA sequence. (b) Genetic changes in (1) the number of binding sites that carry marks; (2) the sequence composition of the binding sites; (3) a gene coding for a binding factor. All changes lead to the gene's activity. (c) Increase elsewhere in the genome in binding sites for one of the binding factors causes permanent activity of the gene.

It is not difficult to imagine changes in hormonal levels occurring and affecting dormant genes in birds or mammals, particularly if they previously lived in large social groups. Stress could change epigenetic marks, and stress-induced genetic changes, including chromosomal changes, could contribute to a genetic revolution. Both genetic and epigenetic changes could contribute to reproductive isolation.

Parapatrically distributed species often have different karyotypes, and this led Mayr (1982*a,b*) to suggest that chromosomal changes have been an important part of founder effect speciation. Other people, notably White (1968, 1978), have argued that chromosome rearrangements can sometimes initiate speciation, even without prior geographic isolation. Whether or not homozygotes for new chromosomal rearrangements are likely to be found without geographical separation of a sub-population is controversial, but there is little doubt that differences in karyotype are commonly found in closely related species, and their origin is something that theories of speciation should be able to explain. We return to the role of chromosome changes in speciation later in this chapter (p. 262).

One instance in which a recent 'domestication' involving a bottleneck in numbers has apparently resulted in speciation has been described by Weinberg and co-workers (1992). In 1964, five or six individuals of the polychaete *Nereis acuminata* were used to start a laboratory culture. The population, which went through three to five generations per year, expanded to thousands. In 1986 four pairs were taken to another laboratory, where again numbers built up to thousands. Between two and four years later, the laboratory strain was tested for reproductive isolation from the field population from which it is thought to have originated. No viable offspring were obtained: post-zygotic isolation was complete. The cause of the post-zygotic isolation is not known, but there is a chromosomal difference between the laboratory and field populations. There is also evidence of pre-zygotic, behavioural isolation. Although other interpretations of the data are possible, this seems to be a case of very rapid allopatric speciation through founder effects.

2. *Sympatric speciation*

Sympatric speciation, the formation of a new species within the range of the ancestral species, is a controversial subject. Following Mayr's forcefully and repeatedly argued case against the possibility of sympatric speciation, it has generally been assumed since the 1950s and 1960s that it is almost impossible for a single interbreeding population to split into two without prior geographical isolation. The only acknowledged exception is the sudden origin of new species through polyploidy. It is believed that the initiation of reproductive isolation by disruptive selection for alternative niches is unlikely, because random mating between individuals from the two niches prevents continued divergence. However, population genetics models have been devised showing that in some circumstances sympatric speciation is feasible.[9] In particular, it is promoted by any tendency for habitat selection and mate selection to be associated. For example, if plant-eating insects mate on their food source, partial reproductive isolation could accompany adaptation to a new host. Gene flow between insects using the old host and those using the new would be reduced because they would meet less frequently. The host specificity of many insect species, and the existence of host races, has been taken as evidence that speciation can occur sympatrically (Tauber and Tauber 1989). There is also experimental evidence showing that disruptive selection on habitat preference can lead to incipient speciation (Rice and Salt 1990).

Could epigenetic inheritance play any role in this mode of speciation? Evidence suggesting that it may comes from studies of host-plant conditioning in insects. It has been found that induced changes in food preference can be passed from mother to offspring.[10] For example, when the parthenogenetic stick-insect *Carausius morosus* was reared on privet, the offspring were less willing to accept ivy as a food source than were the offspring of the same strain reared on ivy. Since parthenogenetic reproduction in this stick-insect does not involve recombination, and since the numbers involved were too small for new mutations to be significant, genetic variation was minimal. Unless stress-induced genomic changes occurred, the inherited differences between the two strains were probably epigenetic rather than genetic.

A similar but even more significant change, which was probably non-genetic, occurred in a clone of the aphid *Dysaphis anthrisci majkopica*: after 15–18 parthenogenetic generations on novel hosts, the clone became reproductively isolated from the original form (Shaposhnikov 1965, 1966, 1985). This species uses apples as its primary host, and umbellifers as its secondary host (see Fig. 9.2 for a typical aphid life cycle). The experimental clone was initially reared on a species of umbellifer that it found less acceptable than its normal host. Once adapted, individuals were transferred to a second umbellifer species, which originally had been totally unacceptable. The result was that the clone rapidly diverged morphologically, and after eight generations it was incapable of living on its original host species. Most of the change occurred during a period of markedly increased variability and intense selection between four and eight generations after transfer to the new host species. When, after 15–18 generations of parthenogenetic reproduction, sexual individuals were produced, the line that had adapted to the new host was almost completely reproductively isolated from the line from which it was derived. Mating occurred normally, but less than 4% of the few eggs produced were viable. Unless stress-induced transposition events occurred, the size of the population used was too small for selection of new mutations to be a likely cause of divergence. However, since aphids methylate their DNA, and inherited changes in methylation associated with altered gene expression are known to occur in aphids (see Chapter 6, p. 145 and Chapter 8, p. 212), it seems quite likely that non-genetic changes in methylation or other types of chromatin marks could have been responsible for the inherited shift in host preference and reproductive isolation.

Divergence in the aphid clone may have been helped by the carry-over of chemicals acquired from the host plant, which influenced chemosensory development in the offspring. Corbet (1985) has suggested that the chemical legacy from parents can modify an insect's chemoresponsiveness, thereby facilitating a shift from one host to another. This type of response could lead to a self-sustaining ecological loop, with induced preferences for a host resulting in induced preference in the offspring, who in turn induce preferences in their offspring, and so on.

One of the most important observations in the experiments with the aphid *D. anthrisci majkopica* was that not only did it acquire a new host preference, morphological differences, and reproductive isolation from its sister clone, it also had adaptive morphological features that made it converge with *D. chaerophyllina*, the species that usually lives on the host to which it had become adapted. Even more remarkably, the newly adapted strain of *D. anthrisci majkopica* produced fertile progeny in crosses with *D. chaerophyllina*. Normally, in spite of breeding sympatrically and synchronously on the same primary host, by all the usual criteria the two species are true species, since they are reproductively isolated: crosses between them produce no progeny. Yet in the experimental situation, after a few generations of selection in the parthenogenetic stage of the life cycle, one species changed into a form that resembled the other in some morphological features, and was capable of interbreeding with it, though many of the progeny were aberrant. The obvious explanation of these extraordinary results is that the experimenters allowed their strains to become contaminated with *D. chaerophyllina*. However, although this was suggested, it was subsequently acknowledged that the strains that showed the new behaviour were indeed *D. anthrisci*, so contamination was not responsible.[11] The results suggest very strongly that the features normally distinguishing the two species are inherited primarily through EISs, and that speciation was an epigenetic event.

Polyphenism and speciation

West-Eberhard (1986, 1989) has suggested another way in which speciation, and perhaps some of the more spectacular evolutionary innovations, could be initiated by essentially epigenetic processes involving very little genetic change. She believes that some evolutionary divergence begins with polyphenism. Polyphenism is 'non-genetic polymorphism': individuals with identical genomes have distinctly different adaptive phenotypes. These alternative, discrete phenotypes are produced in response to environmental cues that determine which genes will be expressed and switch development between alternative pathways. Often there are pronounced differences in the morphology, physiology, and behaviour of the alternative forms.

The presence of alternative adaptive phenotypes in a species is quite common. Aphids provide some of the best examples of polyphenism. Their life cycles vary, but an annual life cycle may begin with a series of wingless parthenogenetic generations on one host plant species, before a winged form is produced and migrates to a second host species, where further parthenogenetic generations of wingless and winged forms occur (see Fig. 9.2). Eventually, another winged generation is produced parthenogenetically and returns to the original host, where it gives rise to sexually reproducing forms. Equally complex life cycles with a variety of polyphenic forms and sexual and asexual reproduction are found in trematodes, some gall midges and gall wasps, and other insect species.

Not all seasonal polyphenisms involve asexual generations. For example, some arthropods having several sexual generations a year at certain times produce offspring that have a period of dormancy, while at other times their offspring develop directly. Some polyphenism is not seasonal at all: different forms can exist together at the same time. The obvious example

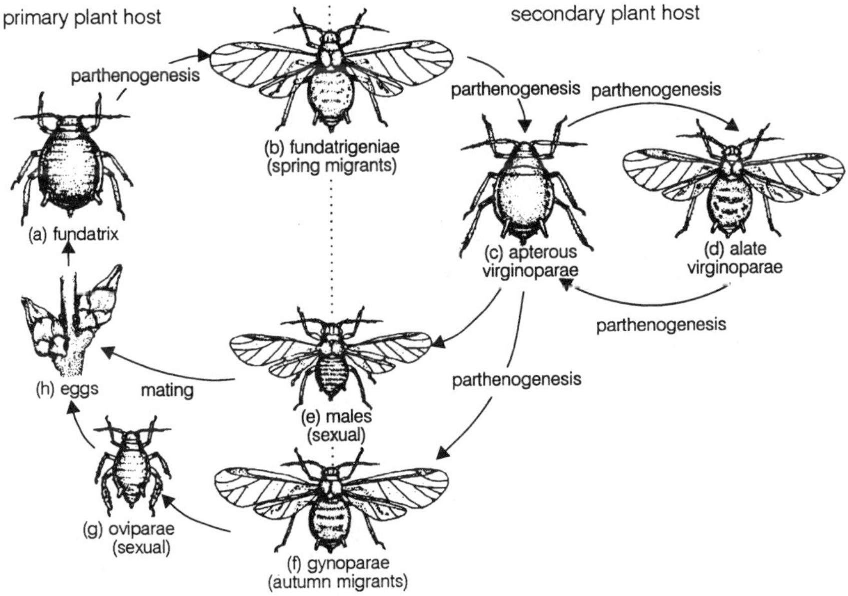

Fig. 9.2 The life cycle of a typical aphid, the bean aphid *Aphis fabae*. (a) Fundatrix, the product of sexual reproduction, which establishes parthenogenetic populations of spring migrants; (b), (c) and (d) are parthenogenetic spring forms, feeding on a variety of secondary host plants. The winged form is produced under crowded conditions. Towards the end of summer, males (e), and gynoparae females (f), which will be parents of sexual females (g), are produced. Mating occurs between males and females leading to the production of eggs (h), from which fundatrix females emerge after a diapause, thus beginning the cycle again. (Modified from Blackman 1974.)

of this is the genotypically identical workers and queens in the social Hymenoptera, but there are other less extreme cases, such as the shell polyphenism of some barnacles, or the carnivorous and omnivorous larval forms of spadefoot toad that live in the same pond (Pfennig 1992).

Polyphenism fascinated and influenced many of the early evolutionists, including Weismann and later Goldschmidt and Waddington. They recognized that environmentally induced switches between different forms could be informative about the origin of evolutionary novelties and the differences between species. As described in Chapter 3 (p. 54), Weismann studied the effects of temperature on butterflies that sometimes had phenotypically different spring and summer generations. His results forced him to acknowledge the existence of a type of Lamarckian inheritance, later called parallel induction: he suggested that the temperature effects seen as polyphenism in wing colour were inherited because parallel changes occurred in the wing determinants in both somatic and germ-line cells.[12]

Goldschmidt used polymorphic and polyphenic forms to back up his belief that mutational change could produce large, adaptive, integrated, phenotypic effects. Superficially, polyphenism does seem to show that a simple switch can produce distinctly different adaptive phenotypes, but of course with polyphenism the response to altered conditions is part of the normal and already canalized phenotypic repertoire of the species. It presumably evolved through selection acting on pre-existing plasticity to shape two well-canalized developmental pathways.[13] It could be initiated by an environmentally-induced change affecting a character such as the size of eggs, which then had repercussions on later development (Chapter 8, p. 221).

West-Eberhard (1986, 1989) has taken a different line from that of the early evolutionists. She argued that the importance of polyphenism in evolution is that it allows innovation to begin without abandoning the existing form. A new adaptation, which enables a population to occupy an additional sympatric niche, can evolve alongside the old one. Old adaptations are not lost when there is environmentally mediated switching between two forms. The contrasting forms are epigenetic alternatives. If conditions change so that only one of the alternative forms is favoured, as might happen if part of the population became geographically isolated, the other form could be lost without any initial genetic change. It would be lost because there is no environmental cue to switch development into the alternative path. Subsequently, West-Eberhard reasoned, rapid genetic changes might occur because the genome is released from the constraint of having to produce two well-canalized phenotypes. She has documented evidence from a variety of organisms that makes the idea of speciation through the fixation of one of the alternative phenotypes found in polyphenic species highly plausible. Her arguments suggest that divergent evolution does not require the isolation of gene pools posited in most models of speciation; on the contrary, sometimes it is the divergent evolution that occurred within a population that makes speciation possible. Initially divergence is a consequence of epigenetic differences, not genetic differences.

Some degree of polyphenism is widespread, but many of the most extreme polyphenisms are found in groups such as aphids, daphnids, some rotifers, and digenetic flukes, all of which have cyclical parthenogenesis. Moran (1992) suggested that the characteristics of groups that enable multiple developmental pathways to be established are first, a short generation time with several generations per year (or per host life cycle in the case of digenetic flukes), and second, pre-natal development. If environmental changes are regular and predictable, environmental stimuli received by the mother can, through their effects on her neuroendocrine system, influence internally developing embryos at a very early stage in their development and determine their adaptive pathway. In other words, polyphenism is favoured if mothers can transmit epigenetic information to their daughters through the environment they provide. The maternal environment induces the genes' phenotypes of daughters.

Species-richness

Many of the groups that show extensive polyphenism, for example daphnids and aphids, are also very species rich. This is consistent with West-Eberhard's suggestion that polyphenism allows experimentation and diversification, and facilitates speciation. However, Buss (1988) has suggested other reasons why groups with parthenogenesis are species-rich. He compared the species-richness of different taxa with the mode of germ-line formation. His results are shown in Fig. 9.3. As Buss stressed, there are many potential sources of error in the estimates of species numbers, and for very few taxa are there more than a handful of studies of the time of germ-line segregation, but the data nevertheless strongly suggest that the species-richness of a taxon is associated with its mode of germ-cell formation. Taxa with early germ-line determination tend to have low species numbers, those with late germ-line determination or somatically derived germ cells have higher species numbers.

How did Buss interpret these non-random patterns of evolutionary divergence? His argument was an extension of those we outlined in Chapter 2 (p. 46): the time of germ-line determination is significant because it dictates both how much of the genetic variation that arises in ontogeny is heritable, and the extent and nature of the selection pressures acting on new genetic variants before they can be transmitted to the next generation. If germ-line cells segregate early, as they do in insects such as

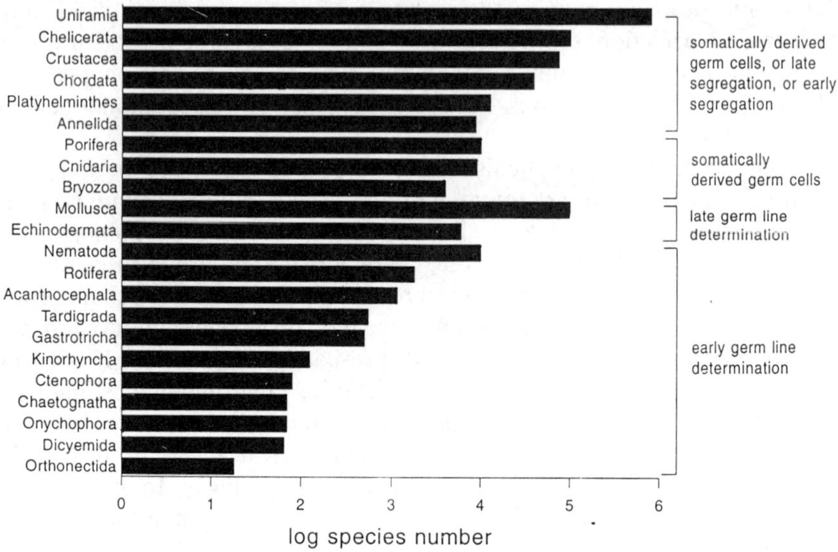

Fig. 9.3 Species richness in different taxa. (Modified from Buss 1988.)

Drosophila and nematodes such as *Caenorhabditis*, the number of cells and cell generations in which heritable genetic variants can arise is limited. Most new variants are not transmitted because they occur in somatic cells, which cannot form part of the germ line. In contrast, in an organism like *Hydra*, which has somatically derived germ cells, and can also reproduce asexually by budding, the number of cell generations between the zygote of one generation and the gametes for the next can be enormous. There are many opportunities for new variants to occur and be screened in the somatic environment before being transmitted to the next generation. Buss believes that variants that compete successfully in the soma and acquire a place in the germ line may initiate new ontogenetic programmes. The opportunities for this to happen are greater in groups in which germ cells are somatically derived or germ-line determination is late, so ontogenetic experiments are more frequent, and this is reflected in the higher species numbers in these groups.

Although the correlation between species number and mode of germ-line determination is good, there are many apparent anomalies. For example, nematodes have early germ-line determination, but many species. At lower taxonomic levels, the digenetic flukes, mites, aphids, and some other insect groups, all have more species than is expected from their early mode of germ-line determination. But some of these are exceptions that prove the rule: in these groups parthenogenesis is common, and amplifies the opportunities for new heritable variants to arise between one sexual generation and the next.[14] New variants arise in the germ line of the asexual generations and are transmitted and screened in the following generations. Consequently, even though the germ line segregates early, there are many cell generations between the zygote and the production of gametes, and extensive competition between variants. These groups therefore have the same potential for evolutionary experimentation as those with late germ-line determination, because their germ line is extended through the asexual generations. Unfortunately, as Buss noted, there are exceptions to the general pattern of association of species-richness and germ-line determination which are not explained by the presence of asexual reproduction. For example, the hypothesis offers no explanation for 'the Creator's inordinate fondness for beetles'.[15]

The arguments just outlined apply to epialleles, as well as to genetic changes. If epigenetic variations initiate new ontogenetic programs, taxa with late-segregating germ lines are expected to be species-rich for the same reasons that Buss suggested. Moreover, if EISs are important, the rate of diversification will be further enhanced when germ-line determination is late or germ cells are somatically derived, because some variants may arise through directed changes and be adaptive. Additional inheritance systems such as EISs provide additional heritable variation; additional heritable variation provides additional opportunities for divergence.

The relationship between polyphenism, parthenogenesis, and species-richness also makes sense in the light of EISs. Polyphenism can evolve when information about the current state of a regularly varying environment can be received and acted on by an individual before it makes critical developmental decisions. One way in which this can happen efficiently is by the parent receiving the information and passing it to their offspring in a form that influences the offspring's developmental pathway. Parents can do this through information provided in the egg cytoplasm, through information carried in chromatin structure, through information provided by hormones and the nature of the nutrients if the embryo develops viviparously or is brooded, and ultimately by behavioural means. Groups that have ameiotic parthenogenesis have unique opportunities to use information provided in chromatin structure, because the asexual generations avoid the epigenetic reprogramming that occurs during meiosis. New epigenetic variants in the chromatin-marking system can be tried out, transmitted, and reinforced through many generations of asexual reproduction. Sometimes chromatin marks in a lineage may become so stubborn the phenotype is fixed. This may be what happens when aphids shift hosts. When parthenogenesis is accompanied by viviparity, the opportunities for information to be passed to offspring are further enhanced.

A series of generations of asexual individuals, each inheriting information about the environment from the previous generation, provides a rather close parallel with ontogenetic development. There are opportunities for heritable changes in lineages, mediated by EISs. In the absence of asexual generations, the opportunities for individuals to change through EISs are more restricted, particularly if the germ line is determined early. With early germ-line determination, only those epigenetic changes that occur in the few cell generations before germ-line segregation, or that occur in the germ line itself, can be inherited and have significance for evolutionary divergence and speciation. Although this may sound as if EISs can have played only a minor role in speciation in groups with early germ-line segregation, in fact the opposite is true. As we discuss later, epigenetic variants influencing germ-line functions may well be one of the most important factors causing reproductive isolation, through their effects on hybrid sterility.

Hybrid viability and inviability

Darwin devoted a chapter of *The origin* to 'hybridism'. He recognized that the failure of species to cross and produce viable and fertile offspring is not something for which there has been direct natural selection; it is an incidental result of selection for other things. He also recognized that the reason why crosses between species sometimes produce no offspring, and the reason why hybrids sometimes produce no offspring, are usually different. Species crosses fail because fertilization cannot occur, or if it does, the embryo dies early in development. Hybrids produce no offspring because, according to Darwin, the whole organization is disturbed and there is failure in the reproductive system, particularly in that of the male.[16]

Darwin's observations have proved to be largely correct, but have remained largely unexplained. Why some hybrids fail to develop successfully, whereas others are somatically vigorous but sterile, is puzzling. As Darwin realized, the viability and sterility of hybrids does not have as much to do with the taxonomic closeness of the parent species as one feels it should, particularly in plants. Why are two morphologically almost identical species incapable of interbreeding, when species belonging to different genera are able to do so? Sometimes morphologically similar species differ in karyotype in ways that might impede interbreeding, but often the karyotypic differences are small, and the two species share many of the same protein polymorphisms, but they cannot interbreed. There seems to be no significant genetic divergence, yet they are reproductively isolated.

We believe that one of the most important factors contributing to hybrid inviability and sterility is differences in the epigenetic variations of the two species (Jablonka and Lamb 1989, 1991). While two parts of a population are geographically separated, they accumulate different epigenetic variations, as well as different genetic variations. Both induced and random changes in epialleles occur, and since the rate at which new epialleles are produced is higher than that for new mutations, the build up of epigenetic differences is initially greater. The same is true during the early stages of adaptation to new hosts during sympatric divergence: the initial changes are induced changes in epigenetic systems.

There is direct evidence that marks in related species diverge. A gene in *Mus musculus musculus* is marked differently from the same gene in the closely related species *M. m. domesticus*: the gene is inactive in *M. m. domesticus* males and active in females, whereas in *M. m. musculus* it is active in both sexes. The difference is associated with an unlinked imprintor gene that has different alleles in the two species (Chapter 5, p. 127). Other evidence suggesting that species carry different heritable marks affecting gene expression in hybrids is summarized in Table 9.1. In some cases the marks are parental-sex specific, with the maternal allele being expressed in both reciprocal crosses. More commonly, the expressed alleles or characters are species-specific, with genes of only one of the parent species being active, regardless of the way the cross is made.[17] Even when both alleles are active in the hybrid, their times of activation are sometimes asynchronous.

In most of the cases in the Table, the reason for the differential expression of the parent genomes is unknown. Chromatin structure in the two parent species could have diverged in any of the ways shown in Fig. 9.1, or genetic changes in genes coding for structural proteins could be responsible. Indications of the types of changes that may accompany the divergence leading to differential gene expression have come from studies of nucleolar dominance. In both plants and animals, hybrids between species commonly have only one of the two parental nucleolar organizer regions (NORs) active. NORs are chromosomal sites that contain many copies of the genes coding for ribosomal RNA; the tandemly repeated copies are separated by spacer regions, several kb long, that are not transcribed. According to Reeder (1984, 1985), nucleolar dominance in hybrids between the toads *Xenopus laevis* and *X. borealis* is associated with differences in the number of copies of a 42 bp repeated sequence in the spacer region; these repeats act as enhancers of gene expression. Flavell and O'Dell (1990) have shown that nucleolar dominance in wheat is also correlated with the number of repeats in an intergenic region, and with the methylation of these sequences. They suggested that the repeats are involved in the cooperative binding of proteins that prevent methylation and heterochromatinization, and consequent silencing, of rRNA genes; competition for the DNA-binding proteins makes NORs with larger blocks of intergenic repeats more active. Different species have NORs that differ in their numbers of repeats, and hence in the likelihood that they will be expressed in hybrids.

The fruitflies *Drosophila melanogaster* and *D. simulans* have their NORs in the heterochromatic regions of the sex chromosomes. Hybrid offspring normally have only the *D. melanogaster* NOR active. In this case,

Table 9.1 Inter-species crosses and gene activity

Parent species		Character studied	Observations in hybrids	References
Female	Male			
Mus mus domesticus (laboratory mouse)	*M. m. musculus* (wild mouse)	Imprinting of *Tme* gene	Not imprinted	Forejt and Gregorová (1992)
M. m. musculus (wild mouse)	*M. m. domesticus* (laboratory mouse)	As above	Imprinted	
Microtus arvalis (vole)	*M. subarvalis* (vole)	X-chromosome inactivation and activity of X-linked enzymes	*M. subarvalis* X inactive in most cells; higher expression of *M. arvalis* enzymes	Zakian *et al*. (1991)
M. subarvalis (vole)	*M. arvalis* (vole)	As above	As above but *M. subarvalis* X preferentially active in extra-embryonic tissues	
Coturnix coturnix japonica (Japanese quail)	*Gallus gallus domesticus* (chicken)	ADH	Maternal form appears first; paternal form absent in some individuals (hybrid produced by artificial insemination)	Castro-Sierra and Ohno (1968)
Ambystoma mexicanum (axolotl)	*Pleurodeles waltlii* (salamander)	DNA-ligase I and II	Expression of axolotl DNA-ligase I and *Pleurodeles* DNA-ligase II	Signoret *et al*. (1983), Signoret and David (1986)
Pleurodeles waltlii (salamander)	*Ambystoma mexicanum* (axolotl)	As above	Expression of *Pleurodeles* DNA-ligase I and axolotl DNA-ligase II	

Xenopus laevis (toad)	*X. borealis* (toad)	NOR	Only the *X. laevis* NOR is active in early development	Reeder and Roan (1984), Reeder (1985)
X. borealis (toad)	*X. laevis* (toad)	NOR	As above	
R. esculenta (hybridogenic European edible frog)	*R. lessonae* (European pool frog)	Meiosis	*Lessonae* chromosomes eliminated in oogonial cells	Bucci *et al*. (1990), Vinogradov *et al*. (1990), Schmidt (1993)
R. lessonae (European pool frog)	*R. esculenta* (hybridogenic European edible frog)	Meiosis	*Lessonae* chromosomes eliminated in spermatogonial cells	
Salmo irideus (rainbow trout)	*S. trutta* (brown trout)	LDH subunit in retina ADH in liver	Maternal form appears long before paternal Paternal form initially suppressed	Hitzeroth *et al*. (1968)
S. trutta (brown trout)	*Salvelinus fontinalis* (brook trout)	α-GPDH	Only paternal form found in early embryo	Schmidtke *et al*. (1976)
S. fontinalis (brook trout)	*S. trutta* (brown trout)	As above	Both parental forms found	
Lepomis gulosus (warmouth)	*L. microlophus* (redear sunfish)	Esterase LDH	Paternal form suppressed Allele found in warmouth heart not detected in hybrid	Whitt *et al*. (1972)
L. microlophus (redear sunfish)	*L. gulosus* (warmouth)		Dies at hatching	
Notemigonus crysoleucas (golden shiner)	*Scardinus erythrophthalmus* (rudd)	NOR	Maternal NOR active in most individuals; remainder with maternal NOR active in most cells, both active in the remaining cells	Gold *et al*. (1991)

Table 9.1 (*continued*)

Parent species		Character studied	Observations in hybrids	References
Female	Male			
S. erythrophthalmus (rudd)	*N. crysoleucas* (golden shiner)	As above	As above	
Poeciliopsis monacha-lucida (hybridogenic Mexican fish)	*P. lucida* (Mexican fish)	Meiosis	Maternal *lucida* chromosome-set eliminated prior to meiotic pairing	Schultz (1977)
Aedes aegypti (mosquito)	*A. mascarensis* (mosquito)	C band on *A. aegypti* chromosome	*A. aegypti* C band suppressed	Motara and Rai (1977)
A. mascarensis (mosquito)	*A. aegypti* (mosquito)	As above	*A. aegypti* C band present	
Bacillus rossius-grandii benazzii (hybridogenic stick-insect)	*B. grandii benazzii* (stick-insect)	Allozymes	Maternal *benazzii* genes not expressed	Mantovani and Scali (1992)
Crepis capillaris (hawk's beard)	*C. neglecta* (hawk's beard)	NOR	Only *C. capillaris* NOR seen	Wallace and Langridge (1971)
C. neglecta (hawk's beard)	*C. capillaris* (hawk's beard)	As above	As above	
Hordeum vulgare (barley)	*H. bulbosum* (barley)	Chromosomes in embryo	*H. bulbosum* chromosomes lost	Davies (1974)
H. bulbosum (barley)	*H. vulgare* (barley)	As above	As above	

NOR: nucleolar organizer region; LDH: lactate dehydrogenase; ADH: alcohol dehydrogenase; α-GPDH: α-glycerophosphate dehydrogenase.

nucleolar dominance is not a function of the different rDNAs themselves, but depends on the presence of a particular region of neighbouring heterochromatin on the X and Y of *D. melanogaster*. When this region is missing, nucleolar dominance is no longer present (Durica and Krider 1978).

Further evidence that divergence in heterochromatin may be responsible for differences in gene expression is seen in Zakian and co-worker's (1991) study of X-inactivation and gene expression in interspecific vole hybrids. The results of this study are summarized in Fig. 9.4. In hybrids between *Microtus subarvalis*, a species with a large block of heterochromatin in the X chromosome, and *M. arvalis*, a species lacking the blocks of hetero-

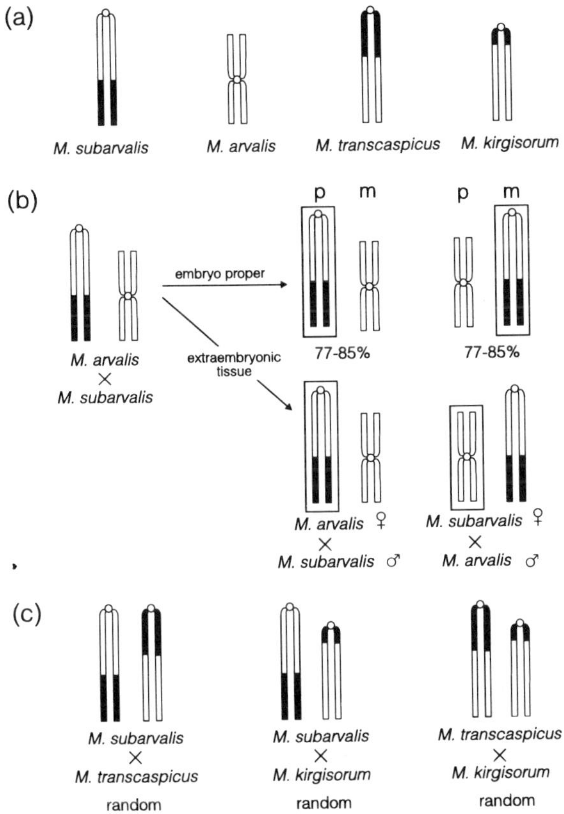

Fig. 9.4 Heterochromatin and X chromosome inactivation in four vole species. (a) The X chromosomes: heterochromatin is shown as solid blocks. (b) Non-random X inactivation in hybrids between *M. arvalis* and *M. subarvalis*; **p**: paternal chromosome; **m**: maternal chromosome; chromosomes in boxes are inactive. (The same pattern was seen when the other species of voles with heterochromatin blocks were crossed with *M. arvalis*.) (c) Random X inactivation in crosses between voles species containing heterochromatin blocks. (Based on data in Zakian *et al.* 1991.)

chromatin, the *M. subarvalis* X is inactive in 77–85% of cells, and most cells show the *M. arvalis* form of the enzymes glucose 6-phosphate dehydrogenase (G6PD) and α-galactosidase (GALA). The exception is in the extra-embryonic tissues where, as in most mammals, the paternal X is inactive, whichever way the cross is made (Fig. 9.4b). In order to see whether the preferential inactivation of the *M. subarvalis* X chromosome is associated with the block of heterochromatin it carries, hybrids between *M. arvalis*, the species lacking heterochromatin blocks, and *M. transcaspicus* and *M. kirgisorum*, two other species with large, although differently located, blocks of heterochromatin, were studied (Fig. 9.4a). The results were similar: the inactive X was the one with the heterochromatin block. Hybrids between two species that both had heterochromatin blocks showed random X inactivation (Fig. 9.4c). The explanation suggested for these data was that since heterochromatin usually replicates after euchromatin, when only one X has a heterochromatic segment, it provokes late replication and inactivation of the whole X chromosome; if both Xs have heterochromatin blocks, inactivation is random.[18]

When crosses are made between distantly related species, which must have many genetic and epigenetic differences, often the hybrids are not, as might be expected, intermediate between the parent species. For example, many intergeneric hybrids in plants show parental dominance, strongly resembling one of the parent species in both morphological and physiological characters.[19] Interestingly, and probably significantly, in hybrids between some cereal plants the chromosomes of the two parent species are not randomly positioned in the cell; those from one species are more peripheral than those from the other. The chromosomes of the two species seem to be marked in ways that cause them to be treated differently: they carry what Heslop-Harrison (1990) calls 'species-specific imprints'.

The most extreme form of parental dominance is seen in animal hybrids in which the genome of one parent species is eliminated completely. This occurs in species showing hybridogenesis: two species cross, a viable and fit hybrid expressing genes from both parents is produced, but one parental genome is selectively eliminated before gamete formation. As Fig. 9.5 shows, the frog species *Rana esculenta* is in fact a hybrid between *R. ridibunda* and *R. lessonae*.[20] It has hemiclonal inheritance: before meiosis, gonial cells eliminate the chromosome set of one of the parent species, and restore diploidy by doubling that of the other. All gametes therefore normally have the complete genome from only one of the parental species. Most populations of *R. esculenta* coexist with *R. lessonae*, and are maintained by matings between *R. lessonae* males and *R. esculenta* females that eliminate the *lessonae* chromosome set. Matings between *R. esculenta* males and females are rare, and usually result in inviable offspring. However there are other populations of *R. esculenta* that coexist with *R. ridibunda*, and eliminate *ridibunda* genomes; yet other populations occur

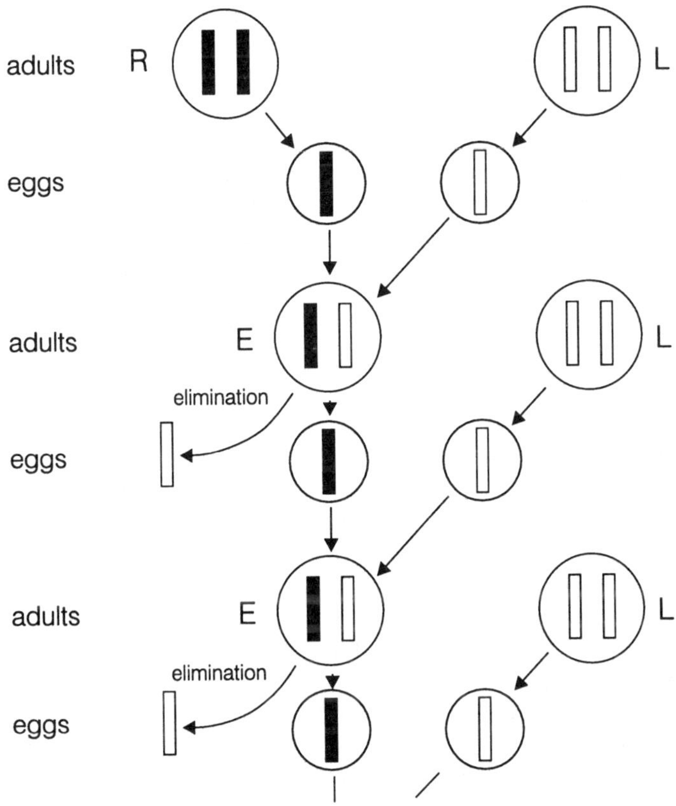

Fig. 9.5 Chromosome elimination in *Rana esculenta*. R: *R. ridibunda*; L: *R. lessonae*; E: the hybrid *R. esculenta*. Hollow bars: lessonae genome; solid bars: ridibunda genome.

in the absence of the parent species. In the latter, some individuals must eliminate *R. ridibunda* genomes, and others *R. lessonae*.

How are the parental chromosome sets distinguished? The chromosomes of *R. lessonae* and *R. ridibunda* are different because the latter have bands of centromeric heterochromatin that are lacking in the former. These heterochromatic blocks are believed to be causally related to the pre-meiotic elimination of one parental genome.

Excluding one of the parental genomes from meiosis is a very effective way of circumventing the meiotic problems found in many interspecific hybrids (see next section). In another hybridogenic species, the Mexican fish *Poeciliopsis monacha-lucida*, a unisexual hybrid between *P. lucida* and *P. monacha*, which is maintained by mating with *P. lucida* males, the paternal *lucida* genome is eliminated during meiosis.[21] The *lucida* chromosomes become heterochromatic and do not take part in pairing, so only the *monacha* chromosomes continue into the second meiotic division. Once again, potential problems associated with meiosis are avoided, in this case by by-passing pairing. In order for chromosomes to be excluded in this way, the chromosome sets from the parents must be marked in different ways, and the marks must be retained in the germ line of the hybrid.

The various lines of evidence just discussed leave no doubt that species diverge in epigenetic marks. In some cases, divergence is clearly associated with DNA divergence, particularly changes in the amount of heterochromatin or number of repetitive sequences, and these changes are probably primary. But can divergence in epigenetic information alone be sufficient to cause inviability in hybrids? We know that it can. The evidence comes from the work on imprinting discussed in Chapter 5. A mouse zygote produced from two haploid mammalian genomes, which is totally adequate genetically, fails to develop normally if the whole, or parts, of both genomes carry imprints from the same sex. To be viable, an embryo needs maternally- and paternally-derived chromosomes with different, but complementary, chromatin marks. If chromatin marks diverge while populations are isolated, the marks of one population and those of the other may no longer be complementary, so regulation of gene expression in hybrid embryos may be impaired. Evidence that inappropriate expression of differently imprinted maternal and paternal genomes can cause early embryonic lethality has been found in a cross between two strains of laboratory mice (Renard and Babinet 1986).

Hybrid inviability is not uncommon, but in most cases the underlying cause of it is unknown. We predict that often epigenetic marks and the DNA sequences that carry them will be found to be important. The hint that hybrid inviability is associated with inappropriate or unbalanced epigenetic marks comes from the frequent association of irregularities in gene expression in hybrids with heterochromatin or repeated sequences. Studies of repeated sequences and methylation patterns in hybrids should prove rewarding.

Hybrid sterility

Commonly species hybrids are viable and vigorous, in spite of having two genomes that evolved in isolation. In many ways this is quite remarkable: why should two genomes work together satisfactorily when they must have diverged in both structural and regulatory sequences? And if they work together satisfactorily in somatic functions, why cannot they do so in reproductive functions? It is now recognized that the sterility or reduced fertility of hybrids is sometimes associated with the production of genetically unbalanced gametes, but it is also frequently caused by problems in the process of gametogenesis itself. But why should there be problems in gametogenesis? Spurway highlighted the enigma in the following way:

> Meiosis, however, is one of the most widespread and constant phenomena in the organic world. Therefore it is curious that its genetic organization should have altered radically during virtually every speciation process so that it fails grossly in hybrids that often show morphological hybrid vigour. Why should the somatic tissues, which may be strikingly different in the parent species, show developmental homeostasis; why should courtship only vary trivially between species, though ethological isolation is the most efficient way to conserve gametes; whereas the gametogenic processes themselves, which have remained fundamentally unaltered since before the divergence of animals and plants, usually show complicated disturbances at several stages in at least one sex of most species hybrids? (Spurway 1955, p. 338)

Spurway's answer to the riddle was that the selection pressures that lead to the normal functioning of somatic tissues are different from those that lead to normality in the germ line. In somatic cells, selection is for developmental homeostasis: for cells to survive and do their job in spite of minor environmental fluctuations or genetic deviations. The germ line is different: there is no advantage in the germ line having developmental homeostasis; in fact there should be as little developmental homeostasis as possible, because if germ cells that are not totally normal survive, they will go on to produce inadequate gametes and inadequate zygotes. Selection for coping with germ-line deviations will not occur. Selection in the germ line should be normalizing, preserving the original by eliminating genetic variants, whereas selection for somatic functions should be canalizing, preserving the original phenotype in spite of genetic and environmental change.[22]

According to this argument, the somatic adequacy of hybrids is the result of selection for a somatic buffering system. Such a buffering system has not evolved in the germ line because there it would be maladaptive. If Spurway's reasoning is taken further, it can be argued that selection should ensure that there is a high degree of quality control in the germ line to prevent the formation and survival of inadequate gametes. Moreover, any processes that check the genome for defects that would impair functioning in the future zygote would also be selected. The enormous wastage of gametocytes or gametes in both sexes in almost all organisms is certainly compatible with the idea that there are quality-control systems that eliminate most defective products before or soon after zygotes are formed.

At what stage or stages is the quality of potential gametes assessed? Quality control may be one of the functions of meiosis, and it may be that it is at this stage that hybrids fail the quality requirements and produce either no gametes or abortive gametes. Meiosis has many functions: it ensures an equal and fair distribution of chromosomes to the gametes; it plays a role in re-setting the epigenetic state before the next generation begins; according to Martin (1977) and Bernstein (1977), during meiotic pairing, damage and defects in the DNA of one chromosome are repaired or eliminated by using the other chromosome as a template. Holliday (1984) extended the repair argument and suggested that epigenetic defects as well as genetic defects can be detected and removed through homologous pairing and recombination, and Ettinger (1986) argued that meiosis enables changes in the overall pattern of chromosome organization, such as those caused by the spread of selfish DNA, to be identified and eliminated. Defence mechanisms such as RIGs, which detect and inactivate foreign DNA, may operate during meiosis.

The various roles of meiosis all seem to depend on the pairing of homologous chromosomes, and it has been recognized for many years that if meiotic pairing is impaired, either gametogenesis is not completed, or an inferior product is produced.[23] Detection of pairing failure, followed by destruction of the problematic gametocyte or gametic product, may be central to the quality control processes of the germ line. Dobzhansky (1951) stressed that failure of meiotic pairing is a basic feature of most sterile hybrids, even when the parental chromosomes are structurally similar. It may be that this pairing failure causes their sterility.

What determines whether or not chromosomes pair? During meiosis, chromosomes are restructured; there are visible changes in the extent of chromatin condensation, and biochemically detectable changes in the associated histones and other proteins. In order for chromosomes to pair, they must have a similar chromatin conformation (Jablonka and Lamb 1988). The chromatin structure of homologous chromosomes in interspecific hybrids could have diverged in ways that prevent adequate restructuring and pairing. This could occur for several different reasons. First, the hybrid gametocyte may have regulatory factors that are inappropriate or insufficient for chromatin restructuring. This may be the situation in hybrids between *Chironomus thummi piger* and *Ch. th. thummi*, where Hägele and Oschmann (1987) showed that during spermatogenesis the two sets of chromosomes differ in their state of condensation. The chromosomes sometimes disintegrate, but even if they do not, the frequency of chromosome aberrations increases and few viable sperm are produced. Second, the two parent species might differ karyotypically; although chromosome rearrangements in themselves need not impair pairing, as is evidenced by the frequency of chromosomal polymorphism within populations, if position effects occur, pairing might be affected. A karyotype change causing the heterochromatinization of formerly euchromatic regions could impair meiosis in the hybrid, because euchromatic segments usually do not pair with heterochromatic segments.

The third reason why hybrids might have inadequate pairing is that even though homologous chromosomes appear morphologically identical, they differ at finer structural levels. Dobzhansky (1951) called the sterility resulting from this 'genic' sterility, but the types of genetic changes that produce it are unknown. Differences in DNA sequences that affect epigenetic marks, for example variations in the number of copies of repeats that compete for binding factors, or epiallelic variations in the same DNA sequence, could cause genic sterility through their effect on the restructuring of chromatin. Divergence in germ-line-specific genes might be particularly important, since they have to change between an active and inactive state during gametogenesis. Defective chromatin restructuring in these genes could upset the whole of gametogenesis.

Haldane's rule

One reason for believing that incompatible chromatin structure underlies many cases of hybrid sterility is Haldane's rule. Haldane (1922) pointed out that, commonly, hybrid males and females differ in viability and fertility, and the sex that is most sensitive to hybridization is the heterogametic sex.[24] In most organisms this is the male, since males are XY, but in a few groups, notably birds and Lepidoptera, where males are XX and females XY, females are more affected. Largely as a result of their genetic studies of *Drosophila* hybrids, Coyne and Orr (1989) suggested that there is a second rule of speciation: the sex chromosomes have a disproportionately large effect on hybrid inviability and sterility.

Many explanations of Haldane's rule have been proposed;[25] usually they are based on the fact that X-linked genes are present in only a single dose in the heterogametic sex. Homogametic hybrids have a complete set of chromosomes from each parent and are balanced, whereas the heterogametic sex, having only a single X from one parent species, is unbalanced, and inappropriate epistatic interactions occur. Coyne and Orr (1989) suggested that genes on the X chromosome are more involved in hybrid sterility than other genes because advantageous recessives are substituted more rapidly on the X chromosome, which is hemizygous in the heterogametic sex, than on the autosomes. Both rules of speciation can be explained by their hypothesis, provided certain assumptions are made about the nature of the mutations. However, some of these assumptions are of doubtful validity, and some observations are not explained by Coyne and Orr's hypothesis. For example, if the divergence of X chromosomes is greater than that of autosomes, why are morphological differences and pre-mating mechanisms not more influenced by sex-linked loci? Sex chromosomes do not have a disproportionately large effect on these characters. Furthermore, as King (1993) pointed out, in their analyses and discussions Coyne and Orr largely ignored chromosomal differences between species, even though X-autosome translocations are a well-known cause of male sterility.

Because of these and other difficulties, we proposed a somewhat

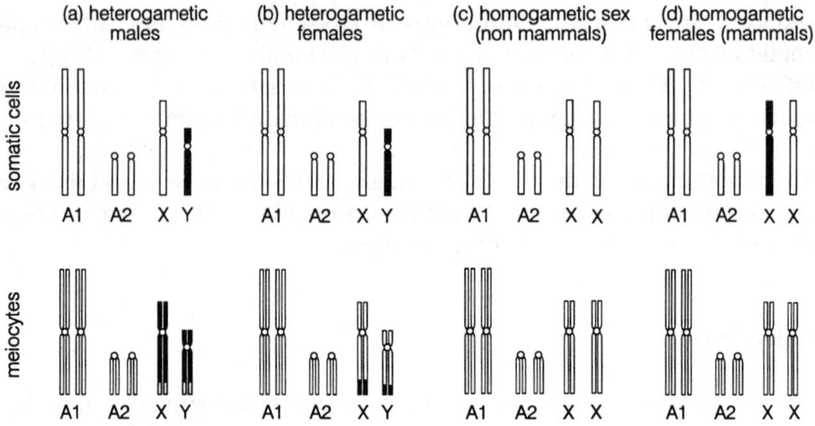

Fig. 9.6 Simplified picture of changes in the conformation of sex chromosomes in animals with two pairs of autosomes (A1 and A2) and a heteromorphic pair of sex chromosomes. (a) Species with heterogametic males: the heterochromatic Y and euchromatic X of somatic cells both become largely heterochromatic in spermatocytes. (b) Species with heterogametic females: the heterochromatic Y and euchromatic X of somatic cells are both largely euchromatic in oocytes. (c) X chromosomes in the homogametic sex of non-mammals are usually euchromatic in both somatic cells and meiocytes. (d) In female mammals one X is heterochromatic in somatic cells, but both are euchromatic in oocytes.

different explanation of the two rules of speciation (Jablonka and Lamb 1991). It was based on the observation that during meiosis changes in the conformation of the sex chromosomes are far greater than those in the autosomes, particularly in the heterogametic sex (Fig. 9.6). For example, in heterogametic males the Y chromosome is normally heterochromatic and the X euchromatic, but before meiotic pairing, unpaired regions of the X chromosome also become heterochromatic. Similarly, in heterogametic females, the normally heterochromatic Y chromosomes become euchromatic and provide the Xs with a pairing partner at meiosis. Pairing failure is thus avoided by altering the state of condensation of the sex chromosomes so that there is conformational matching between them, and no euchromatic region remains unpaired. When pairing is impossible, heterochromatinization of the unpaired segments protects them from the adverse effects of pairing failure. If these processes do not occur properly, as is sometimes the case with chromosome rearrangements which impair the normal processes of chromatinization, pairing is inadequate and the result is partial or total sterility. Haldane's rule is a consequences of the greater remodelling of chromatin structure that occurs in the heterogametic sex, and the consequent greater opportunities for defects to occur in the chromatinization processes. Sex chromosomes are more involved in sterility than are autosomes because the sex chromosomes undergo more remodelling in gametogenesis. This extensive remodelling also means that sex chromosomes are likely to be more strongly imprinted than autosomes in zygotes. Consequently, incompatibility between imprints on the sex chromosomes may contribute more than that on the autosomes to the inviability of hybrids.[26]

As with other explanations of Haldane's rule, evidence for the one just outlined is difficult to find. However, it is compatible with the observation that X-autosome translocations usually lead to male, but not female, sterility. Direct evidence that differences between the sex chromosomes affect fertility more than similar differences between the autosomes comes from a study of hybrids between *Drosophila serido* and *D. buzzatii*. Using a sophisticated breeding programme, Naveira and Fontdevila (1986) produced flies with segments of *D. serido* chromosomes in an otherwise *D. buzzatii* genome. They could identify the chromosomes by studying the banding patterns in polytene chromosomes, where normally homologues are paired. They found that segments of *D. serido* autosomes, identified by their failure to pair with their homologous region, did not affect fertility until they were above a certain size, whereas any segment of *D. serido* X chromosome, however small, produced male sterility.

The study by Matsuda and colleagues (1991, 1992) of hybrids between the laboratory mouse and *M. spretus*, a distantly related European mouse, also points to failure of the sex chromosomes to pair as a cause of sterility. Hybrid males from this cross are sterile, but females are fertile and can be backcrossed to the parent species. Analysis of spermatogenesis in the progeny of such crosses showed a strong correlation between pairing of the X and Y chromosomes and fertility: mice in which they failed to pair were sterile, those in which they paired were fertile. An association between the X and Y was a prerequisite for normal spermatogenesis; when it failed, meiosis was disrupted, and both sex chromosomes and autosomes showed atypical condensation at metaphase.

Although at present relevant experimental evidence is limited, the little that there is suggests that Haldane's rule and the greater involvement of the sex chromosomes in hybrid sterility and inviability are associated with the conformational changes needed for the pairing of the non-homologous sex chromosomes in the heterogametic sex. Unlike some other explanations of Haldane's rule, this hypothesis explains why the sex chromosomes are not particularly involved in the morphological differences between species and in pre-zygotic isolation, and also why the germ line is so sensitive to hybridization.

Meiotic drive

One recent explanation of Haldane's rule suggests that the sterility of heterogametic hybrids is a consequence of divergence of sex-linked meiotic drive systems in the parent species.[27] 'Meiotic drive' is a term used to describe the preferential recovery of one of the allelic alternatives in the progeny of heterozygous parents. The Mendelian law of equal segregation is violated because one of a pair of homologous chromosomes is represented in more than half of the functional gametes. Chromosomes showing meiotic drive have been found in natural populations of many different organisms, from fungi to mammals, and both sex chromosome and autosomes can show drive.

In some cases of meiotic drive, a drive gene (a distorter gene) distorts normal segregation by causing dysfunction or loss of gametes that have a homologous chromosome carrying a sensitive locus. The genetic elements responsible for drive have been called ultra-selfish, because they promote their own transmission at the expense of alternative genes. If they are present on the sex chromosomes, they are potentially capable of distorting segregation so that all gametes carry the X (or Y), and all progeny are therefore of the same sex. If unchecked, sex-linked drive loci could cause the extinction of a population.

One of the most well-studied meiotic drive systems is Segregation Distorter (SD) in *D. melanogaster*.[28] It involves a group of tightly linked loci in, or near, the centromeric heterochromatin of an autosome. Males heterozygous for an *SD* chromosome and a normal, SD^+, chromosome transmit the *SD* chromosome to more than 98% of their progeny. Meiotic products carrying SD^+ show failures of chromatin condensation and histone replacement during maturation. The basic elements of *SD* complexes are *Segregation distorter* (*Sd*), the selfish locus responsible for producing distortion, which is in a euchromatic region, and *Responder* (*Rsp*), which is the target for the action of *Sd*. *Responder* can be either *sensitive* to *Sd* action (Rsp^s) or *insensitive* (Rsp^i), although there are different levels of sensitivity. In heterozygotes SD/SD^+, the *Sd* locus on the *SD* chromosome acts on Rsp^s on the homologue to cause dysfunction of the sperm that receive Rsp^s. Figure 9.7 shows the basic action of genes in the *SD* complex.

Molecular analysis has shown that *Sd* is a duplication, and that it produces protein products. *Rsp*, by contrast, is not transcribed, and consists of variable numbers of copies of a 120 bp satellite DNA sequence. There is a good correlation between the number of copies of the repeat and the sensitivity to distortion: the *Rsp* allele on the *SD* chromosome has fewer than 20 copies, an insensitive wild type allele (Rsp^i) has 10–200 copies, a sensitive allele (Rsp^s) has about 700 copies, and supersensitive alleles (Rsp^{ss}) have as many as 2500 copies. *Rsp* is thought to be a site for the binding of a product that is directly or indirectly produced by *Sd*. Binding of this protein causes the failure of spermiogenesis in meiotic products carrying the Rsp^s chromosome, probably because it results in improper chromatin condensation.

An important feature of meiotic drive systems such as *SD* in *Drosophila*

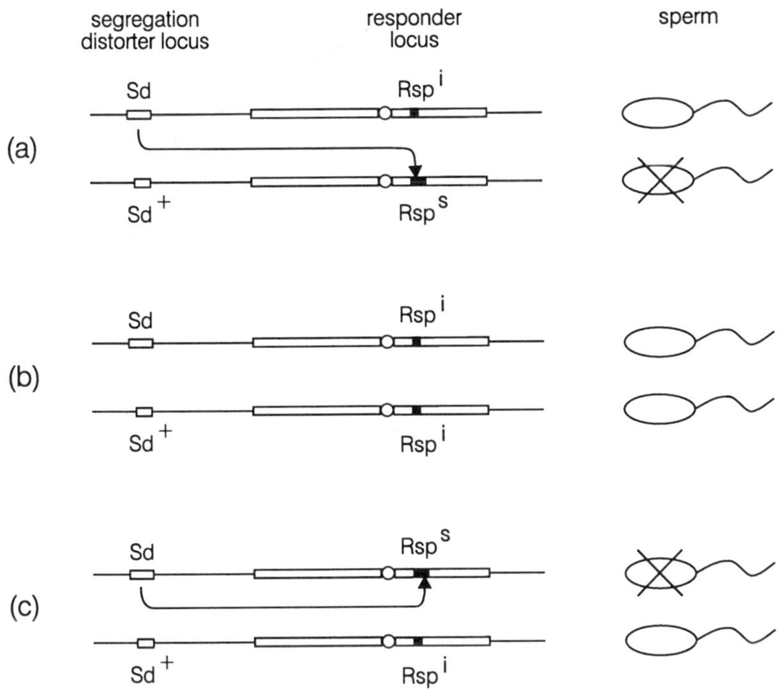

Fig. 9.7 A simplified picture of the Segregation Distorter system in *Drosophila*. (a) The effect of *Sd* when the homologous chromosome carries a sensitive *Rsp* locus (*Sd* Rsp^i/Sd^+ Rsp^s). (b) The ineffectiveness of *Sd* when the homologue carries an insensitive *Rsp* locus (*Sd* Rsp^i/Sd^+Rsp^i). (c) A suicide chromosome with *Sd* linked to a sensitive *Rsp* locus (*Sd* Rsp^s/Sd^+ Rsp^i).

and the autosomal *t*-complex in the mouse is that the distorter locus and the target locus are tightly linked: distorter loci are linked to insensitive targets, and sensitive targets to non-distorter loci. Recombination between distorter and target loci would create a suicide chromosome that would drive itself out of existence (Fig. 9.7c).[29] This does not occur because drive systems are often associated with heterochromatin, and always with a series of inversions that effectively preclude recombination. Linked enhancers and repressors of drive may be included within the inversion complex.

Most known meiotic drive systems are associated with sex chromosomes rather than autosomes, and their effect is to distort the normal sex ratio. For example, several *Drosophila* species have X-linked drivers that affect a sensitive target on the Y, and therefore produce almost entirely female progeny. It has been suggested that sex-linked drive is common because since most regions of the X and Y are non-homologous, distorter and target loci can be on different chromosomes with no danger of forming a suicide chromosome through recombination. It is therefore of interest that most sex chromosome meiotic drive systems are nevertheless associated with chromosome rearrangements, even though rearrangements are not necessary to reduce recombination. This may be a clue to the mechanisms responsible for drive. McKee (1991) found that in laboratory strains of *D. melanogaster* with various chromosomal rearrangements, meiotic drive causing sex-ratio distortion was strongly correlated with chromosome non-disjunction and male sterility. He suggested that there is a common mechanism underlying all three: reduced fertility and meiotic drive, as well as non-disjunction, are caused by disruption of pairing between the sex chromosomes. Inactivation of the X chromosome in the germ line of males is initiated in heterochromatin, and spreads into euchromatin. When a chromosomal change moves X euchromatin away from X heterochromatin, it leads to drive and reduced fertility because inadequate X-inactivation upsets pairing.

Because drive chromosomes favour their own transmission, even if they carry genes with deleterious effects on fitness, they can rapidly become fixed in a population by driving out responder chromosomes. Obviously if this occurs, segregation distortion is no longer seen. When segregation distortion is detected in natural populations, it is because both the drive chromosome and its sensitive target are still present. Modifiers that counteract the effects of drive loci accumulate in natural populations, and the chromosomes that carry the drive locus accumulate enhancers. The result is a delicate balance between enhancers and suppressors of drive. Frank (1991*a*) and Hurst and Pomiankowski (1991) suggested that it is the upsetting of this balance that underlies the sterility of the heterogametic sex described by Haldane's rule. The drive systems of the parent species evolve independently, and in hybrids the countermeasures against sex chromosome drive fail: X chromosomes drive against Ys, and Ys drive against Xs. The result is sterility.

The meiotic drive interpretation of Haldane's rule has been challenged on both theoretical and empirical grounds.[30] On the basis of the arguments developed earlier, we believe that although there is a link between meiotic drive and hybrid sterility, it is somewhat different from that which Hurst and Pomiankowski and Frank suggested. The link is the stringent requirements of chromatin restructuring during gametogenesis. Like hybrid sterility, meiotic drive is associated with changes affecting chromatin structure: molecular studies of *SD* chromosomes show that the target locus is an amplified block of heterochromatin repeats, and genetic analyses of sex ratio distortion strongly suggest that it results from inappropriate chromatin condensation during meiosis. It may not be coincidence that the *t*-locus in the mouse, another well-studied segregation distorter, is in a chromosome region that is known to be imprinted. All drive systems seem to be associated with changes in DNA packaging and reconstruction. With sex chromosomes, drive may be the result of a type of position effect, in which a rearrangement associated with the driving chromosome allows heterochromatinization to spread into normally euchromatic regions. This leaves the corresponding region of the homologous chromosome without a pairing partner. This unpaired region fails to reconstruct at subsequent stages of gametogenesis, and the gamete harbouring it is aborted. The cause of drive is therefore similar to that of many cases of hybrid sterility, except that in the latter the chromatin restructuring necessary for pairing is inadequate in both homologues, so all gametes are aborted.

The association of drive with chromatin structure is also seen with B chromosomes. These are supernumerary, non-essential chromosomes which are not homologous with those of the normal complement. They are found in variable numbers in populations of many different species of animals and plants.[31] Commonly B chromosomes are largely heterochromatic and lack genes with major effects, although often they reduce fitness and have effects on recombination frequency. Usually they show drive, frequently accumulating preferentially in one sex. Sometimes drive results from mitotic non-disjunction in the germ line or, in plants, from non-disjunction in the first pollen-grain or egg-cell mitosis. Non-disjunction leads to one daughter cell having both copies of the chromosome and the other none.

In rye, B chromosome transmission shows the effects of parental imprinting (Puertas *et al.* 1990). In crosses between plants with two Bs and those with none, plants inheriting the Bs from their female parent transmit them at a higher frequency than those inheriting them from the male parent. Plants with maternally transmitted Bs also accept 2B pollen more readily the 0B pollen. In view of this imprinting, it is interesting that non-disjunction of rye B chromosomes can be influenced by the de-methylating agent 5-azacytidine. Normally non-disjunction of B chromosomes occurs only in the first pollen grain mitosis, but root cells treated with 5-azacytidine also show B chromosome non-disjunction, although other chromosomes behave normally (Neves *et al.* 1992). This suggests that B chromosomes carry methylation marks that influence their segregation in mitosis, and since they also show parental imprinting, some of these marks are not totally erased at meiosis.

In some species B chromosomes show drive during meiosis; the chromosomes sometimes preferentially segregate into the egg nucleus rather than the polar bodies. In the mealy-bug *Pseudococcus affinis*, in which the paternal set of chromosomes becomes heterochromatic and is eliminated during spermatogenesis, B chromosomes exploit the asymmetry of transmission: they decondense and become euchromatic, thereby ensuring that they segregate with the euchromatic maternal chromosome set that is transmitted to the next generation. The drive of these B chromosomes is suppressed by various modifiers that affect their condensation during spermatogenesis (Nur and Brett 1988).

The most extreme form of drive, the ultimate selfish element, is found in the parasitic wasp *Nasonia vitripennis*. Like other hymenopterans, males in this species are haploid, normally developing from unfertilized eggs, whereas females develop from fertilized eggs and are diploid. Werren and his colleagues found that a small B chromosome, *Paternal Sex Ratio* (*PSR*), carried by the sperm, causes the condensation and elimination of the rest of the sperm's chromosome complement; the originally diploid female zygote is converted into a haploid male (Werren 1991). The chromosomal element *PSR* consists of several blocks of tandemly repeated sequences, and Beukeboom and Werren (1993) suggested that it may act as a sink for proteins that are required for the normal processing of paternal chromosomes. The defective processing that *PSR* induces in the paternal chromosome set could occur either during spermatogenesis, perhaps by interfering with the replacement of histones by protamines, or during the short period between fertilization and the first meiotic division.

There is still a lot to be learnt about drive, but so far all of the well-studied drive systems seem to be associated with heterochromatin and the way chromatin is restructured; all are associated with chromosome rearrangements or additional chromosomes. As with hybrid sterility, the data point to the conclusion that chromatin restructuring is a vital and vulnerable part of gametogenesis. Any divergence in chromatin structure occurring in isolated populations could lead to incompatible drive systems in hybrids. Within a population, chromosome changes such as rearrangements that have position effects, or the addition of some types of repetitive sequences, may affect chromatin structure in ways that distort Mendelian ratios by preventing some chromosomes from undergoing the chromatin changes necessary for the survival of gametes. Consequently, one chromosome may replace another in the population. How widespread and important such replacement is in natural populations is unknown, but drive systems are common. Drive could be an important cause of divergence during periods of population isolation, with one type of chromosome becoming fixed in one subpopulation, and a different one in another subpopulation.

Chromosomes and speciation

Meiotic drive is one of the mechanisms that White (1978) suggested could be important in speciation. He argued that some types of chromosome rearrangements initiate sympatric speciation when individuals become homozygous, because the new arrangement produces an immediate partial reproductive barrier. Heterozygotes for the old and new chromosome arrangements have reduced fertility because recombination between rearranged chromosomes, and the segregation of translocations, result in gametes with incomplete or unbalanced sets of chromosomes. Meiotic drive, White suggested, is one way in which a new arrangement can become common enough for homozygotes to be produced.

The rapid route to new species that White proposed has not been widely accepted.[32] The objection that is usually made to White's model of sympatric speciation, which he called 'stasipatric' speciation, is that the initiating event, a chromosomal change, would be found first in a heterozygote. For the rearrangement to become homozygous, in spite of its effects on fertility, some kind of inbreeding would have to occur. For this reason, it is argued that this type of speciation could occur only in the relatively few species in which the population structure is such that mating with relatives is likely.[33]

White believed that many types of chromosomal change could be important in speciation. He interpreted karyotype changes quite widely, including gains and losses of heterochromatin, and general changes in the amounts of DNA, as well as chromosome fissions, fusions, translocations, and inversions. In his book *Modes of speciation* he wrote:

> One of the main conclusions of this book is that over 90 percent (and perhaps over 98 percent) of all speciation events are accompanied by karyotypic changes, and that in the majority of these cases the structural chromosomal rearrangements have played a primary role in initiating divergence. (White 1978, p. 324)

King (1993) also strongly advocates the primacy of chromosomal change in speciation, although he rejects the notion that the changes often occur sympatrically. He emphasizes the importance of the reduced fertility of individuals heterozygous for some chromosome rearrangements, while stressing that not all types of chromosome change have adverse effects on the fecundity of heterozygotes. Many populations are polymorphic for chromosomal rearrangements, and there is no reason to think that heterozygotes in these populations suffer reductions in fecundity. It is also unlikely that small differences in the amount of heterochromatin have serious effects on the viability or fertility of heterozygotes. Yet many of the karyotypic differences between species are of this type.

Wilson, Bush and others assembled data suggesting that there is a relationship between rates of karyotype change and rates of speciation.[34] Vertebrates such as mammals, which have a high tendency to speciate, also have a high rate of karyotype evolution, and small mammals have higher rates of karyotype evolution than large. The explanation Wilson and his colleagues offered for these patterns is that mammals are more likely to have a population structure, such as small colonies or herds, that allows inbreeding and consequent fixation of chromosome rearrangements. Karyotypes of small mammals change more than those of large because small mammals are less mobile and therefore more likely to inbreed, or alternatively because their higher rate of reproduction makes them more tolerant of the reduction in fecundity occurring when they are heterozygous for a new rearrangement. Whatever the cause, the association between rate of karyotype change and rate of speciation shows that conditions that are favourable for karyotype change are also favourable for speciation.

The observation that the karyotypes of related species often show striking differences demands some explanation. Is it simply the result of fixation of chance chromosomal mutations in small populations, or are there ways in which chromosome structure can change and bring about divergence? As with most questions in evolutionary biology, there is probably no single and simple answer to this question. New linkage groups created by a rearrangement might be beneficial, and aid its spread. Spread of a new rearrangement in an incipient species might also be aided by meiotic drive, and if the new rearrangement causes position effects that influence gene expression, this could give it a selective advantage that aids its spread. There is evidence that chromosome structure *per se* can have phenotypic effects that are subject to selection. For example, Groeters and Shaw (1992) found gradual and systematic changes in the position of the centromere in the chromosomes of the grasshopper *Caledia captiva* across 11° of latitude in Australia. The change was closely correlated with development time, suggesting that the chromosome rearrangements provided the heritable basis of the change.

One way in which variation important for evolution may be generated is through processes that Dover (1982, 1986) has called molecular drive. Despite the too similar name, molecular drive processes have nothing to do with meiotic drive. They are processes that can cause substantial divergence in the DNA sequences of different populations in the absence of selection. The sequences involved are those present in the haploid genome in more than one copy; this includes many genes coding for structural proteins, such as those coding for histones and elements of the cytoskeleton, and the genes for ribosomal and transfer RNA. It also includes repetitive sequences found throughout the genome in both heterochromatin and euchromatin. Dover realized that mutations in one member of a family of repeats can spread throughout that family, and through the population, as a consequence of various types of non-reciprocal transfer between chromosome regions. The mechanisms of transfer are transposition, slippage replication, gene conversion, and unequal crossing over. Details of these mechanisms are unimportant for present purposes, but Fig. 9.8 illustrates how one of them, unequal crossing over, could generate different variants in different populations.

The consequences of molecular drive are what Dover calls 'concerted evolution': within a species the sequences of members of a gene family become very similar, but there are marked differences between species. The average number of repeats in some repeated sequences may also

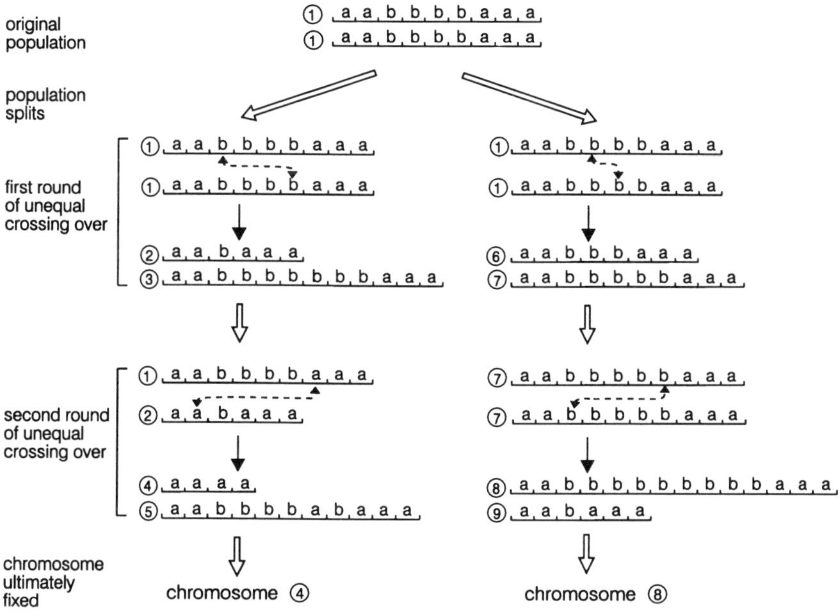

Fig. 9.8 Molecular drive causing divergence in two populations. A population splits into two isolated sub-populations, initially having chromosomes with identical blocks of repeats of sequences **a** and **b** (chromosome type 1). Rounds of unequal crossing-over produce different chromosome types (2–9) in the two sub-populations, and eventually different chromosomes are fixed.

change, expanding or contracting regions of the chromosome. If the repeated sequences have functions such as being binding sites for control proteins, co-evolution of the binding proteins and repeated sequences is likely, and may lead to incompatibility in hybrids between two populations (Dover and Tautz 1986). Molecular drive may therefore generate the variation in size and composition of sequences that are important for carrying the protein and methylation marks that control gene activity. The changes spread without selection, but may be an important source of variation on which selection can act. Sometimes they lead to visible karyotype changes, but chromosomes can be substantially reorganized while remaining morphologically similar.[35]

What about other karyotypic differences between species? Are they all the result of random changes occurring in small isolates? An alternative possibility is that not all karyotypic changes are random. According to Groves (1989), diversification of some of the major primate groups is associated with characteristic types of chromosomal change: in lemurs Robertsonian translocations are common, whereas in old world monkeys there are more fissions, and in hominoids more than 70% of the changes are pericentric inversions. Many spontaneous chromosomal aberrations seen in new-born humans give karyotypes resembling those of ancestral forms or of related species. King (1993) has argued that the non-randomness of chromosomal evolution stems from structural characteristics of the genome which restrict the types of rearrangement that are produced. He pointed to the recurrence of particular types of rearrangements in intraspecific hybrids, and suggested that factors such as the amount and type of heterochromatin present in different regions, and the presence of molecular structures making some sites hot spots at which transposable elements can cause breaks, produces directed karyotypic change. King also highlighted evidence suggesting that sometimes multiple chromosome rearrangements occur simultaneously, indicating that chromosomal evolution and consequent reproductive isolation could be very rapid.

Some karyotypic changes may be generated as part of the response to stressful conditions that reduce population size, or are found in newly colonized areas. If, as suggested in Chapter 7 (p. 171), directed changes involving chromosome structure are part of the evolved response to extreme stress, it is not surprising that speciation is usually accompanied by chromosomal changes. By altering the amounts of heterochromatin, changing the relationship between blocks of heterochromatin and euchromatin, moving genes around the genome, and creating new chromosome domains, stress responses could generate changes in the epigenetic control of gene activity, which would interact with those induced directly by the environment, to produce morphological and physiological changes on which selection can act. In this case adaptive divergence and speciation are initiated by chromosomal variation, but not for the reasons White suggested.

The arguments developed in this section suggest that chromosomal speciation may be the consequence of several distinct, but usually associated, processes. Rearrangements that reduce fertility in heterozygotes may become fixed through inbreeding in isolates, and molecular drive may enhance the genetic divergence between isolated populations until they no longer interbreed. Induced chromosomal changes, such as breaks, non-reciprocal recombination, and transposition events, many of which may be highly localized, are likely to occur in isolates where individuals are exposed to various types of stress. All of these may contribute to the genetic revolutions of speciation.

Epigenetic inheritance and the origin of species

Since species usually show heritable differences, speciation is generally assumed to be a genetically based and genetically initiated process, but as we have shown is this chapter, there is an important epigenetic dimension. It is impossible to make a sharp distinction between genetic and epigenetic effects, because epigenetic modifications bias DNA sequence changes, which in turn alter the potential for epiallelic variation. The feedback between the genetic and epigenetic systems means that it is often impossible and misleading to assign primacy to one system rather than the other. Sometimes speciation is initiated by purely epigenetic processes: by directed changes in marks that are inherited and lead to sterility in crosses with members of other populations. Adaptation and reproductive isolation would in this case be directly linked. Speciation may also begin with genetic changes, although these may not be random mutations. Often there will be an epigenetic basis to the genetic changes: stressful conditions, which are likely to be found in isolated populations near the borders of a species range, or in conditions where populations have crashed, or in new colonizers, may induce bursts of genetic variation, including chromosome variation. The variations that are of particular importance are those that affect the stability of epigenetic memory by determining the ease with which marks are acquired or lost, and those that affect gene expression and chromatin restructuring during gametogenesis.

Most discussions of modes of speciation in animals reflect the present emphasis in genetical research on a few groups of animals, notably rodents and dipterans, which have fairly early germ-line determination. Geographic variation also tends to be studied more in vertebrates and insects than in other animal groups. This may well bias our perception of the nature of speciation processes. The route to new species for groups with somatically derived germ cells may be very different from that in higher vertebrates and insects. When germ cells are somatically derived, and new epialleles are tested somatically, the opportunities for somatic divergence are greater. In hybrids divergence of epigenetic marks is therefore likely to have strong effects on viability and on all pre-zygotic isolating mechanisms. In contrast, in groups with early germ-line determination, epialleles that can be transmitted to progeny occur only in the germ line, and most of the incompatibilities between marks are expected to be associated with the germ-line specific functions of meiosis and gamete differentiation, and therefore lead to sterility. For many years there has been a suspicion that speciation in plants may be somewhat different from that in animals; part of the reason for this may be associated with their different modes of germ-cell formation and their abilities to reproduce asexually.

Summary

Darwin believed that an inevitable by-product of the evolution of adaptations in divergent populations is the formation of distinct species. He used domestication as a model for the origin, as well as the selection, of variations, and collected data showing that conditions of domestication induced variations. He believed that, in nature, evolution occurs under circumstances similar to domestication, which promote variability. The idea that domestication induces heritable variations by imposing a severe stress has been developed by several biologists, notably Belyaev and his group. They emphasized the effect that the stressful conditions of domestication have on gene expression. Domestication, and domestication-like processes, which couple induced heritable variations with novel selection pressures, may sometimes initiate speciation events.

The effect of a new, and often stressful, environment probably commonly leads to directed changes both in epialleles and DNA sequences. Such environmental conditions are likely to be found at the margins of a population's distribution, or in small isolates. Induced heritable variations may also occur when a subpopulation uses a novel resource that leads to physiological adaptations and heritable changes in gene expression. The formation of new epialleles can therefore facilitate sympatric speciation. Directed epigenetic and genetic changes probably play an important role in chromosomal evolution, and may be one of the causes of the differences in karyotype shown by closely related species.

Divergence in chromatin structure is of central importance in hybrid inviability, hybrid sterility, and meiotic drive. Inviability can be caused by incompatible chromatin marks in a hybrid zygote; for example, incompatible marks could result from differences in parental imprinting, leading to inappropriate gene expression in hybrid embryos. Often chromatin remodelling during gametogenesis in hybrids is impaired, because chromatin cannot be restructured properly in the hybrid cytoplasm. Commonly the effects of impaired restructuring are manifest at the pairing stage of meiosis, which is an important, although not unique, 'quality control' stage. Defective restructuring leads to inadequate gametes: usually all products of meiosis fail, leading to sterility, but in more circumscribed situations, only half of the meiotic products are inadequate, and the result is meiotic drive. Haldane's rule can be interpreted as a consequence of the peculiar requirements for restructuring of the sex chromosomes in gametogenesis in the heterogametic sex: the chromosome-wide changes that the X or Y have to undergo are particularly vulnerable to hybridization. Impaired restructuring in gametogenesis leads to sterility, and incompatibility of parental imprints can lead to reduced viability.

Epigenetic variations that are tested somatically and are transmitted through the germ line may allow new developmental programmes to be established, and contribute to the species-richness observed in organisms with late germ-line segregation. But heritable epigenetic variations affect speciation in all groups of organisms, through their effects on post-zygotic isolation.

Notes

1. It has been suggested that a better name for what is usually referred to as the 'biological' species concept would be the 'isolation' species concept, since other equally 'biological' definitions have been proposed. Templeton (1989) discusses and evaluates various alternative definitions of species. Mayr (1992) gives a detailed rebuttal of many of the objections to the biological species concept.
2. This lecture is reproduced in Belyaev (1979). See Belyaev and Borodin (1982), Belyaev *et al.* (1981*b*), and Trut (1987) for more detailed accounts of the work on foxes.
3. It might be thought that some of the phenotypic effects were pleiotropic effects of genes selected for their role in producing tame behaviour since, as has been known for many years, genes influencing behaviour can affect characters such as coat colour, because pigments and neurotransmitters share a common synthesis pathway (see Hemmer 1990, Chapter 8). However, although they considered this possibility, the Russian workers rejected this explanation of their data.
4. Bottema (1989) reviews this work.
5. See McClintock (1984), Wills (1984), and the reviews by Parsons (1988) and Kohane and Parsons (1988).
6. For a discussion and description of the possible modes of speciation see Bush (1975) and White (1978).
7. See Mayr (1942, p. 237), (1963, pp. 529–535), and (1982*b*, pp. 3–5).
8. For example, see Barton (1989). As King (1993, p. 65) has pointed out, over the years Mayr's original concept has been modified, and this confuses discussion of founder effects.
9. Maynard Smith (1962) showed that theoretically sympatric speciation can occur, but only under certain rather unlikely conditions. More recent models with different assumptions suggest that sympatric speciation is quite plausible (e.g. Diehl and Bush 1989).
10. For a critical review of the data, see Mackenzie (1992). The work on *Carausius* is described in Sladden and Hewer (1938).
11. Blackman (1979) suggested that Shaposhnikov's observations were probably the result of contamination, but subsequently conceded that there was little doubt that the reported changes had indeed occurred (Shaposhnikov 1985, pp. 76–77).
12. Weismann (1893) pp. 399–409.
13. See Stearns (1989) for a discussion of the role of plasticity in evolution.
14. Not all parthenogenesis in these groups involves clonal inheritance. Only if it is ameiotic can parthenogenesis be regarded as equivalent to extended clonal lineages within an organism. Many parthenogenetic nematodes are produced by fusion of two meiotically produced products, so parents and offspring are not genetically uniform, although they may be very similar (Bell 1982).
15. Beetles are an extremely species-rich group. Haldane is said to have replied to a question about what could be inferred about the Creator from a study of the works of creation with 'an inordinate fondness for beetles' (Fisher 1988).
16. Darwin believed that hybrid sterility was causally similar to the sterility encountered in domestication, and that understanding the latter would provide

the key to understanding hybrid sterility. There is little reason for thinking this suggestion is correct.

17. Whitt (1981) briefly reviews the many studies of enzymes in species hybrids, particularly in fish.
18. Zakian *et al.*'s interpretation suggesting that the heterochromatic block initiates heterochromatinization of the chromosome carrying it, is supported by Jablonka *et al.*'s (1987) finding that in a female cell line of another species of vole, *M. agrestis*, in which one X had a larger block of heterochromatin than the other, the X with the larger block was always inactive. Although this could be coincidence, it would be interesting to know how other cell lines with unequally sized X chromosomes behave.
19. See Heslop-Harrison (1990) for a brief review of work on parental dominance.
20. For details of the population structure and peculiar chromosome behaviour in this group of species, see Vinogradov *et al.* (1990) who describe genome exclusion in male *R. esculenta*, Bucci *et al.* (1990) who describe it in females, and Schmidt (1993) who discusses the outcomes of crosses between the species involved.
21. Work on the ecology and evolution of *Poeciliopsis* was reviewed by Schultz (1977).
22. Waddington (1957) made the useful distinction between two forms of stabilizing selection: canalizing selection and normalizing selection.
23. Details of the evidence for this are given in Miklos (1974), Burgoyne and Baker (1984), Jablonka and Lamb (1988), and Burgoyne and Mahadevaiah (1993).
24. Read and Nee (1991) have challenged the validity of Haldane's rule, claiming that the data are insufficient to rule out the possibility that the apparent correlation of sterility with the sex of hybrids is coincidence.
25. Reviewed by Wu and Davis (1993).
26. Jablonka and Lamb (1991) predicted that Haldane's rule would not apply to the viability of mammalian hybrids, because X-inactivation means that extensive chromatin restructuring occurs during meiosis in homogametic females as well as heterogametic males. A comparison based on data from mammals and *Drosophila* species showed that, as predicted, in mammalian hybrids there are no sex differences in viability, whereas in *Drosophila* hybrids there are.
27. This hypothesis was put forward independently by Frank (1991*a*) and Hurst and Pomiankowski (1991). It generated a lot of discussion, and some quite fierce opposition (see, for example, Coyne *et al.* 1991; Coyne and Orr 1993; Frank 1991*b*; Pomiankowski and Hurst 1993).
28. Segregation Distorter and other drive systems are reviewed by Crow (1991), Lyttle (1991, 1993), and Wu and Hammer (1991). The *American Naturalist*, Vol. 137, part 3 (pp. 281–456), 1991, carries a series of papers given at a symposium on the genetics and evolutionary biology of meiotic drive.
29. Eshel (1985) and Haig and Grafen (1991) have argued that recombination is essential as a defence against meiotic drive, and the latter authors suggest that this may be one of the most important short-term advantages of sex.
30. See for example, Coyne *et al.* (1991) and Charlesworth *et al.* (1993). The experimental finding that there is little evidence of sex-ratio distortion in semi-sterile hybrids has been used as an argument against the validity of the meiotic drive hypothesis (Johnson and Wu 1992; Coyne and Orr 1993).
31. Jones and Rees (1982) review the nature, properties and distribution of B chromosomes.

32. For an example of arguments leading to the rejection of White's ideas, see Futuyma and Mayer (1980).
33. Lande (1979, 1985) has shown that chromosomal rearrangements can spread if they first become fixed in small isolates though random drift, and then spread throughout the population as a result of local extinctions and colonizations.
34. Details of the work can be found in Wilson *et al.* (1975), Bush *et al.* (1977), and Bengtsson (1980). King (1993, p. 88) queried the adequacy of the methods used to determine the role of karyotype change in evolution, pointing out that many of the rearrangements used for the analysis were unlikely to have played a causative role in speciation because they would not act as a sterility barrier.
35. The curious, coordinated, expansion or contraction of the centromeric heterochromatin of different chromosomes within a karyotype could have evolved by molecular drive: association of centromeric regions in chromocentres in interphase nuclei may promote non-reciprocal exchange between non-homologous, heterochromatic, centromeric regions, and lead to homogenization of chromosome morphology in those regions.

10

Multiple inheritance systems

> The multiplicity that does not reduce to unity is confusion; the unity that is not dependent on multiplicity is tyranny.
>
> Pascal: *Pensées*, 871

In 1958, the centenary of the first publication of Darwin and Wallace's theory of evolution by natural selection, Donald Michie expressed the opinion that the 'third stage' in genetics had arrived. The first stage was that of classical Mendelian genetics, dominated by Johannsen's genotype concept and the chromosomal theory. The second stage, which began in the post-war period, was the incorporation of the inheritance of cytoplasmic factors into genetics; it was recognized that there are cytoplasmic inheritance systems which can transmit epigenetic information germinally. The third stage, which, according to Michie, was just dawning in 1958, was a synthesis of these two: the acceptance that DNA and the cytoplasm are not independent, and that there are persistent interactions between them (Fig. 10.1). The dogmatic assumption that transmission of genes is independent of their expression would, he argued, be rejected:

> . . . the new genetics will consist in the rejection of the former dualism and the search for pathways by which the chromosomal genes may be reached and modified by cytoplasmic action. (Michie 1958, p. 70)

The third stage in genetics revisited

Michie's paper anticipated important advances in genetics, particularly in molecular biology and developmental genetics. Since 1958, and especially in the past decade, much of genetical research has been concerned with the ways in which genes are controlled and modified by developmental and environmental stimuli. It is now realized that DNA sequences can be altered developmentally, and cells can be infected by 'foreign' DNA and acquire new genes; cellular inheritance and EISs are beginning to be important areas of research in developmental biology. But Michie's optimism that these discoveries would lead to a more complex and mature concept of biological heredity and, necessarily, of evolution, was premature. Their assimilation into a coherent theory of heredity and evolution still has not taken place.

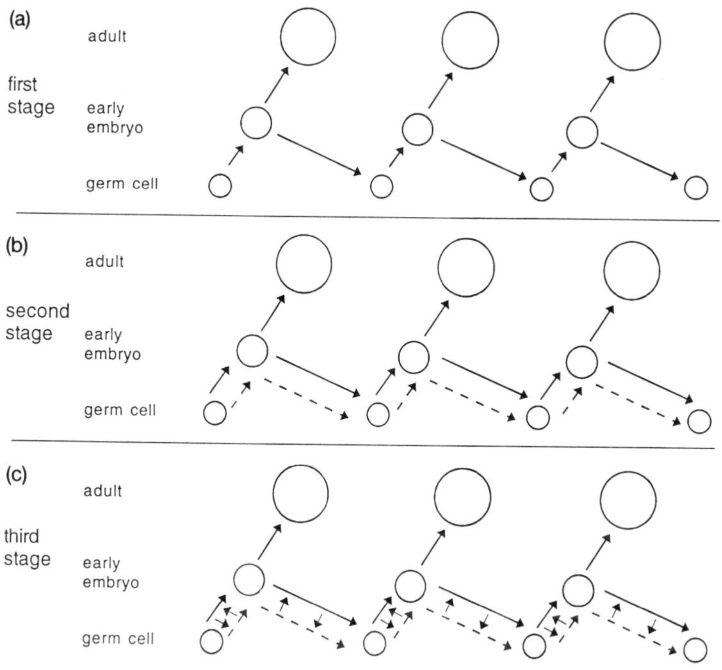

Fig. 10.1 Stages in the history of genetics. (a) The first stage, in which only well behaved Mendelian genes are transmitted between generations. (b) The second stage, incorporating non-Mendelian inheritance. (c) The third stage, in which there are interactions between the Mendelian and non-Mendelian systems. In all cases it is assumed that there is a rigid and fairly early segregation of the germ line and soma. Solid lines: Mendelian genes; broken lines: non-Mendelian factors.

The neo-Darwinian 'Modern Synthesis' incorporated only the first stage of genetics, that is, only the point of view of classical genetics, and it imposed an ideological strait-jacket on evolutionary theory. Mayr's campaign for 'hard inheritance' as the basis of evolutionary change led to the belief that any variation that can be directed by the environment has infinite malleability, and therefore lacks the stability required for a transmissible unit of variation in evolution.[1] But as we have argued throughout this book, the 'softness' and 'hardness' of inheritance are not absolutes. The evidence in Chapters 3 and 7 suggests that there are internal mechanisms for producing directed variations in DNA. Organisms seem to be capable of manufacturing variations in response to environmental changes using intracellular genetic-engineering machinery. In addition, epigenetic systems, which are very sensitive to the environment, produce cellular phenotypes that can be transmitted from one generation to the next, and even between generations of organisms. Induced, or 'acquired', characters can be inherited. Yet, even in the genetics textbooks of the early 1990s, let alone in textbooks of evolution, little is said about the fact that variation in DNA sequence may be directed; similarly, apart from a brief look at genomic imprinting, transmission of epigenetic information through the germ line is completely ignored. The view of inheritance depicted in even the best books is completely DNA-centric, and the notion that variations in germinal DNA are 'random' is taken for granted.

In spite of the adherence of most geneticists and evolutionary biologists to such a restricted view of heredity, the time is now ripe for the unification of the 'third stage in genetics' with evolutionary theory. The third stage in genetics today entails going beyond the DNA-centric view of inheritance, and beyond the view that hereditary variations, both in primary DNA sequence and in epigenetic marks, must be blind. Inevitably, the picture of biological heredity that emerges from this is complex and somewhat messy. The elegance of Mendelian analysis is compromised; Mendelian generalizations are violated. We can no longer dogmatically assert that hereditary information transmitted through the germ line is immune from any influence of the environment, and that it is indifferent to the organism's developmental history. The third stage in genetics demands that additional transmission mechanisms are considered: several transmission mechanisms, including several EISs, underlie the diverse hereditary phenomena that are observed.

At present, because so little research is carried out in this area, it is impossible to know how common transgenerationally-transmitted epigenetic information is, in what groups of organisms it is most important, and in what types of loci. It has been detected most frequently in organisms that reproduce clonally or have late segregation of the germ line, so it may be in these groups that investigating the underlying mechanisms of epigenetic inheritance will be most profitable. However, animals such as mammals and *Drosophila* also transmit some epigenetic information to the next generation, and it is these genetically well-understood groups that have provided hints about the nature of the loci that carry this information. For example, it seems that in higher organisms epigenetic information in loci near heterochromatin, or in loci containing repeated motifs or CpG clusters, tends to be transmitted germinally; the loci that carry epigenetic information to the next generation may be similar to imprinted loci and be expressed during early development. However, these are only very preliminary hints, and the picture that will emerge will no doubt be more interesting and less simple. Unfortunately, the research programme of genetics, and the assumptions underlying it, have largely precluded a direct approach to the problems of epigenetic inheritance, let alone provided answers. Nevertheless, our understanding of epigenetic inheritance is increasing, and cherished assumptions about biological heredity are beginning to be challenged. There is no doubt that heredity is a Pandora's box still holding many surprises.

Recognition of the significance of epigenetic inheritance could affect epidemiological research, social medicine, and plant and animal breeding. The possibility that environmentally induced epigenetic diseases can be transmitted through several generations will have to be incorporated into epidemiological research programmes. Understanding epigenetic inheritance and directed variation may even make epigenetic engineering possible: detailed knowledge of the role of epigenetic systems in development will ultimately enable environmental and developmental conditions to be manipulated in ways that induce transmissible phenotypes. For example, a particular chemical agent, applied at a particular developmental stage, might alter the heritable epigenetic state of germ cells, and this change, which could involve several genes rather than a single gene, might become a new heritable phenotype. Such epigenetic engineering could improve plant and animal breeding strategies.

For those who are concerned with measuring heritability, recognition of the existence of epigenetic inheritance and directed variations will produce an unwelcome complication. Measures of heritability are notoriously difficult to interpret, particularly because of genotype-environment interactions.[2] Epigenetic inheritance of environmentally induced variations obviously contributes to these interactions, making the distinction between the genetic and environmental components of variance impossible to disentangle, even in theory, and rendering the interpretation of heritability even more obscure.

In the immediate future, recognition that the concept of heredity has to be extended to incorporate directed variations in DNA sequence and EISs is likely to have its major impact on evolutionary biology. The theory of evolution by natural selection is based on the existence of heritable variations that affect fitness. A theory of variation is therefore a fundamental part of the theory of evolution, and will determine its content, its predictions, and its limits. The present theory is based largely on the assumption that heritable variations are random and involve changes in DNA sequences. If, as we have argued in the first part of this book, some variation is not random and is not based on sequence change, but rather is 'acquired' variation induced by the environment, this must alter many parts of evolutionary theory. In the second half of this book we have tried to show how, by including inherited epigenetic information and directed DNA changes as sources of variation, the interpretation of some evolutionary events is different, and often simpler. In what follows we first briefly summarize some of our main conclusions, before discussing other, more general, implications of the broader concept of heredity that we have presented.

The evolutionary importance of epigenetic inheritance: a summary

Epigenetic inheritance affects evolutionary change in two ways. The first is direct: the existence of epialleles means that selection, migration, drift, epimutation pressure, and epiallele-induction alter the frequencies of epialleles in a population and contribute to evolutionary change. The second type of evolutionary effect is indirect: once cellular inheritance systems exist, they open up new evolutionary opportunities, and impose certain constraints on the organisms possessing them. Organisms with EISs can evolve complex multicellular organizations, but possession of EISs endangers them because readily induced epigenetic states can destroy the harmonious development of descendants. Strategies such as chromatin restructuring during gametogenesis and early embryogenesis, and early segregation of the germ line and soma, are thought to be evolved responses to the hazards of EISs.

We have argued that epigenetic memory systems evolved very early in the history of life, in early unicellular organisms, and their elaboration and sophistication probably occurred in environments that fluctuated in a regular way. Certain DNA sequence organizations, such as the number and composition of dispersed and clustered repeats, determined the evolution of mutation and epimutation rates, as well as the evolution of inducible genomic and epigenomic responses to environmental challenges. Changes in DNA sequences that carry epigenetic marks cause changes in the memory span of loci because they influence the stability of epialleles and determine the ease with which loci can be turned on and off by environmental stimuli. Not only do DNA changes alter the epiallelic form of a locus, but both random and induced changes in a gene's phenotype alter the probability of DNA changes. The conclusion that follows from this is that the rate of molecular evolution of certain genes is influenced by the environmental conditions that affected their activity. In other words, rates of molecular evolution may be affected by environmentally-induced mutation bias.

In Chapter 7 we showed how persistent interactions between epigenetic information and genetic information provide an evolutionary explanation of some aspects of genome organization and of regional chromosomal regulation. We suggested that:

(1) Repeated sequences of DNA, heterochromatin blocks, and CpG sites in control regions are all likely carriers of heritable epigenetic marks; their number, composition, and organization determine the memory span of individual loci and chromosomal domains, both within a cell lineage in a multicellular organism, and between generations of individuals.

(2) The organization of the vertebrate genome into bands is a consequence of the evolutionary sophistication of cell memory. Bands are the result of selection for the long-term activity or inactivity of tissue-specific and stage-specific genes, which is necessary in any organism with many cell types and complex development. The sequence composition of large genomic units such as G bands, and the type of dispersed repeats they contain, affect the regulation of chromosomal regions and help determine the stability of the repressed or active state of the region.
(3) Constitutive heterochromatin may play a role in determining cellular memory. The multiple DNA repeats in heterochromatin may act as a sink for the DNA-binding proteins that determine the stability of chromatin structure in many euchromatic loci, and affect the average range of memory spans in a species.
(4) At the level of small genomic units, such as single loci or looped domains, neighbouring control regions have been selected both to respond to particular regulators, and to determine certain memory spans. Changes in memory carriers, such as CpG sequences or clusters of repeats, may cause subtle, but potentially important, alterations in development.

The modulation of cell memory through alterations of genome organization has been an essential part of the evolution.

Cellular inheritance systems and chromatin marking have played a central role in the evolution of complex multicellular organisms. They underlie the processes of determination and differentiation, without which development is unthinkable. In Chapter 8 we argued that EISs were probably crucial for the transition from the unicellular to multicellular mode of organization. They increased the efficiency of group selection between primitive cell colonies, and facilitated the evolution of group adaptations and hence the evolution of multicellular 'individuals'.

One of the main conclusions of this book is that the inducibility of alternative epigenetic states and the transmissibility of epialleles have allowed environmentally directed evolution, especially in organisms without a segregated germ line. The most straightforward consequences of the inheritance of induced variations are a high rate of adaptive evolution and an efficient evolutionary response to novel conditions. Obviously, the organism's developmental plasticity and the range of epiallelic forms that the DNA sequence composition can carry constrain and limit the response to the environment, but the limits are not always easy to define, and how much the range can be stretched is unknown. A new epiallelic form that affects the phenotype may also alter the probability of a DNA base change, and if the epiallelic form is advantageous, there will be selection for any DNA change that stabilizes it. The new phenotype may become genetically assimilated. If, as we argued, the vehicles of epigenetic memory are dynamic sequence motifs with very high rates of change, and if the direction of change is guided by the epiallele's conformation, this process will be rapid and its cost to the population small.

The responsiveness of epigenetic systems to environmental changes, and the transmission of such changes to the next generation, can lead to rapid adaptive evolution, but it can also be maladaptive. EISs, the architects of multicellularity, also endanger it. The inheritance of epialleles, and the potential for competition between cells with different epialleles to achieve germ-line status, threaten the integrity of the multicellular individual. Developmental strategies such as beginning development from a single cell, maternal control of development, early segregation of the germ line, irreversible differentiation, and extensive reprogramming of epigenetic information during gametogenesis, can be interpreted as evolutionary responses to the potential dangers of transmitting epialleles and mutations to the next generation.

In Chapter 9 we argued that in most groups of organisms, even those with strategies that minimize the transmission of variants to the next generation, epigenetic inheritance may be a key factor in speciation. Even when the germ line is segregated, both random and induced changes in epigenetic marks can occur in the germ-line cells themselves. Sometimes the effects of such changes may be a fairly direct, with the changed marks contributing to reproductive isolation through pre-zygotic or post-zygotic isolating mechanisms. For example, induced changes could lead to different times of activity in animals, or different times of flowering in plants, and thereby prevent the formation of hybrids between individuals from different populations. If hybrids are formed, incompatibilities between the parental epigenetic marks could reduce their viability and fertility.

Epigenetic inheritance could also have indirect effects on speciation, by triggering genetic changes. In particular, stressful conditions, which alter gene expression and chromatin structure, can trigger chromatin breaks and rearrangements and thus facilitate chromosomally-based speciation. Although epigenetic systems are probably important for speciation in all groups, they may have somewhat different effects in different taxa. As we discussed in Chapter 9, it is probable that the rate of speciation in groups with late germ-line determination can be greatly enhanced by epigenetic variation.

Further implications of the epigenetic perspective

The precise effects that epigenetic inheritance and directed variations can have on evolutionary changes will be revealed only when empirical and theoretical research is conducted in this field. It is clear that at present we are barely scratching the tip of a large and fascinating iceberg. Although we cannot predict the outcome of future investigations, we can explore some of the general consequences of the epigenetic view that we have presented.

We want to emphasize once again that we fully accept the basic theoretical framework of evolution by natural selection, but we contend that the conventional version of the theory is based on an incomplete theory of variation. It is the version of the theory of evolution by natural selection as originally portrayed by Darwin, rather than the neo-Darwinian version, that is in agreement with current knowledge of heredity. The new developments in molecular biology do not alter the basic structure of the Darwinian theory that heritable variations affecting fitness are the basis of adaptive evolution, but they do alter the content of the theory and its implications. Recognition that some heritable variations are directed, and that there are information carriers additional to DNA, makes it necessary to re-think some fundamental aspects of evolutionary biology. It also allows a more satisfactory unification of some parts of biology, in particular the integration of developmental biology with evolutionary biology.

1. *The implications of directed heritable variation*

If directed mutations and epimutations play a significant role in evolution, the environment is both an agent of variation and an agent of selection. This suggests different interpretations of some evolutionary phenomena. We may have to reconsider the interpretation of microevolutionary rates and trends, and the evolution of morphological novelties; we may also have to amend some practices in theoretical population genetics and in phylogenetic analysis.

The most straightforward effect of directed epimutations and mutations is the effect on rates of microevolution. If the environment acts both as the agent of selection and the agent of variation, evolutionary changes within populations may be rapid. New environmental conditions will induce new patterns of gene expression during development. New patterns of gene activity, and the alleles and epialleles underlying them, will be selected. The induced epiallelic forms may change mutation rates in a locus-specific manner, and hence alter the fixation rate of genetic variants. As we have shown in the model on page 168, the rate of fixation of a new allele following the induction of a new epiallelic form that enhances the mutation rate can be significantly increased. In a sexually reproducing population, this type of mutation-based assimilation of a new variant will assist genetic assimilation based on segregation and recombination. If environmental conditions are stressful, in addition to local changes, a burst of variations, reshaping the genome and enabling rapid selection of new ranges of adaptability, can be generated.

A persistent environment, which both imposes a selection pressure and induces epimutations and mutations, can lead to an evolutionary trend. A trend may result solely, or mainly, from epimutational and mutational bias in the set of genes activated in the new conditions. One of the suggested explanations for the evolution of isochores in vertebrates assumes that the driving force is mutational bias mediated by the epigenetic state of active and inactive chromatin (Chapter 7, p. 184).

Examples of environmentally related trends are geographical clines in morphological characters. These were interpreted by the Lamarckians at the beginning of the century as being the result of environmentally induced non-adaptive hereditary changes. They were re-interpreted by the neo-Darwinians as being due solely to the selection exerted by the environmental gradient along the cline. The clines may now have to be re-interpreted once more to include both explanations. A detailed study at both the ecological and molecular levels may tell us the relative importance of selection and biased epimutation and mutation.

Since the rate of induced change may be substantial, if directed heritable variations are to be incorporated into theoretical population genetics, the usual practice of neglecting mutation pressure is no longer appropriate. Moreover, when variations are both directed and adaptive, and are related to the extent of environmental change, the selection coefficient and the mutation pressure are not independent. An additional problem is that the number of generations for which the ancestors of an individual have been exposed to the inducing environmental conditions needs to be considered, because the accumulation of marks can be progressive, and the reliability of transmission may be enhanced in a progressive manner. Both phenotypic expression and transmissibility can depend on past history. Evolution often may not be a simple Markovian process.

If locus-specific, environment-dependent, mutational bias occurs, convergence at the molecular level may be much more common than is acknowledged today, especially when aspects of genome organization (e.g. the number of repeats, or the organization of different sequence motifs in a locus), rather than base substitutions, is compared between species. The 'molecular clock', which is based on the randomness and neutrality of mutations, may be applicable only to base substitutions.[3]

Most DNA base substitutions are slightly deleterious or neutral, and sequence composition seems to evolve mainly by drift.[4] Sequence organization, however, probably changes mainly through selection: chromosomal evolution involving gross karyotypic changes appears to be correlated more with morphological evolution than with molecular evolution (Wilson *et al.* 1977). We expect that morphological evolution will also often be found to be associated with evolutionary changes in smaller, more subtle features in genome organization, which can affect cell memory and through it alter ontogenetic pathways. They can also affect the interaction between chromosomes during meiosis and impair gametogenesis in hybrids. Divergence of small elements of chromosomal organization can affect both adaptation and reproductive isolation.

2. *Directed variation and the origin of novelty*

The possibility that novelties can arise through directed variations seems at first sight to pose a theoretical problem. When Bateson (1979) compared biological and cultural evolution, he talked about their fundamentally stochastic nature, and following Ross Ashby, stressed that no system can produce a novelty if it does not contain some source of the random. Since the evolutionary process obviously does produce novelties, a view such as ours, that emphasizes the importance of directed variations, which are seemingly part of a pre-existing repertoire of responses, appears to be paradoxical.

In order to see why there is no paradox, it is essential to realize that 'directed' does not mean that induced variations are 'uniform', and it also does not mean that they are 'predictable' or 'adaptive'. If heritable variations are influenced and even controlled by environmental cues, it does not mean that all individuals in the population have identical epialleles. Even if an environmental stimulus affects a few loci in a highly specific manner, not all affected genes will be identical with respect to the chromatin marks or induced DNA changes that they carry. For example, assume that a stimulus activates a gene and this activation is associated with the removal of methyl groups from CpG sites; some epialleles will have more methyl groups removed, some less (Fig. 10.2). The response to the stimulus, although locus-specific and therefore directed, is not uniform. Locus-specific induction of epialleles followed by selection enables evolutionary fine-tuning of genetic responses. As already noted, the targeting of variations to a few loci may both increase the rate of evolution, and determine its direction. The main point is that even with directed adaptive changes,

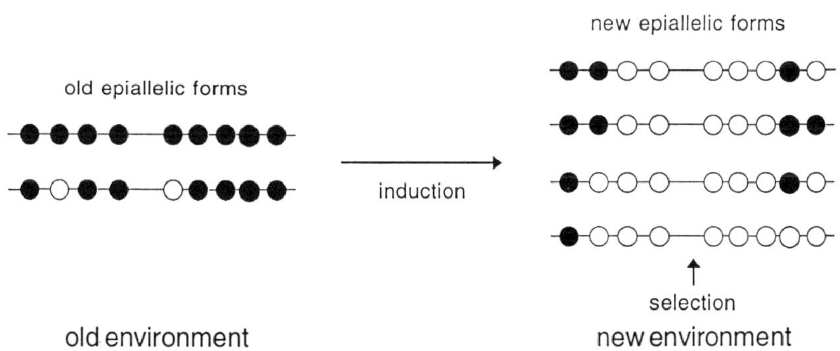

Fig. 10.2 Induced variability in epiallelic forms in a new environment. The DNA sequence is identical in all cases. Solid circles: methylated CpG sites; open circles unmethylated CpG sites.

not all epialleles in the population are identical, and natural selection can act on this variability.

More important for understanding why there is no paradox in the evolution of novelties through directed changes, is the realization that new conditions may alter gene expression in unanticipated ways. The particular combination of active and repressed loci that is induced at a particular stage of development may be novel. The number of combinations of regulatory genetic networks is astronomical, and although only a small subset of these will be stable, it is difficult to believe that all stable networks have been realized in the ecological history of the population. The 'reaction range' of the individual—the range of its possible responses to environmental challenges—can be defined only a posteriori, because the environmental conditions may have unique aspects. A population exposed to a new environmental challenge will show a range of novel responses, and this variability is the material on which selection can operate. It is this range of novel responses which could underlie the induction of hopeful monsters, or merely hopeful freaks.

3. *Multiple inheritance systems*

Following the publication in 1962 of Wynne-Edwards's book advocating evolution through group selection, the question of the unit on which natural selection acts, and the significance of the unit of selection in evolutionary history, became topics of heated debate among biologists and philosophers of biology.[5] According to the neo-Darwinian view, the gene is the unit of heritable variation, the individual is the unit of selection, and the population is the unit of evolution. The arguments centred on the importance in adaptive evolution and in speciation of other entities as units of selection, particularly the gene, the group, and the species. Eventually, the debate yielded a clearer picture of the requirements for selection at the various levels. One of the most interesting outcomes of this discussion was the definition of questions about the evolutionary transitions to new levels of complexity and individuality.[6] Maynard Smith defined eight levels of complexity from replicating molecules through cells, multicellular organisms, groups, demes, to groups with cultural inheritance. He asked a series of questions about these stages, the answers to which would constitute, in his opinion, 'a solution to most of the outstanding problems in evolutionary biology'. These questions were:

> (i) What is the nature of the genetic information that is passed from generation to generation at each stage? (ii) How is the integrity of that information protected against selection at lower levels? (iii) How did natural selection bring about the transition from one stage to another, since, at each transition, selection for 'selfishness' between entities at the lower level would tend to counteract the change. (Maynard Smith 1988*a*, pp. 222–223)

These three questions are interrelated, and any attempt to examine the transition from a lower to a higher level of organization or individuality (the third question), requires an answer to the first two. 'Individual' is difficult to define, but a definition that satisfies the requirements of the present discussion is that an individual is an entity that is systematically the target of selection, and functions as a cohesive whole as it interacts with the environment. As Buss, Maynard Smith, and others who have discussed the evolutionary transitions between levels of individuality have stressed, the emergence of a hierarchically higher unit depends on the evolution of systems that ensure the cohesiveness of the new individual, and protect it from its selfish, potentially competitive, components, which attempt to assume a direct reproductive role. We discussed this problem within the framework of the transition from unicellular to multicellular organization in Chapter 8 (p. 205), suggesting that the hereditary variations produced by EISs played an essential role in this major evolutionary transition. We believe that as with the transition from unicellularity to multicellularity, the transitions to other levels of individuality also involved the mobilization of an inheritance system that evolved at a previous level, and assumed central importance in the new context.

The possibility that additional units of variation transmitted by non-DNA inheritance systems have a role in the transition to a new level of individuality has received little attention.[7] While the nature of the units of selection and the units of evolution have been examined meticulously, the nature of the units of variation has been almost completely ignored, because it is assumed that the unit of variation has been defined once and for all as the DNA sequence. We believe that this restricted view needs to be remedied.

The mobilization of additional types of heritable variations may be a way of escaping from stasis in some groups. The opportunities for evolutionary change become increasingly restricted as canalization becomes more effective, because genetic variation cannot be 'seen' by natural selection. Consider DNA variations in a genetic system that has evolved into a highly complex and stable network of interactions, able to compensate efficiently for variations. Each individual allele in such a system is on the average selectively neutral: its selection coefficient is very small and fluctuating, and depends on the genetic background and environment in which the allele finds itself. The phenomenon of stasis—the lack of morphological change in a group for millions of years—is usually attributed to such genetic and physiological homeostasis.[8] Natural selection, acting on sequence composition, seems to be fairly ineffective unless there is a major genomic change that breaks homeostasis. This could happen through a change in ploidy, the introduction of a new genome through symbiosis, a transition to asexual reproduction, or a mutation with a very large effect. But there are other ways in which the organism can be released from the straitjacket of genetic homeostasis. When the environment changes drastically and induces new phenotypes, evolution through genetic assimilation is possible. The store of previously 'neutral' variation becomes visible to natural selection and new combinations of genes can become fixed. For example, variations in the number of repeated sequences can assume selective importance in the new environment if they affect the ease with which the phenotype is induced.

Even without changes in DNA composition, induced heritable epigenetic variations provide a way out of stasis, especially in groups without a segregated germ line. They provide an additional source of variation on which natural selection can act.

The evolution of evolutionary systems

As cells evolved, natural selection favoured having reserve copies of parts and information, having repair and defence systems, and alternative biochemical pathways. However, an almost inevitable consequence of the evolution of DNA repair and defence systems, which were needed for genetic homeostasis, was the evolution of internal genetic-engineering systems that could overcome this genetic homeostasis. For example, repair enzymes, so very important for correcting damage and for defence against genetic parasites, can also perform other tasks: they can recombine DNA motifs and rearrange the chromosome. The homeostatic mechanisms of repair and defence became a kind of Trojan horse, enabling the creation of a new kind of heritable variation. With the genetic engineering-kit, mutations can become directed if the engineering enzymes evolve to respond to developmental and environmental signals, and combine DNA sequence motifs according to cellular 'rules of grammar'. The major unit of variation becomes the DNA motif, not the base composition of that motif.

Another example of the evolution of new rules of evolution, which has been discussed at some length in this book, is the exploitation of epigenetic inheritance. This type of inheritance is vital in organisms that need flexible, reliable cellular inheritance systems, with adaptive variations that are mainly induction-based rather than selection-based. Once the epigenetic inheritance systems evolved, they imposed their own rules on the evolutionary game. A further stage in the evolution of evolution is behavioural transmission. Behavioural transmission can occur by various routes: behavioural phenotypes can be transmitted from parents to offspring, and in cultural evolution, which depends on group structure, there is transmission of behavioural information both from parents to offspring (vertical transmission), and to and from other individuals in the group (horizontal and oblique transmission). Finally there is cultural, language-based, specifically human evolution, which has shaped the history of our species.[9] Since there is more than one channel of transmission, and there are several types of inherited variations, the 'evolutionary game' is played simultaneously in several interacting dimensions.

The approach to heredity that a consideration of multiple inheritance systems demands is an epidemiological approach. Mendelian transmission is only one way, albeit a very important way, of passing on information, and other routes are open for the transmission and spread of random and directed variations. With DNA-based inheritance and epigenetic inheritance, the transmission is mainly vertical, from parents to progeny; with behavioural and cultural inheritance systems, variations are also transmitted horizontally and obliquely, between relatives and non-relatives, in a manner most appropriately described by epidemiological models.

The two great transitions to new levels of individuality, the transition from unicellularity to multicellularity, and the transition from the individual to the behaviourally or culturally integrated group, have both been associated with the novel exploitation of inheritance systems that evolved at previous levels.[10] These inheritance systems provided a new context for selection: the focus of selection was changed. Probably the most important consequence of the utilization of new inheritance systems is that they impart group-identity on an assembly of interacting components such as a lineage of cells, or a society of individuals in a cultural system. This is essential for the formation of enduring new levels of individuality and complexity.

Evolutionary biology and developmental biology

As acknowledged by the architects of the Modern Synthesis, embryology was not integrated within the framework of the evolutionary synthesis.[11] It is customary to blame this deficiency on the very different research traditions in embryology and genetics. Yet palaeontology and genetics also had different traditions and philosophies, and this did not stop Simpson from unifying the two disciplines to the satisfaction of the other architects.[12] There was also no lack of developmental research, both theoretical and experimental, directly relevant to genetics and evolution. The work of Garstang and de Beer on the effects of changes in early development on phylogeny, was well known and well respected;[13] Richard Goldschmidt integrated development and his version of genetics within an evolutionary framework, and Schmalhausen and Waddington published their work on the evolution of genetic adaptability and autoregulation, during the 1940s.[14] As Harwood (1993) has documented, in Germany in the interwar years there were many attempts to bridge the gap between embryology and genetics, and there was a general recognition of the importance of non-genic factors in heredity and evolution. In America there were definite, conscious beginnings at the end of the 1930s of a synthesis between embryology and genetics, in part influenced by the work of German immigrants such as Gluecksohn-Schoenheimer,[15] but also as a result of the studies of maternal effects and lethal factors. The chasm between embryology and evolutionary biology is therefore surprising.

We suspect that a major problem in bringing developmental biology into the unifying evolutionary framework developed fifty years ago was that its incorporation at that time might have endangered the whole Modern Synthesis. The focus of many embryologists on macromutations that change the timing of early development and dramatically modify the adult phenotype, threatened gradualism. Richard Goldschmidt, at that time the leading authority in the field of animal physiological genetics, was a firm believer in the importance of macromutation in evolution, and was against gradualism. Schmalhausen and Waddington concentrated their discussion on adaptability, which emphasizes individual development, and this had a smell of Lamarckism about it, despite the fact that their ideas were well within the confines of neo-Darwinism. Many of the insights offered by these biologists are now being resurrected. The main difference today is that their insights can be translated into a molecular language, so that the developmental effects of some mutations can be defined biochemically. Moreover, strict gradualism is no longer adhered to as religiously as in the past, and interest in the evolution of developmental processes is fastening the bond between studies of ontogeny and evolutionary biology. One of the principal European research programmes, the structuralist approach, which concentrates on the search for general laws of form and development, and emphasizes the role of developmental constraints, is assuming increasing importance in both developmental and evolutionary biology.[16]

Many developmental and evolutionary biologists believe that the hoped-for synthesis between evolutionary biology and developmental biology is at long last under way. Gilbert expressed this view very clearly:

> We are at a remarkable point in our understanding of nature, for a synthesis of developmental genetics with evolutionary biology may transform our appreciation of the mechanisms underlying evolutionary change and animal diversity. Such a synthesis is actually a return to a broader-based evolutionary theory that fragmented at the turn of the past century. . . . During the mid-twentieth century, population genetics merged with evolutionary biology to produce the evolutionary genetics of the modern synthesis, while molecular genetics merged with developmental biology to produce developmental genetics. These two vast areas, developmental genetics and evolutionary genetics, are on the verge of a merger that may unite these long-separated strands of biology and that may produce a developmental genetic theory capable of explaining macroevolution. (Gilbert 1991*a*, pp. 855–856)

An inevitable part of the merger that Gilbert anticipates will be a return to a broader concept of heredity, which was narrowed and restricted at the turn of the century. To illustrate the importance of the study of epigenetics in the unification, in Fig. 10.3 we have added epigenetics to Gilbert's picture of the history of developmental biology, genetics, and evolutionary biology.

We believe that acceptance of a broader concept of heredity, which includes multiple inheritance systems, will have far reaching consequences for our understanding of evolutionary processes. Epigenetic inheritance raises some problems for neo-Darwinian evolutionary theory because it allows directed mutations and epimutations, and therefore the inheritance of acquired characters. It also complicates the analysis of evolutionary change by introducing an additional inheritance system. But epigenetic inheritance exists, and has to be assimilated into evolutionary biology.

Some of the evidence showing that epigenetic inheritance can lead to

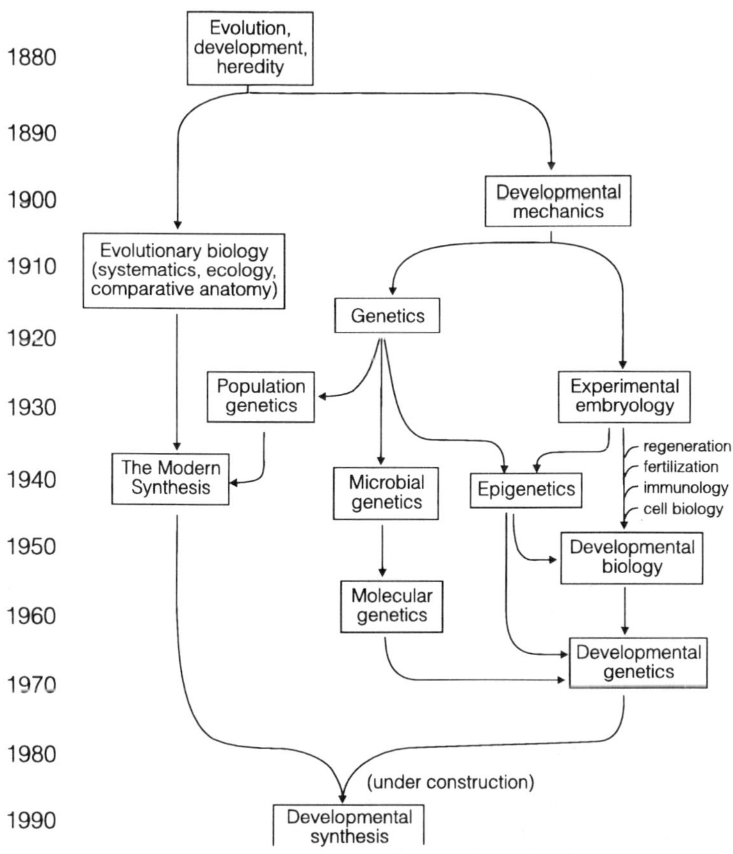

Fig. 10.3 Simplified scheme of the evolution of genetics, embryology, and evolutionary biology during the twentieth century. (Modified from Gilbert 1991*a*.)

evolutionary change is old. We have disinterred it from the graveyard of forgotten research, where it was buried because the observations were unexplained or inconvenient. However, it is the new molecular biology that has provided the strongest evidence for epigenetic inheritance, albeit as a spin-off of research into other things. The importance of epigenetic inheritance in development is beyond doubt. Its accidental discovery in transgenerational inheritance in so many animals and plants makes it clear that it also has importance in heredity and evolution. Understandably, with the present distortion of academic research programmes by the need to solicit research funding, few people have been able to centre a research programme on germinally-transmitted non-DNA variations. The current dogma is almost totally DNA-centric.

The achievements of molecular genetics, which reflect the prevailing DNA-centric view of heredity, may be used as an argument against some of the dissenting views expressed in this book. We do not argue with the importance of molecular genetics, nor do we want to belittle its achievements. What we do claim is that it presents a partial and incomplete view of biological heredity, and therefore of evolution. Molecular genetics has greatly expanded our understanding of heredity, but as with other disciplines, what is astonishing is not how much we know, but how much we have managed to do with the little that we know.

Notes

1. Mayr's terms were discussed in Chapter 1 (p. 13). The demise of soft inheritance is discussed by Mayr in *The growth of biological thought*, (1982*a*, pp. 793–794). He concluded (p. 828): 'the genetic material (DNA) is completely constant ("hard") from generation to generation, except for a very occasional (about one in 100,000) "mutation" (that is, an error in replication)'. The idea that inheritance has to be very hard to be effective in evolution was developed by Dawkins in his books *The selfish gene* (1976), *The extended phenotype* (1982), and *The blind watchmaker* (1986).
2. See the discussion in Rose *et al.* (1984).
3. In Chapter 7 (p. 184) we explained how either some sort of mutational bias, or selection, can lead to the evolution of isochores. When large parts of the genome are compared between species differing in genome organization, the changes in DNA composition will be biased by the existing differences in organization and will not obey the molecular clock. However, when the rate of molecular evolution is compared between species that have a similar compartmentalized, banded, genome organization, the rate of evolution should follow the molecular clock.
4. A detailed discussion of this point of view and evidence for neutral molecular evolution are given by Kimura (1983) and Li and Grauer (1991).
5. Williams argued against the significance of group selection in his 1966 classic *Adaptation and natural selection*; Lewontin (1970) gave a more balanced discussion of the topic; Dawkins (1976) argued strongly for the gene as the unit of

selection; Wade (1978) reviewed models of group selection that showed that it could occur under some conditions. Sober (1984) provided a discussion from a philosophical point of view. Wilson and Sober (1989) and Williams (1992) present recent views on units of selection.

6. Buss's book *The evolution of individuality* was very important in stimulating these discussions.
7. The only exception is the consideration of cultural transmission in humans.
8. See for example, Wake *et al.* (1983) and Turner (1988).
9. Studies of behavioural transmission in mammals and birds have been accumulating; see for example, Zentall and Galef's *Social Learning* (1988). Several books on cultural evolution in humans have appeared since the beginning of the 1980s: Cavalli-Sforza and Feldman (1981), Lumsden and Wilson (1981), Boyd and Richerson (1985), Durham (1991).
10. For further discussion, see Jablonka (1994).
11. See Mayr and Provine (1980).
12. Simpson's book *Tempo and mode in evolution* (1944) is one of the classics of the Modern Synthesis.
13. For example, de Beer published *Embryology and evolution* in 1930, and updated it in *Embryos and ancestors* in 1940.
14. Gottlieb (1992) gives a general and popular review of these approaches.
15. Gilbert (1991*b*) describes the importance of the genetic work by Gluecksohn-Schoenheimer and by Waddington for the emergence of developmental genetics. Gluecksohn-Schoenheimer summarized her ideas on the relationship between genetics and embryology in several publications, for example in 1949.
16. Riedl (1978) in *Order in living organisms* represents the structuralist approach to the origin and evolution of form; in the *The origins of order*, Kauffman (1993) reviews the structuralist approach to biology, and investigates the order imposed on a biological system by cybernetic constraints.

Glossary

(Words in italic are defined elsewhere in the glossary.)

allopatric speciation The evolution of new species through divergence of populations that are geographically separated.

***Alu* sequence** The predominant *SINE* family (with nearly a million copies) in the human genome.

amplification Selective increase in the number of copies of one region of DNA.

aneuploid Having a chromosome number that is one or more greater than, or less than, that of the regular chromosome set.

5-azacytidine An analogue of cytidine in which the carbon in position 5 of the base ring is replaced by nitrogen.

B chromosomes Supernumary chromosomes, present in some members of some animal and plant species, that are frequently heterochromatic, may have no pairing partner, and lack genes with major effects.

Baldwin effect The effect seen when the environmental induction of a physiological or behavioural adaptation allows a population to survive long enough for the accumulation by selection of similar *constitutive* hereditary changes.

basal body Cytoplasmic organelle made up of microtubules which is found at the base of eukaryotic cilia and flagella.

biophores Weismann's term for the fundamental vital hereditary units of living matter.

blastocyst An embryonic stage in mammals that consists of an outer layer of cells, the trophectoderm, and an inner cell mass from which the embryo is formed.

blastula A stage in the development of animals in which the embryo consists of a sheet of epithelial cells around a cavity.

boundary element A DNA sequence that isolates one domain of DNA from the regulatory influences of a neighbouring domain.

C bands Blocks of constitutive *heterochromatin* which stain densely with most dyes.

canalization The adjustment of developmental pathways by natural selection so as to bring about a uniform result in spite of genetic and environmental variations.

canalizing selection Selection for a well-buffered developmental pathway that ensures the production of a standard phenotype in spite of genetic and environmental variations.

cell memory The retention of functional or structural states in cell lineages.

cellular inheritance Transmission of structural or functional states in cell lineages.

central dogma The belief, enunciated by Francis Crick, that the flow of hereditary information is unidirectional: it passes from nucleic acids to proteins, but never in the reverse direction.

centrioles Cytoplasmic organelles, containing nine triplets of microtubules, that are found in most animal cells.

centromere A short segment of the eukaryotic chromosome with which the spindle fibres become associated during cell division, and at which chromatids remain attached until anaphase.

chiasma A cytologically visible region of contact between two non-sister chromatids which is seen during meiotic prophase and reflects crossing-over.
chromatin The complex of DNA, RNA, and proteins that makes up chromosomes.
chromatin mark The non-DNA part of a chromosomal locus that affects the nature and stability of gene expression.
chromatin diminution Developmentally regulated loss of the whole or parts of chromosomes from somatic cells during ontogeny.
chromocentre The aggregation of *heterochromatin* from different chromosomes that is visible in some interphase cells, particularly in dipteran cells containing *polytene chromosomes*.
chromosome domain A region of chromosome, thought to be a unit of function and replication, containing 30–300 kb of DNA anchored at each end to the *nuclear matrix*.
chromosome puff A visibly swollen region of a *polytene chromosome* which is characteristic of sites of active transcription.
cis-acting factor A chromosomal element that affects the regulation of regions of the chromosome carrying it.
cline Continuous variation in a character across the geographic range of a species.
clonal memory span The average number of cell divisions through which a particular epigenetic state is faithfully transmitted to daughter cells.
concerted evolution A term now usually used for the tendency for members of a gene family to evolve together and become very similar, even though they are found in different chromosomal locations.
constitutive activity Continuous gene activity that does not depend on external stimuli.
copia-like element One of a family of common *transposable elements* in *Drosophila*.
CpG island A GC-rich region of DNA, usually at the 5′ end of a gene, that is relatively rich in unmethylated CpG dinucleotides.
CpG site A DNA site at which C (cytosine) is followed by G (guanine); p denotes the phosphate group, so the C is at the 5′ position relative to the G.
cultivar A variety of plant produced and maintained by cultivation and not normally found in wild populations.
cytoskeleton A network of protein filaments in the cytoplasm that gives the cell its shape and is responsible for cell movements.
cytotaxis A term coined by Sonneborn to describe the processes whereby new cell structures are ordered and arranged under the influence of existing cell structures.
dark G bands Dark chromosome bands, revealed by staining procedures using Giemsa, which replicate late in *S phase* and contain few genes (cf. *light G bands*).
Darwinism Darwin's theory of evolution by descent with modification; usually limited to that part of the theory dealing with the mechanism of evolutionary change, i.e. natural selection of heritable variations.
Dauermodifikation (lingering modification) A term coined by Jollos for an induced heritable variation that is not as stable as a classical mutation, but is transmitted to descendants for several generations.
determinant Weismann's term for a component of an *id* (a complete set of hereditary information) which is responsible for the properties of a particular cell type.
determination The process of commitment of cells to a particular developmental pathway.

DH site (DNase-I hypersensitive site) A *nucleosome*-free region of DNA, extremely sensitive to degradation by *DNase-I*, which is associated with transcriptional activity.

differentiation The processes of change in cell shape and physiology that lead to the production of a specialized cell type.

diplogenesis The theory, developed by Cope around 1900, proposing that an environmental agent alters simultaneously and in the same way both the soma and germ cells.

directed assembly The assembly of cell structures under the influence of existing cell structures.

directed mutation Non-random, environmentally-induced genomic changes.

DNA modification The addition of covalently bonded small chemical groups (e.g. the methyl group $-CH_3$) to specific bases of DNA, e.g. the change of cytosine to 5-methyl cytosine.

DNA methylation Modification of DNA by the addition of methyl groups ($-CH_3$) to some bases; in eukaryotes the bases modified are usually cytosines.

DNase-I An endonuclease: an enzyme that degrades DNA to nucleotides.

DNase-I sensitive region A chromatin region showing greater sensitivity to degradation by *DNase-I* than adjacent regions; sensitivity is associated with transcriptional activity or potential activity.

dormant gene A gene that is inactive because there is no tissue in which intracellular conditions are suitable for its expression.

early germ-line determination Irreversible segregation of the germ cell lineage from somatic lineages early in development (cf. *late germ-line determination, somatic germ-cell determination*).

ectopic pairing Pairing between non-homologous chromosome regions.

EIS Abbreviation for *epigenetic inheritance system.*

enhancer A regulatory DNA sequence that is a *cis-acting factor* that increases the level of transcription of neighbouring or distant genes.

epiallele One of the alternative heritable chromatin forms of a gene with an unchanged DNA sequence.

epigenetic Pertaining to the developmental processes through which the genetic information in an individual is expressed.

epigenetic inheritance system (EIS) A system that enables the phenotypic expression of the information in a cell or individual to be transmitted to the next generation.

epimutation A heritable abnormality in phenotype which is not the result of an altered DNA base sequence.

epistasis An interaction between genes at different loci that produces a phenotype different from that expected from the independent expression of the genes.

euchromatin Less compact chromatin, usually staining in a characteristic way at metaphase (less intensely with many dyes), which contains most normal genes (cf. *heterochromatin*).

exons The sequences of the eukaryotic gene that are ultimately translated into proteins (cf. *introns*).

expressivity The extent to which a particular genotype is expressed in the phenotype.

extracelluar matrix The proteins and polysaccharides that form a matrix outside cells.

extraembryonic tissues Tissues derived from the fertilized egg that lie outside the embryo itself.
facultative heterochromatin *Chromatin* that contains normal genes but in certain cell types becomes condensed and inactive.
fluctuation test A test, based on the analysis of the distribution of mutants in replicate cultures, that is used to measure mutation rates or to determine whether mutations are random or directed.
founder effect Mayr's term for the drastic genetic reorganization that could occur in a population derived from a few founder individuals which represent a very small sample of the gene pool to which they formerly belonged.
frameshift mutation A mutation that upsets the triplet-codon reading frame of DNA through the insertion or deletion of one or more bases.
G bands Giemsa bands: light and dark bands seen in euchromatic regions of the metaphase chromosomes of vertebrates after staining procedures involving the dye Giemsa (see *light G bands* and *dark G bands*).
gemmules Darwin's term for hypothetical small hereditary particles that are sent from each part of the body to the reproductive organs where they form the 'germ' that gives rise to the next generation.
gene conversion A process occurring during meiosis in which one allele converts the other allele to its own sequence.
gene phenotype The chromatin structure of a gene.
genetic assimilation Inheritance of an acquired character arising through the replacement by natural selection of a response originally dependent on an environmental stimulus by a stimulus-independent response.
genetic load The reduction in the average fitness of individuals due to the presence of deleterious alleles in the population.
genetic revolution A major reorganization of the genome that can bring about a shift to a new co-adapted combination of alleles.
genomic imprinting The process that causes the expression of genetic information to depend on the sex of the parent from which it was inherited; also used for the result of this process.
germ line The cell lineage that in normal development gives rise to gametes.
germ plasm The cellular material containing the hereditary information for the production of the next generation. Weismann identified the germ plasm with the nuclear material.
germinal selection Weismann's concept that selection occurs between homologous *determinants* in germ line cells.
Haldane's rule The generalization that in the offspring of interspecific crosses, when one sex is absent, rare, or sterile, it is the *heterogametic sex*.
hard inheritance Mayr's term for inheritance in which the genetic material is constant from generation to generation and is not modified by environmental factors (cf. *soft inheritance*).
heterochromatin Condensed *chromatin*, which stains darkly with most dyes, is late replicating, and is largely transcriptionally inactive (cf. *euchromatin*).
heterochrony An evolutionary change in the time or rate at which the different body parts develop relative to each other.
heterogametic sex The sex that has two different sex chromosomes and therefore produces two types of gamete differing in their sex chromosomes (cf. *homogametic sex*).
homeobox A DNA sequence, characteristic of genes that influence segmenta-

tion in animals, that encodes a polypeptide sequence associated with DNA binding.

homeostasis The maintenance of relatively steady states in an organism through internal regulatory mechanisms, despite variations in internal and external conditions.

homeotic gene A gene determining characteristic structures associated with body segments in *Drosophila* and some other organisms.

homogametic sex The sex in which the two sex chromosomes are morphologically the same and which therefore produce gametes all having the same type of sex chromosome (cf. *heterogametic sex*).

hopeful monster A term used by Goldschmidt to describe a mutant organism that differs dramatically from its immediate ancestors in ways that may have adaptive significance.

housekeeping genes Genes that are active in all cell types and code for the proteins that are essential for cell maintenance (cf. *tissue-specific* and *stage-specific genes*).

hybridogenesis Maintenance of a hybrid species by the selective elimination of one of the parental genomes during gametogenesis (to produce gametes of one parental type) and breeding with the other parental type.

id The term Weismann used for a unit of heredity that contains all the information necessary for the development of the entire organism. (In effect, an id is a haploid genome, but Weismann assumed each cell contained many ids.)

ILC environment Abbreviation for *i*ntermediate *l*ength *c*ycle environment, which is an environment that fluctuates regularly with a cycle length greater than the generation time of an individual, but shorter than the time needed to fix classical mutations.

imprint A *chromatin mark* that is determined by the sex of the parent.

introns Intervening sequences of DNA bases that interrupt the coding sequences and are not represented in the mature RNA because they are spliced out of the primary transcript (cf. *exons*).

isochores Stretches of DNA, more than 300 kb long, of homogeneous base composition, which are interspersed with other regions of different base composition.

Lamarckian inheritance The concept that characters acquired during the lifetime of an organism can be transmitted to its offspring; a synonym for the inheritance of acquired characters.

Lamarckism The theory of evolution, developed by Lamarck in the early nineteenth century, that proposes that evolutionary change results from an inherent tendency for self-complication and from acquired adaptations that are passed to descendants and become evolutionary adaptations. Often used for evolutionary theories based on the inheritance of acquired characters.

lampbrush chromosome A type of chromosome found at the diplotene stage of meiosis (particularly associated with the primary oocytes of amphibians) that is characterized by large numbers of lateral loops containing DNA; the loops are associated with transcriptional activity.

Lansing effect The cumulative but reversible effect of parental age on the longevity of the next generations.

late germ-line determination The segregation of a distinct germ cell lineage relatively late in development (cf. *early germ-line determination, somatic germ-cell determination*).

light G bands Light chromosome bands, revealed by staining procedures using Giemsa, which contain most genes and replicate early in *S phase* (cf. *dark G bands*).

LINEs *L*ong *in*terspersed repeated *e*lements, found throughout the genome, that result from the movement of *transposable elements*.

macronucleus The large somatic nucleus of ciliated protozoa.

Markovian process A stochastic process where future development is determined only by the present state, and does not depend on the way the present state was reached.

meiotic drive A process that produces non-Mendelian ratios through causing one allele or chromosome rather than its alternative to be incorporated into gametes.

meristematic tissue Plant tissue in which active division of multipotent cells occurs; some meristematic tissue gives rise to germ cells.

methyl-transferase (methylase) The enzyme catalyzing the addition of methyl groups to DNA bases.

methylation The attachment of a methyl group to some other molecule (see *DNA methylation*).

micronucleus The small germ-line nucleus of ciliated protozoa.

microtubules Elements of the eukaryote cytoskeleton, also forming parts of cilia and flagella, that are made of tubulin proteins.

Modern Synthesis The theory of evolution that resulted from the unification of Darwin's theory of natural selection with Mendelian genetics, systematics, and palaeontology.

molecular clock The concept that the rate at which mutational changes in a gene accumulate is constant over time, so the amount of molecular divergence can be used to date the time at which lineages split.

molecular drive The spread and fixation of different variants in multi-copy gene families and non-coding repeated sequences, which result from intragenomic processes such as slippage, gene conversion, and unequal crossing-over; it causes all members of the family to change in the same way and homogenizes gene families.

mosaic development Embryonic development in which the fate of cell lineages is determined mainly by localized determinants in the fertilized egg (cf. *regulative development*).

MTOC Abbreviation for the *m*icro*t*ubule *o*rganizing *c*entre, a site in the cell that is involved with the organization of the *microtubules* of the *cytoskeleton*.

neo-Darwinism A term originally used by Romanes (1896) for a Darwinian evolutionary theory that excludes the possibility of the inheritance of acquired variations; now more commonly used for the evolutionary theory that results from the unification of Darwinian natural selection with Mendelian genetics.

neo-Lamarckism A term used for the various versions of Lamarck's ideas that emphasize the role of the transmission of acquired characters in evolution.

non-disjunction A failure of homologous chromosomes or sister chromatids to separate in meiosis or mitosis, which results in daughter cells with too many or too few chromosomes.

NOR Acronym for *n*ucleolar *o*rganizer *r*egion, a chromosomal region containing multiple copies of the ribosomal RNA sequences.

normalizing selection Selection that eliminates deviants.

nuclear matrix Structural elements in the nucleus that are associated with the organization and activity of chromosomes.

nucleolar dominance The expression of only one of the nucleolar organizers (*NOR*s) in a heterozygote.

nucleosome A fundamental unit of eukaryotic *chromatin*, consisting of a core particle made up of a histone octamer, around which about 140 bp of DNA are wound.

oncogene A gene that is associated with the production of cancer.

operon A region of DNA that acts as a single transcription unit and is made up of one or more structural genes and their regulatory sequences.

orthogenesis The theory that there are directional trends in evolution that affect a group of related species; commonly used with the implication that the trends are non-adaptive and result from internal causes.

P element A type of *transposable element* found in *Drosophila*.

pairing failure The failure of the whole or part of euchromatic chromosomes to pair with a homologous region during meiotic prophase; often results in defects later in gametogenesis.

pangenesis Darwin's theory of heredity; now used generally for all of the many theories, including Darwin's, that suggest that some kind of 'seed material' coming from all parts of the body contributes to the germinal material that forms the next generation.

parallel induction A theory suggesting that environmental agents can cause the same change in both the soma and germ line.

paramutation A term coined by Brink for the process whereby one allele in a heterozygote alters the heritable properties of the other allele.

parapatric Used for populations with ranges that are contiguous but not overlapping, so that gene flow between them is possible.

parental-age effect An influence of the age of the parent on the phenotype of the offspring.

penetrance The proportion of individuals with a specific genotype that express the genotype at the phenotypic level.

peripatric speciation Mayr's term for the speciation that occurs following geographic isolation of a small peripheral population.

PEV Acronym for *position effect variegation*.

phenocopy An environmentally-induced phenotypic variation that resembles the effect of a known gene mutation.

plasmagene An archaic concept referring to cytoplasmic hereditary particles.

polite DNA A term coined by Zuckerkandl for the preferential insertion of transposable elements and other sequences into regions of DNA having a base composition similar to themselves.

polyploid Having more than two sets of homologous chromosomes (e.g. triploid, with three sets, tetraploid, with four sets).

polyphenism The existence of distinctly different adaptive phenotypes in a population of individuals with similar genotypes.

polytene chromosomes Giant multistranded chromosomes, characteristic especially of some dipteran tissues, that are formed by repeated replication of the chromosomes without separation of the daughter chromatids.

position effect A change in the phenotypic expression of a gene that results from a change in its position in the genome, or a change in the position of sequences that are normally adjacent to it.

position effect variegation (PEV) The mosaic phenotype resulting from the inactivation in some cells of a gene that has been relocated within the genome.

post-zygotic isolation Reproductive isolation in which two populations do not interbreed successfully because although a zygote is formed, it either fails to develop or develops into a sterile or reproductively inadequate adult.
preformationism The concept that all the structures of an organism are present in miniature in the germ cell.
pre-zygotic isolation Reproductive isolation in which two populations do not interbreed because ecological, behavioural, or mechanical factors prevent a zygote from being formed.
promoter A control region 5′ to a gene that binds RNA polymerase and is involved in the initiation of transcription.
pseudogene A DNA sequence that has significant homology with a functional gene, but cannot be transcribed.
reduction division The meiotic division in which the chromosome number is halved.
regulative development Embryonic development in which the fate of cell lineages is determined mainly by cell–cell interactions (cf. *mosaic development*).
replica plating A technique that allows simultaneous transfer of colonies of cells from one plate of growth medium to identical positions on additional plates.
replicative transposition Movement of a *transposable element* that occurs when an existing element is replicated and the new copy is moved to an additional site.
retrotransposon A *transposon* that moves via an RNA intermediate.
retrovirus A virus that encodes its genetic information in single-stranded RNA and produces a reverse transcriptase that transcribes this RNA into DNA during the replicative cycle.
reverse transcription A process in which RNA is used as a template for the synthesis of DNA.
RIGS Acronym for *r*epeat-*i*nduced *g*ene *s*ilencing.
RIP A process, first identified in fungi, whereby repeated sequences of DNA above a certain length are either excised through rearrangements (*r*earrangements *i*nduced *p*re-meiotically), or are modified and mutated (*r*epeat-*i*nduced *p*oint-mutation).
ripping A synonym for *RIP*.
satellite DNA Highly repeated DNA sequences having a base composition sufficiently different from the bulk of DNA to form a satellite band in caesium chloride gradient centrifugation.
Segregation Distorter (SD) A gene complex in the *centromere* region of a *Drosophila* autosome which is responsible for a system of male *meiotic drive*.
SINEs *S*hort *in*terspersed repeated *e*lements found throughout the genome.
soft inheritance A term coined by Mayr who defined it as 'inheritance during which the genetic material is not constant from generation to generation but may be modified by the effects of the environment, by use or disuse, or other factors' (Mayr 1982*a*, p. 959), (cf. *hard inheritance*).
soma All the cells of the body other than the germ line.
somatic germ-cell determination The production of germ cells from somatically-derived cells (cf. *early germ-line determination* and *late germ-line determination*).
somatic induction A theory suggesting that a somatic change induced by an environmental agent can induce a corresponding change in germ cells.
somatic mutation A mutation occurring in a somatic cell.
somatic selection Selection among cells within a tissue or organ.
S phase The phase in the cell cycle during which DNA is replicated.

splicing The removal from RNA of *intron* sequences, and joining together of *exon* sequences, to form uninterrupted protein-coding mRNA.

stabilizing selection Selection for the average or standard phenotype that removes extreme variants and reduces the phenotypic and genotypic variance in future generations.

stage-specific genes Genes that are expressed at only certain stages of development (cf. *housekeeping genes*).

stasipatric speciation Speciation resulting from novel chromosomal changes that cause homozygotes to have increased fitness and be reproductively isolated in part of the ancestral species range.

stasis Lack of morphological change in an evolutionary lineage.

steady-state system A self-perpetuating metabolic pattern that is maintained by positive feedback.

stubborn mark A *chromatin mark* that is not readily erased during early embryogenesis and gametogenesis.

sympatric speciation The evolution of new species within populations without geographical isolation.

synaptonemal complex A structure formed between homologous chromosomes during the pairing phase of meiosis.

T-DNA A region of the Ti plasmid of *Agrobacterium tumefaciens* used to transform plant cells.

tandem repeats Repeated DNA sequences that are organized in an uninterrupted linear array.

telomere The tip of a chromosome.

tissue-specific genes Genes that are expressed in only some tissues (cf. *housekeeping genes*).

trans-acting factor A diffusible gene-product that affects gene regulation.

transgene The newly integrated DNA within a *transgenic organism*.

transgenic organism An organism whose genome includes new DNA introduced by experimental manipulation.

transposable element (transposon) A general term for a genetic unit that can move to and from different sites in the genome.

trophectoderm The external epithelial layer of the mammalian *blastocyst* that later develops into *extraembryonic tissues*.

Weismann's doctrine The belief, attributed to Weismann, that very early in development a group of cells is set aside for the eventual formation of germ cells, while the remaining cells become irreversibly committed to somatic functions.

X-inactivation The process occurring during embryogenesis that results in one of the two X chromosomes of female mammals becoming transcriptionally silent.

References

Achwal, C.W., Ganguly, P., and Chandra, H.S. (1984). Estimation of the amount of 5-methylcytosine in *Drosophila melanogaster* DNA by amplified ELISA and photoacoustic spectroscopy. *EMBO Journal*, **3**, 263–266.

Adams, R.L.P. (1990). DNA methylation. *Biochemical Journal*, **265**, 309–320.

Adams, R.L.P. and Burdon, R.H. (1985). *Molecular biology of DNA methylation*. Springer-Verlag, New York.

Alberts, B., Worcel, A., and Weintraub, H. (1977). On the biological implications of chromatin structure. In *The organization and expression of the eukaryotic genome*, (ed. E.M. Bradbury and K. Javaherian), pp. 165–191. Academic Press, London.

Alberts, B., Bray, D., Lewis, J., Raff, M., Roberts, K., and Watson, J.D. (1989). *Molecular biology of the cell*, (2nd edn). Garland, New York.

Albrecht-Buehler, G. (1977). Daughter 3T3 cells: are they mirror images of each other? *Journal of Cell Biology*, **72**, 595–603.

Allen, G.E. (1979). Naturalists and experimentalists: the genotype and phenotype. *Studies in History of Biology*, **3**, 179–209.

Allen, N.D., Norris, M.L., and Surani, M.A. (1990). Epigenetic control of transgene expression and imprinting by genotype-specific modifiers. *Cell*, **61**, 853–861.

Allison, C. and Hughes, C. (1991). Bacterial swarming: an example of prokaryotic differentiation and multicellular behaviour. *Science Progress*, **75**, 403–422.

Ashley, T. (1988). G-band position effects on meiotic synapsis and crossing over. *Genetics*, **118**, 307–317.

Ashley, T. (1990). G-bands and chromosomal meiotic behavior. In *Chromosomes Today*, Vol. 10, (ed. K. Fredga, B.A. Kihlman, and M.D. Bennett), pp. 311–320. Unwin Hyman, London.

Badino, G. and Robotti, C. (1975). Selection in parthenogenetic lines of *Asplanchna sieboldi* (Leydig) 1854 (Rotatoria). *Experientia*, **31**, 298–299.

Baldwin, J.M. (1896). A new factor in evolution. *American Naturalist*, **30**, 441–451, 536–553.

Bander, S.A.A., Watson, S.C., and Shire, J.G.M. (1989). Paternal inheritance of egg traits in mice: a case of genomic imprinting. *Genetical Research*, **54**, 213–219.

Banks, J.A. and Fedoroff, N. (1989). Patterns of developmental and heritable change in methylation of the *Suppressor-mutator* transposable element. *Developmental Genetics*, **10**, 425–437.

Banks, J.A., Masson, P., and Fedoroff, N. (1988). Molecular mechanisms in the developmental regulation of the maize *Suppressor-mutator* transposable element. *Genes and Development*, **2**, 1364–1380.

Barlow, D.P. (1993). Methylation and imprinting: from host defense to gene regulation? *Science*, **260**, 309–310.

Barlow, D.P., Stöger, R., Herrmann, B.G., Saito, K., and Schweifer, N. (1991). The mouse insulin-like growth factor type-2 receptor is imprinted and closely linked to the Tme locus. *Nature*, **349**, 84–87.

Bartolomei, M.S., Zemel, S., and Tilghman, S.M. (1991). Parental imprinting of the mouse H19 gene. *Nature*, **351**, 153–155.

Barton, N.H. (1989). Founder effect speciation. In *Speciation and its consequences*, (ed. D. Otte and J.A. Endler), pp. 229–256. Sinauer, Sunderland, Mass.

Bateman, K.G. (1959). The genetic assimilation of four venation phenocopies. *Journal of Genetics*, **56**, 443–474.

Bateson, G. (1979). *Mind and nature: a necessary unity*. Wildwood House, London.

Beale, G.H. (1954). *The genetics of Paramecium aurelia*. Cambridge University Press.

Beardmore, J.A. and Shami, S.A. (1985). The Lansing effect and age-mediated changes in genetic parameters. *Genetica*, **68**, 37–46.

Beisson, J. and Sonneborn, T.M. (1965). Cytoplasmic inheritance of the organization of the cell cortex in *Paramecium aurelia*. *Proceedings of the National Academy of Sciences, USA*, **53**, 275–282.

Bell, G. (1982). *The masterpiece of nature: the evolution and genetics of sexuality*. Croom Helm, London.

Bell, G. (1985). The origin and early evolution of germ cells as illustrated by the Volvocales. In *The origin and evolution of sex*, (ed. H. Halvorson and A. Monroy), pp. 221–256. Alan R. Liss, New York.

Bell, G. (1988*a*). Uniformity and diversity in the evolution of sex. In *The evolution of sex*, (ed. R.E. Michod and B.R. Levin), pp. 126–138. Sinauer, Sunderland, Mass.

Bell, G. (1988*b*). *Sex and death in protozoa*. Cambridge University Press.

Bell, G. (1989). Darwin and biology. *Journal of Heredity*, **80**, 417–421.

Bell, G. and Koufopanou, V. (1991). The architecture of the life cycle in small organisms. *Philosophical Transactions of the Royal Society of London Series B*, **332**, 81–89.

Bell, L.R., Horabin, J.I., Schedl, P., and Cline, T.W. (1991). Positive autoregulation of *Sex-lethal* by alternative splicing maintains the female determined state in Drosophila. *Cell*, **65**, 229–239.

Belyaev, D.K. (1979). Destabilizing selection as a factor in domestication. *Journal of Heredity*, **70**, 301–308.

Belyaev, D.K. and Borodin, P.M. (1982). The influence of stress on variation and its role in evolution. *Biologisches Zentralblatt*, **100**, 705–714.

Belyaev, D.K., Ruvinsky, A.O., and Borodin, P.M. (1981*a*). Inheritance of alternative states of the fused gene in mice. *Journal of Heredity*, **72**, 107–112.

Belyaev, D.K., Ruvinsky, A.O., and Trut, L.N. (1981*b*). Inherited activation-inactivation of the star gene in foxes. *Journal of Heredity*, **72**, 267–274.

Belyaev, D.K., Ruvinsky, A.O., Agulnik, A.I., and Agulnik, S.I. (1983). Effect of hydrocortisone on the phenotypic expression and inheritance of the *Fused* gene in mice. *Theoretical and Applied Genetics*, **64**, 275–281.

Ben-Jacob, E., Shmueli, H., Shochet, O., and Tenenbaum, A. (1992). Adaptive self-organization during growth of bacterial colonies. *Physica* A, **187**, 378–424.

Bengtsson, B.O. (1980). Rates of karyotype evolution in placental mammals. *Hereditas*, **92**, 37–47.

Bernardi, G. (1989). The isochore organization of the human genome. *Annual Review of Genetics*, **23**, 637–661.

Bernardi, G. (1993). The vertebrate genome: isochores and evolution. *Molecular Biology and Evolution*, **10**, 186–204.

Bernstein, H. (1977). Germ line recombination may be primarily a manifestation of DNA repair processes. *Journal of Theoretical Biology*, **69**, 371–380.

Bernstein, H., Hopf, F.A., and Michod, R.E. (1987). The molecular basis of the evolution of sex. *Advances in Genetics*, **24**, 323–370.

Bernstein, H., Hopf, F.A., and Michod, R.E. (1988). Is meiotic recombination an adaptation for repairing DNA, producing genetic variation, or both? In *The evolution of sex*, (ed. R.E. Michod and B.R. Levin), pp. 139–160. Sinauer, Sunderland, Mass.

Berrill, N.J. and Liu, C.K. (1948). Germplasm, Weismann, and Hydrozoa. *Quarterly Review of Biology*, **23**, 124–132.

Bestor, T.H. (1990). DNA methylation: evolution of a bacterial immune function into a regulator of gene expression and genome structure in higher eukaryotes. *Philosophical Transactions of the Royal Society of London Series B*, **326**, 179–187.

Beukeboom, L.W. and Werren, J.H. (1993). Deletion analysis of the selfish *B* chromosome, *Paternal Sex Ratio* (*PSR*), in the parasitic wasp *Nasonia vitripennis*. *Genetics*, **133**, 637–648.

Bickmore, W.A. and Sumner A.T. (1989). Mammalian chromosome banding—an expression of genome organization. *Trends in Genetics*, **5**, 144–148.

Biémont, C. (1991). Are imprinting and inbreeding two related phenomena? *Genetics Selection Evolution*, **23**, 85–102.

Bird, A. (1992). The essentials of DNA methylation. *Cell*, **70**, 5–8.

Blacher, L.I. (1982). *The problem of the inheritance of acquired characters. A history of a priori and empirical methods used to find a solution*, (trans. F.B. Churchill from the Russian 1971 edn). Amerind, New Delhi.

Blackman, R. (1974). *Aphids*. Ginn, London.

Blackman, R.L. (1979). Stability and variation in aphid clonal lineages. *Biological Journal of the Linnean Society*, **11**, 259–277.

Blau, H.M. (1992). Differentiation requires continuous active control. *Annual Review of Biochemistry*, **61**, 1213–1230.

Blau, H.M. and Baltimore, D. (1991). Differentiation requires continuous regulation. *Journal of Cell Biology*, **112**, 781–783.

Bodnar, J.W. and Ward, D.C. (1987). Highly recurring sequence elements identified in eukaryotic DNAs by computer analysis are often homologous to regulatory sequences or protein binding sites. *Nucleic Acids Research*, **15**, 1835–1851.

Boe, L. (1990). Mechanism for induction of adaptive mutations in *Escherichia coli*. *Molecular Microbiology*, **4**, 597–601.

Bohr, V.A. (1988). DNA repair and transcriptional activity in genes. *Journal of Cell Science*, **91**, 175–178.

Bohr, V.A. and Wassermann, K. (1988). DNA repair at the level of the gene. *Trends in Biochemical Sciences*, **13**, 429–433.

Bonner, J.T. (1958). *The evolution of development*. Cambridge University Press.

Bonner, J.T. (1974). *On development*. Harvard University Press, Cambridge, Mass.

Bonner, J.T. (1988). *The evolution of complexity*. Princeton University Press, New Jersey.

Borst, P. and Greaves, D.R. (1987). Programmed gene rearrangements altering gene expression. *Science*, **235**, 658–667.

Bottema, S. (1989). Some observations on the modern domestication processes. In *The walking larder*, (ed. J. Clutton-Brock), pp. 31–45. Unwin Hyman, London.

Boulikas, T. (1992). Evolutionary consequences of nonrandom damage and repair of chromatin domains. *Journal of Molecular Evolution*, **35**, 156–180.

Bowler, P.J. (1983). *The eclipse of Darwinism*. John Hopkins University Press, Baltimore.

Bowler, P.J. (1988). *The non-Darwinian revolution*. John Hopkins University Press, Baltimore.

Bowler, P.J. (1989). *Evolution: the history of an idea*, (revised edn). University of California Press, Berkeley.

Bownes, M. (1990). Preferential insertion of P elements into genes expressed in the germ-line of *Drosophila melanogaster*. *Molecular and General Genetics*, **222**, 457–460.

Boyd, R. and Richerson, P.J. (1985). *Culture and the evolutionary process*. University of Chicago Press.

Boyes, J. and Bird, A. (1992). Repression of genes by DNA methylation depends on CpG density and promoter strength: evidence for involvement of a methyl-CpG binding protein. *EMBO Journal*, **11**, 327–333.

Brandeis, M., Kafri, T., Ariel, M., Chaillet, J.R., McCarrey, J., Razin, A., and Cedar, H. (1993). The ontogeny of allele-specific methylation associated with imprinted genes in the mouse. *EMBO Journal*, **12**, 3669–3677.

Breese, E.L., Hayward, M.D., and Thomas, A.C. (1965). Somatic selection in perennial ryegrass. *Heredity*, **20**, 367–379.

Brenner, S. (1992). Dicing with Darwin. *Current Biology*, **2**, 167–168.

Brent, L., Rayfield, L.S., Chandler, P., Fierz, W., Medawar, P.B., and Simpson, E. (1981). Supposed lamarckian inheritance of immunological tolerance. *Nature*, **290**, 508–512.

Brent, L., Chandler, P., Fierz, W., Medawar, P.B., Rayfield, L.S., and Simpson, E. (1982). Further studies on supposed lamarckian inheritance of immunological tolerance. *Nature*, **295**, 242–244.

Brink, R.A. (1960). Paramutation and chromosome organization. *Quarterly Review of Biology*, **35**, 120–137.

Brink, R.A. (1962). Phase change in higher plants and somatic cell heredity. *Quarterly Review of Biology*, **37**, 1–22.

Brink, R.A. (1973). Paramutation. *Annual Review of Genetics*, **7**, 129–152.

Brockdorff, N., Ashworth, A., Kay, G.F., McCabe, V.M., Norris, D.P., Cooper, P.J., Swift, S., and Rastan, S. (1992). The product of the mouse *Xist* gene is a 15 kb inactive X-specific transcript containing no conserved ORF and located in the nucleus. *Cell*, **71**, 515–526.

Brown, C.J., Hendrich, B.D., Rupert, J.L., Lafrenière, R.G., Xing, Y., Lawrence, J., and Willard, H.F. (1992). The human *XIST* gene: analysis of a 17kb inactive X-specific RNA that contains conserved repeats and is highly localized within the nucleus. *Cell*, **71**, 527–542.

Brown, D.P. (1984). The role of stable complexes that repress and activate eucaryotic genes. *Cell*, **37**, 359–365.

Brown, S.W. and Chandra, H.S. (1977). Chromosome imprinting and the differential regulation of homologous chromosomes. In *Cell biology: a comprehensive treatise*, Vol. 1, *Genetic mechanisms of cells*, (ed. L. Goldstein and D.M. Prescott), pp. 109–189. Academic Press, New York.

Brown, S.W. and Nelson-Rees, W.A. (1961). Radiation analysis of a lecanoid genetic system. *Genetics*, **46**, 983–1007.

Brun, J. (1965). Genetic adaptations of *Caenorhabditis elegans* (Nematoda) to high temperatures. *Science*, **150**, 1467.

Bucci, S., Ragghianti M., Mancino, G., Berger, L., Host, H., and Uzzell, T. (1990). Lampbrush and mitotic chromosomes of the hemiclonally reproducing hybrid *Rana esculenta* and its parental species. *Journal of Experimental Zoology*, **255**, 37–56.

Burgoyne, P.S. and Baker, T.G. (1984). Meiotic pairing and gametogenic failure. In *Controlling events in meiosis* (ed. C.W. Evans and H.G. Dickinson), pp. 349–362. Company of Biologists, Cambridge.

Burgoyne, P.S. and Mahadevaiah, S.K. (1993). Unpaired sex chromosomes and gametogenic failure. *Chromosomes Today*, Vol. 11, (ed. A.T. Sumner and A.C. Chandley), pp. 243–263. Chapman and Hall, London.

Burt, A., Bell, G., and Harvey, P.H. (1991). Sex differences in recombination. *Journal of Evolutionary Biology*, **4**, 259–277.

Bush, G.L. (1975). Modes of animal speciation. *Annual Review of Ecology and Systematics*, **6**, 339–364.

Bush, G.L., Case, S.M., Wilson, A.C., and Patton, J.L. (1977). Rapid speciation and chromosomal evolution in mammals. *Proceedings of the National Academy of Sciences, USA*, **74**, 3942–3946.

Buss, L.W. (1982). Somatic cell parasitism and the evolution of somatic tissue compatibility. *Proceedings of the National Academy of Sciences, USA*, **79**, 5337–5341.

Buss, L.W. (1983). Evolution, development, and the units of selection. *Proceedings of the National Academy of Sciences, USA*, **80**, 1387–1391.

Buss, L.W. (1987). *The evolution of individuality*. Princeton University Press, New Jersey.

Buss, L.W. (1988). Diversification and germ-line determination. *Palaeobiology*, **14**, 313–321.

Bussey, H. and Fieldes, M.A. (1974). A model for stably inherited environmentally induced changes in plants. *Nature*, **251**, 708–710.

Cairns, J. (1990). Causes of mutation and Mu excision. *Nature*, **345**, 213.

Cairns, J. and Foster, P.L. (1991). Adaptive reversion of a frameshift mutation in *Escherichia coli*. *Genetics*, **128**, 695–701.

Cairns, J., Overbaugh, J., and Miller, S. (1988). The origin of mutants. *Nature*, **335**, 142–145.

Cairns-Smith, A.G. (1985). *Seven clues to the origin of life*. Cambridge University Press.

Campbell, J.H. (1982). Autonomy in evolution. In *Perspectives on evolution*, (ed. R. Milkman), pp. 190–201. Sinauer, Sunderland, Mass.

Campbell, J.H. (1985). An organizational interpretation of evolution. In *Evolution at a crossroads: the new biology and the new philosophy of science*, (ed. D.J. Depew and B.H. Weber), pp. 133–167. MIT Press, Cambridge, Mass.

Campbell, J.H. (1987). The new gene and its evolution. In *Rates of evolution*, (ed. K.S.W. Campbell and M.F. Day), pp. 283–309. Allen and Unwin, London.

Campbell, J.H. and Perkins, P. (1988). Transgenerational effects of drug and hormonal treatments in mammals: a review of observations and ideas. *Progress in Brain Research*, **73**, 535–553.

Cannon, H.G. (1959). *Lamarck and modern genetics*. Manchester University Press.

Capdevila, M.P. and Garcia-Bellido, A. (1974). Development and genetic analysis of *bithorax* phenocopies in *Drosophila. Nature*, **250**, 500–502.

Castro-Sierra, E. and Ohno, S. (1968). Allelic inhibition at the autosomally inherited gene locus for liver alcohol dehydrogenase in chicken-quail hybrids. *Biochemical Genetics*, **1**, 323–335.

Cattanach, B.M. (1974). Position effect variegation in the mouse. *Genetical Research*, **23**, 291–306.

Cattanach, B.M. (1986). Parental origin effects in mice. *Journal of Embryology and Experimental Morphology*, **97** (Supplement), 137–150.

Cavalli-Sforza, L.L. and Feldman, M.W. (1981). *Cultural transmission and evolution: a quantitative approach*. Princeton University Press.

Chandler, V.L. and Walbot, V. (1986). DNA modification of a maize transposable element correlates with loss of activity. *Proceedings of the National Academy of Sciences, USA*, **83**, 1767–1771.

Chandley, A.C. (1986). A model for effective pairing and recombination at meiosis based on early replicating sites (R-bands) along chromosomes. *Human Genetics*, **72**, 50–57.

Chandra, H.S. and Nanjundiah, V. (1990). The evolution of genomic imprinting. *Development*, 1990 Supplement, 47–53.

Charlesworth, D., Charlesworth, B., and Bull, J.J. (1988). Origin of mutants disputed. *Nature*, **336**, 525.

Charlesworth, B., Coyne, J.A., and Orr, H.A. (1993). Meiotic drive and unisexual hybrid sterility: a comment. *Genetics*, **133**, 421–424.

Chomet, P.S., Wessler, S., and Dellaporta, S.L. (1987). Inactivation of the maize transposable element *Activator* (*Ac*) is associated with its DNA modification. *EMBO Journal*, **6**, 295–302.

Churchill, F.B. (1974). William Johannsen and the genotype concept. *Journal of the History of Biology*, **7**, 5–30.

Conklin, K.F. and Groudine, M. (1984). Chromatin structure and gene expression. In *DNA Methylation: biochemistry and biological significance*, (ed. A. Razin, H. Cedar, and A.D. Riggs), pp. 293–351. Springer-Verlag, New York.

Cook, P.R. (1974). On the inheritance of differentiated traits. *Biological Reviews*, **49**, 51–84.

Cooper, D.W., VandeBerg, J.L., Sharman, G.B., and Poole, W.E. (1971). Phosphoglycerate kinase polymorphism in kangaroos provides further evidence for paternal *X* inactivation. *Nature New Biology*, **230**, 155–157.

Cooper, D.W., Johnston, P.G., Sharman, G.B., and VanderBerg, J.L. (1977). The control of gene activity on eutherian and metatherian X chromosomes: a comparison. In *Reproduction and evolution*, (ed. J.H. Calaby and C.H. Tyndale-Biscoe), pp. 81–87. Australian Academy of Sciences, Canberra.

Cope, E.D. (1904). *The primary factors of organic evolution*. Open Court Company, London.

Corbet, S.A. (1985). Insect chemosensory responses: a chemical legacy hypothesis. *Ecological Entomology*, **10**, 143–153.

Coyne, J.A. and Orr, H.A. (1989). Two rules of speciation. In *Speciation and its consequences*, (ed. J. Otte and J.A. Endler), pp. 180–207. Sinauer, Sunderland, Mass.

Coyne, J.A. and Orr, H.A. (1993). Further evidence against meiotic-drive models of hybrid sterility. *Evolution*, **47**, 685–687.

Coyne, J.A., Charlesworth, B., and Orr, H.A. (1991). Haldane's rule revisited. *Evolution*, **45**, 1710–1714.

Crick, F.H.C. (1958). On protein synthesis. *Symposia of the Society for Experimental Biology*, **12**, 138–163.

Crosby, J.L. (1956). A suggestion concerning the possible role of plasmagenes in the inheritance of acquired adaptations. *Journal of Genetics*, **54**, 1–8.

Crouse, H.V. (1960). The controlling element in sex chromosome behavior in Sciara. *Genetics*, **45**, 1429–1443.

Crow, J.F. (1991). Why is Mendelian segregation so exact? *BioEssays*, **13**, 305–312.

Cullis, C.A. (1984). Environmentally induced DNA changes. In *Evolutionary theory: paths into the future*, (ed. J.W. Pollard), pp. 203–216. Wiley, Chichester.

Darlington, C.D. (1953). *The facts of life*. Allen and Unwin, London.

Darwin, C. (1859). *On the origin of species by means of natural selection, or the preservation of favoured races in the struggle for life*. (6th edn, 1872). John Murray, London. Republished Watts and Co., 1929, London.

Darwin, C. (1868). *The variation of animals and plants under domestication*, 2 vols. John Murray, London.

Darwin, C. (1871). *The descent of man, and selection in relation to sex*. John Murray, London.

Davies, D.R. (1974). Chromosome elimination in inter-specific hybrids. *Heredity*, **32**, 267–270.

Davies, K. (1993). Triplet repeats on the rise. *Nature*, **364**, 88.

Davis, B.D. (1989). Transcriptional bias: a non-Lamarckian mechanism for substrate-induced mutations. *Proceedings of the National Academy of Sciences, USA*, **86**, 5005–5009.

Dawkins, R. (1976). *The selfish gene*. Oxford University Press.

Dawkins, R. (1982). *The extended phenotype*. Freeman, Oxford.

Dawkins, R. (1986). *The blind watchmaker*. Longman, Essex.

Dean, A.C.R. and Hinshelwood C. (1963). Integration of cell reactions. *Nature*, **199**, 7–11.

de Beer, G.R. (1930). *Embryology and evolution*. Clarendon Press, Oxford.

de Beer, G.R. (1940). *Embryos and ancestors*. Clarendon Press, Oxford.

DeChiara, T.M., Robertson, E.J., and Efstratiadis, A. (1991). Parental imprinting of the mouse insulin-like growth factor II gene. *Cell*, **64**, 849–859.

Delage, Y. and Goldsmith, M. (1912). *The theories of evolution*, (trans. A. Tridon from the 1909 French edn). Frank Palmer, London.

Delbrück, M. (1949). In the discussion of T.M. Sonneborn and G.H. Beale: Influence des gènes, des plasmagènes et du milieu dans le déterminisme des caractères antigéniques chez *Paramecium aurelia* (variété 4). In *Unités Biologiques Douées de Continuité Génétique*, Colloques Internationaux du Centre National de la Recherche Scientifique, VIII, pp. 33–34.

Dennis, E.S. and Brettell, R.I.S. (1990). DNA methylation of maize transposable elements is correlated with activity. *Philosophical Transactions of the Royal Society of London Series B*, **326**, 217–229.

Detlefsen, J.A. (1925). The inheritance of acquired characters. *Physiological Reviews*, **5**, 244–278.

Deumling, B. and Clermont, L. (1989). Changes in DNA content and chromosomal size during cell culture and plant regeneration of *Scilla siberica*: selective chromatin diminution in response to environmental conditions. *Chromosoma*, **97**, 439–448.

Diehl, S.R. and Bush, G.L. (1989). The role of habitat preference in adaptation and speciation. In *Speciation and its consequences*, (ed. D. Otte and J.A. Endler), pp. 345–365. Sinauer, Sunderland, Mass.

Dittrich, B., Buiting, K., Groß, S., and Horsthemke, B. (1993). Characterization of a methylation imprint in the Prader-Willi syndrome chromosome region. *Human Molecular Genetics*, **2**, 1995–1999.

Dobzhansky, T. (1951). *Genetics and the origin of species*, (3rd edn). Columbia University Press, New York.

Doerfler, W. (1991). Patterns of DNA methylation—evolutionary vestiges of foreign DNA inactivation as a host defense mechanism. *Biological Chemistry Hoppe-Seyler*, **372**, 557–564.

Dooner, H.K., Robbins, T.P., and Jorgensen, R.A. (1991). Genetic and developmental control of anthocyanin biosynthesis. *Annual Review of Genetics*, **25**, 173–199.

Dorn, R., Krauss, V., Reuter, G., and Saumweber, H. (1993). The enhancer of position-effect variegation of *Drosphila, E(var)3–93D*, codes for a chromatin protein containing a conserved domain common to several transcriptional regulators. *Proceedings of the National Academy of Sciences, USA*, **90**, 11376–11380.

Dougherty, E.C. (1955). Comparative evolution and origin of sexuality. *Systematic Zoology*, **4**, 145–169, 190.

Dover, G. (1982). Molecular drive: a cohesive mode of species evolution. *Nature*, **299**, 111–117.

Dover, G.A. (1986). Molecular drive in multigene families: how biological novelties arise, spread and are assimilated. *Trends in Genetics*, **2**, 159–165.

Dover, G.A. and Tautz, D. (1986). Conservation and divergence in multigene families: alternatives to selection and drift. *Philosophical Transactions of the Royal Society Series B*, **312**, 275–289.

Drake, J.W. (1991). Spontaneous mutation. *Annual Review of Genetics*, **25**, 125–146.

Durham, W.H. (1991). *Coevolution: genes, culture and human diversity*. Stanford University Press, California.

Durica, D.S. and Krider, H.M. (1978). Studies on the ribosomal RNA cistrons in interspecific Drosophila hybrids. II. Heterochromatic regions mediating nucleolar dominance. *Genetics*, **89**, 37–64.

Durrant, A. (1971). Induction and growth of flax genotrophs. *Heredity*, **27**, 277–298.

Dyson, F.J. (1985). *Origins of life*. Cambridge University Press.

Eissenberg, J.C. and Elgin, S.C.R. (1991). Boundary functions in the control of gene expression. *Trends in Genetics*, **7**, 335–340.

Engler, P., Haasch, D., Pinkert, C.A., Doglio, L., Glymour, M., Brinster, R., and Storb, U. (1991). A strain-specific modifier on mouse chromosome 4 controls the methylation of independent transgene loci. *Cell*, **65**, 939–947.

Ephrussi, B. (1958). The cytoplasm and somatic cell variation. *Journal of Cellular and Comparative Physiology*, **52**, Supplement 1, 35–53.

Epstein, H., James, T.C., and Singh, P.B. (1992). Cloning and expression of *Drosophila* HP1 homologs from a mealybug, *Planococcus citri*. *Journal of Cell Science*, **101**, 463–474.

Eshel, I. (1985). Evolutionary genetic stability of Mendelian segregation and the role of free recombination in the chromosomal system. *American Naturalist*, **125**, 412–420.

Ettinger, L. (1986). Meiosis: a selection stage preserving the genome's pattern of organization. *Evolutionary Theory*, **8**, 17–26.

Ettinger, L. and Doljanski, F. (1992). On the generation of form by the continuous interactions between cells and their extracellular matrix. *Biological Reviews*, **67**, 459–489.

Evenson, D.P. and Prescott, D.M. (1970). Disruption of DNA synthesis in *Euplotes* by heat shock. *Experimental Cell Research*, **63**, 245–252.

Falk, R. (1994). The struggle for independence in genetics. *Journal of the History of Biology* (in press).

Fedor, M.J. (1992). Chromatin structure and gene expression. *Current Opinion in Cell Biology*, **4**, 436–443.

Fedoroff, N.V. (1989). About maize transposable elements and development. *Cell*, **56**, 181–191.

Fedoroff, N., Masson, P., and Banks, J.A. (1989). Mutations, epimutations, and the developmental programming of the maize *Suppressor-mutator* transposable element. *BioEssays*, **10**, 139–144.

Field, L.M., Devonshire, A.L., Ffrench-Constant, R.H., and Forde, B.G. (1989). Changes in DNA methylation are associated with loss of insecticide resistance in the peach-potato aphid *Myzus persicae* (Sulz.). *FEBS Letters*, **243**, 323–327.

Filipski, J. (1988). Why the rate of silent codon substitutions is variable within a vertebrate's genome. *Journal of Theoretical Biology*, **134**, 159–164.

Finch, C.E. (1990). *Longevity, senescence, and the genome*. University of Chicago Press.

Fincham, J.R.S., Connerton, I.F., Notarianni, E., and Harrington, K. (1989). Premeiotic disruption of duplicated and triplicated copies of the *Neurospora crassa am* (glutamate dehydrogenase) gene. *Current Genetics*, **15**, 327–334.

Fisher, R.C. (1988). An inordinate fondness for beetles. *Biological Journal of the Linnean Society*, **35**, 313–319.

Fitch, W.M. (1982). The challenges to Darwinism since the last centennial and the impact of molecular studies. *Evolution*, **36**, 1133–1143.

Fitch, W.M. and Atchley, W.R. (1985*a*). Evolution in inbred strains of mice appears rapid. *Science*, **228**, 1169–1175.

Fitch, W.M. and Atchley, W.R. (1985*b*). Rapid mutations in mice? *Science*, **230**, 1408–1409.

Flavell, R.B. and O'Dell, M. (1990). Variation and inheritance of cytosine methylation patterns in wheat at the high molecular weight glutenin and ribosomal RNA loci. *Development*, 1990 Supplement, 15–20.

Forejt, J. and Gregorová, S. (1992). Genetic analysis of genomic imprinting: an *Imprintor-1* gene controls inactivation of the paternal copy of the mouse *Tme* locus. *Cell*, **70**, 443–450.

Foster, P.L. (1992). Directed mutation: between unicorns and goats. *Journal of Bacteriology*, **174**, 1711–1716.

Foster, P.L. and Cairns, J. (1992). Mechanisms of directed mutation. *Genetics*, **131**, 783–789.

Fothergill, P.G. (1952). *Historical aspects of organic evolution*. Hollis and Carter, London.

Francis, A. and Jones, R.N. (1989). Heritable nature of colchicine induced variation in diploid *Lolium perenne*. *Heredity*, **62**, 407–410.

Frank, S.A. (1991*a*). Divergence of meiotic drive-suppression systems as an explanation for sex-biased hybrid sterility and inviability. *Evolution*, **45**, 262–267.

Frank, S.A. (1991*b*). Haldane's rule: a defense of the meiotic drive theory. *Evolution*, **45**, 1714–1717.

Frankel, J. (1983). What are the developmental underpinnings of evolutionary changes in protozoan evolution? In *Development and Evolution*, (ed. B.C. Goodwin, N. Holder, and C.C. Wylie), pp. 279–314. Cambridge University Press.

Frankel, J. (1989). *Pattern formation: ciliate studies and models*. Oxford University Press, New York.

Frankel, J. (1990). Positional order and cellular handedness. *Journal of Cell Science*, **97**, 205–211.

Futuyma, D.J. and Mayer, G.C. (1980). Non-allopatric speciation in animals. *Systematic Zoology*, **29**, 254–271.

Garrard, W.T. (1991). Histone H1 and the conformation of transcriptionally active chromatin. *BioEssays*, **13**, 87–88.

Gaunt, S.J. and Singh, P.B. (1990). Homeogene expression patterns and chromosomal imprinting. *Trends in Genetics*, **6**, 208–212.

Gerbi, S.A. (1986). Unusual chromosome movements in Sciarid flies. In *Germ line—soma differentiation. Results and problems in cell differentiation*, Vol. 13, (ed. W. Hennig), pp. 71–104. Springer-Verlag, Berlin.

Gilbert, S.F. (1991*a*). *Developmental biology*, (3rd edn). Sinauer, Sunderland, Mass.

Gilbert, S.F. (1991*b*). Induction and the origins of developmental genetics. In *Developmental biology: a comprehensive synthesis*, Vol. 7, *A conceptual history of modern embryology*, (ed. S.F. Gilbert), pp. 181–206. Plenum, New York.

Gluecksohn-Schoenheimer, S. (1949). Causal analysis of mouse development by the study of mutational effects. *Growth*, **13** (Supplement), 163–176.

Gold, J.R., Li, Y., Schmidt, T.R., and Tave, D. (1991). Nucleolar dominance in interspecific hybrids of cyprinid fishes. *Cytobios*, **65**, 139–147.

Goldman, M.A. (1988). The chromatin domain as a unit of gene regulation. *BioEssays*, **9**, 50–55.

Goldschmidt, R. (1940). *The material basis of evolution*. Yale University Press, New Haven, Conn.

Gottlieb, G. (1992). *Individual development and evolution: the genesis of novel behavior*. Oxford University Press, New York.

Gottschling, D.E., Aparicio, O.M., Billington, B.L., and Zakian, V.A. (1990). Position effect at *S. cerevisiae* telomeres: reversible repression of Pol II transcription. *Cell*, **63**, 751–762.

Gould, J.L. and Gould, C.G. (1989). *Sexual selection*. Scientific American Library, New York.

Gould, S.J. (1977). *Ontogeny and phylogeny*. Belknap Press, Cambridge, Mass.

Gould, S.J. (1982). Darwinism and the expansion of evolutionary theory. *Science*, **216**, 380–387.

Gould, S.J. (1989). *Wonderful life*. Norton, New York.

Grafen, A. (1988). Origin of mutants disputed. *Nature*, **336**, 525–526.

Green, M.C., Bailey, D.W., Green, E.L., Roderick, T.H., and Russell, E.S. (1985). Rapid mutations in mice? *Science*, **230**, 1407–1408.

Grell, R.F. (1971). Heat-induced exchange in the fourth chromosome of diploid females of *Drosophila melanogaster*. *Genetics*, **69**, 523–527.

Grell, R.F. (1978*a*). A comparison of heat and interchromosomal effects on recombination and interference in *Drosophila melanogaster*. *Genetics*, **89**, 65–77.

Grell, R.F. (1978*b*). High frequency recombination in centromeric and histone regions of *Drosophila* genomes. *Nature*, **272**, 78–80.

Griesemer, J.R. and Wimsatt, W.C. (1989). Picturing Weismannism: a case study of conceptual evolution. In *What the philosophy of biology is*, (ed. M. Ruse), pp. 75–137. Kluwer Academic Publishers, Dordrecht.

Grimes, S.R. (1986). Nuclear proteins in spermatogenesis. *Comparative Biochemistry and Physiology*, **83B**, 495–500.

Grimes, G.W. and Aufderheide, K.J. (1991). *Cellular aspects of pattern formation: the problem of assembly. Monographs in Developmental Biology*, Vol. 22. Karger, Basel.

Groeters, F.R. and Shaw, D.D. (1992). Association between latitudinal variation for embryonic development time and chromosome structure in the grasshopper *Caledia captiva* (Orthoptera: Acrididae). *Evolution*, **46**, 245–257.

Gross, D.S. and Garrard, W.T. (1987). Poising chromatin for transcription. *Trends in Biochemical Sciences*, **12**, 293–297.

Gross, D.S. and Garrard, W.T. (1988). Nuclease hypersensitive sites in chromatin. *Annual Review of Biochemistry*, **57**, 159–197.

Groudine, M. and Conklin, K.F. (1985). Chromatin structure and de novo methylation of sperm DNA: implications for activation of the paternal genome. *Science*, **228**, 1061–1068.

Groudine, M. and Weintraub, H. (1982). Propagation of globin DNAase I-hypersensitive sites in absence of factors required for induction: a possible mechanism for determination. *Cell*, **30**, 131–139.

Groves, C.P. (1989). *A theory of human and primate evolution*. Oxford University Press.

Grüneberg, H. (1970). Is there a viral component in the genetic background? *Nature*, **225**, 39–41.

Haack, H. and Hodgkin, J. (1991). Tests for parental imprinting in the nematode *Caenorhabditis elegans. Molecular and General Genetics*, **228**, 482–485.

Hadchouel, M., Farza, H., Simon, D., Tiollais, P., and Pourcel, C. (1987). Maternal inhibition of hepatitis B surface antigen gene expression in transgenic mice correlates with *de novo* methylation. *Nature*, **329**, 454–456.

Hägele, K. and Oschmann, B. (1987). Non-reciprocal gonadal dysgenesis in hybrids of the chironomid midge *Chironomus thummi*. III. Germ line specific abnormalities. *Chromosoma*, **96**, 50–54.

Haig, D. and Grafen, A. (1991). Genetic scrambling as a defence against meiotic drive. *Journal of Theoretical Biology*, **153**, 531–558.

Haig, D. and Graham, C. (1991). Genomic imprinting and the strange case of the insulin-like growth factor II receptor. *Cell*, **64**, 1045–1046.

Haig, D. and Westoby, M. (1989). Parent-specific gene expression and the triploid endosperm. *American Naturalist*, **134**, 147–155.

Haigh, L.S., Owens, B.B., Hellewell, S., and Ingram, V.M. (1982). DNA methylation in chicken α-globin gene expression. *Proceedings of the National Academy of Sciences, USA*, **79**, 5332–5336.

Haldane, J.B.S. (1922). Sex ratio and unisexual sterility in hybrid animals. *Journal of Genetics*, **12**, 101–109.

Haldane, J.B.S. (1932). *The causes of evolution*. Longmans, Green and Company, London. (Reprinted by Cornell University Press, New York, 1966).

Hall, B.G. (1988). Adaptive evolution that requires multiple spontaneous mutations. I. Mutations involving an insertion sequence. *Genetics*, **120**, 887–897.

Hall, B.G. (1990). Spontaneous point mutations that occur more often when advantageous than when neutral. *Genetics*, **126**, 5–16.

Hall, B.G. (1991). Adaptive evolution that requires multiple spontaneous mutations: mutations involving base substitutions. *Proceedings of the National Academy of Sciences, USA*, **88**, 5882–5886.

Hall, B.K. (1992a). Waddington's legacy in development and evolution. *American Zoologist*, **32**, 113–122.

Hall, B.K. (1992b). *Evolutionary developmental biology*. Chapman and Hall, London.

Hall, J.G. (1990). Genomic imprinting: review and relevance to human diseases. *American Journal of Human Genetics*, **46**, 857–873.

Hamburger, V. (1980). Embryology and the modern synthesis in evolutionary theory. In *The evolutionary synthesis: perspectives on the unification of biology*, (ed. E. Mayr and W.B. Provine), pp. 97–112. Harvard University Press, Cambridge, Mass.

Hämmerling, J. (1929). Dauermodifikationen. *Handbuch der Vererbungswissenschaft*, **1**(E), 1–69.

Hanawalt, P.C. (1987). Preferential DNA repair in expressed genes. *Environmental Health Perspectives*, **76**, 9–14.

Hanawalt, P.C. (1989). Preferential repair of damage in actively transcribed DNA sequences *in vivo*. *Genome*, **31**, 605–611.

Hansen, J.C. and Ausio, J. (1992). Chromatin dynamics and the modulation of genetic activity. *Trends in Biochemical Sciences*, **17**, 187–191.

Harris, M. (1989). Phenotypic changes in cell culture. In *Developmental biology: a comprehensive synthesis, Vol. 6, Genomic adaptability in somatic cell specialization*, (ed. M.A. DiBerardino and L.D. Etkin), pp. 79–95. Plenum Press, New York.

Harwood, J. (1993). *Styles of scientific thought: the German genetics community 1900–1933*. University of Chicago Press.

Hayes, J.J. and Wolffe, A.P. (1992). The interaction of transcription factors with nucleosomal DNA. *BioEssays*, **14**, 597–603.

Hayes, W. (1968). *The genetics of bacteria and their viruses*, (2nd edn). Blackwell, Oxford.

Hayman, D.L., Moore, H.D.M., and Evans, E.P. (1988). Further evidence of novel sex differences in chiasma distribution in marsupials. *Heredity*, **61**, 455–458.

Hayward, M.D. and Breese, E.L. (1968). The genetic organisation of natural populations of *Lolium perenne* L. III. Productivity. *Heredity*, **23**, 357–368.

Hemmer, H. (1990). *Domestication: the decline of environmental appreciation*. Cambridge University Press.

Henikoff, S. (1992). Position effect and related phenomena. *Current Opinion in Genetics and Development*, **2**, 907–912.

Hergersberg, M. (1991). Biological aspects of cytosine methylation in eukaryotic cells. *Experientia*, **47**, 1171–1185.

Herman, R.K. and Dworkin, N.B. (1971). Effect of gene induction on the rate of mutagenesis by ICR-191 in *Escherichia coli*. *Journal of Bacteriology*, **106**, 543–550.

Heslop-Harrison, J.S. (1990). Gene expression and parental dominance in hybrid plants. *Development*, 1990 Supplement, 21–28.

Highkin, H.R. (1958*a*). Transmission of phenotypic variability within a pure line. *Nature*, **182**, 1460.

Highkin, H.R. (1958*b*). Temperature-induced variability in peas. *American Journal of Botany*, **45**, 626–631.

Hinshelwood, C.N. (1946). *The chemical kinetics of the bacterial cell.* Clarendon Press, Oxford.

Hinshelwood, C.N. (1953). Adaptation in micro-organisms and its relation to evolution. *Symposia of the Society for Experimental Biology*, **7**, 31–42.

Hitzeroth, H., Klose, J., Ohno, S., and Wolf, U. (1968). Asynchronous activation of parental alleles at the tissue-specific gene loci observed on hybrid trout during early development. *Biochemical Genetics*, **1**, 287–300.

Hjelm, K.K. (1986). Is non-genic inheritance involved in carcinogenesis? A cytotactic model of transformation. *Journal of Theoretical Biology*, **119**, 89–101.

Ho, L., Bohr, V.A., and Hanawalt, P.C. (1989). Demethylation enhances removal of pyrimidine dimers from the overall genome and from specific DNA sequences in Chinese hamster ovary cells. *Molecular and Cellular Biology*, **9**, 1594–1603.

Ho, M.W., Tucker, C., Keeley, D., and Saunders, P.T. (1983). Effects of successive generations of ether treatment on penetrance and expression of the *Bithorax* phenocopy in *Drosophila melanogaster. Journal of Experimental Zoology*, **225**, 357–368.

Hoffmann, A.A. and Parsons, P.A. (1991). *Evolutionary genetics and environmental stress.* Oxford University Press.

Holliday, R. (1984). The biological significance of meiosis. In *Controlling events in meiosis*, (ed. C.W. Evans and H.G. Dickinson), pp. 381–394. Company of Biologists, Cambridge.

Holliday, R. (1987). The inheritance of epigenetic defects. *Science*, **238**, 163–170.

Holliday, R. (1988). A possible role for meiotic recombination in germ line reprogramming and maintenance. In *The evolution of sex*, (ed. R.E. Michod and B.R. Levin), pp. 45–55. Sinauer, Sunderland, Mass.

Holliday, R. (1990*a*). Mechanisms for the control of gene activity during development. *Biological Reviews*, **65**, 431–471.

Holliday, R. (1990*b*). Genomic imprinting and allelic exclusion. *Development*, 1990 Supplement, 125–129.

Holliday, R. and Rosenberger, R.F. (1988). Origin of mutants disputed. *Nature*, **336**, 526.

Holmquist, G. (1988). DNA sequences in G-bands and R-bands. In *Chromosomes and chromatin*, Vol. 2, (ed. K.W. Adolph), pp. 75–121. CRC Press, Boca Raton, Florida.

Holmquist, G.P. (1989). Evolution of chromosome bands: molecular ecology of noncoding DNA. *Journal of Molecular Evolution*, **28**, 469–486.

Holmquist, G.P. (1992). Chromosome bands, their chromatin flavors, and their functional features. *American Journal of Human Genetics*, **51**, 17–37.

Hsieh, C.L. and Lieber, M.R. (1992). CpG methylated minichromosomes become inaccessible for V(D)J recombination after undergoing replication. *EMBO Journal*, **11**, 315–325.

Hull, D.L. (1984). Lamarck among the Anglos. Introductory essay in *Zoological Philosophy*, (J.B. Lamarck, trans. H. Elliot), pp. xl–lxvi. Chicago University Press.

Hull, D.L. (1988). Interactors versus vehicles. In *The role of behavior in evolution*, (ed. H.C. Plotkin), pp. 19–50. MIT Press, Cambridge, Mass.

Hulten, M.A. and Hall, J.G. (1990). Proposed meiotic mechanism of genomic imprinting. *Chromosomes Today*, Vol. 10, (ed. K. Fredga, B.A. Kihlman, and M.D. Bennett), pp. 157–162. Unwin Hyman, London.

Hurst, L.D. (1992). Intragenomic conflict as an evolutionary force. *Proceedings of the Royal Society of London Series B*, **248**, 135–140.

Hurst, L.D. and Pomiankowski, A. (1991). Causes of sex ratio bias may account for unisexual sterility in hybrids: a new explanation of Haldane's rule and related phenomena. *Genetics*, **128**, 841–858.

Huxley, J. (1942). *Evolution: the modern synthesis*. Allen and Unwin, London.

Jablonka, E. (1994). Inheritance systems and the evolution of new levels of individuality. *Journal of Theoretical Biology*, **170**, 301–309.

Jablonka, E. and Lamb, M.J. (1988). Meiotic pairing constraints and the activity of sex chromosomes. *Journal of Theoretical Biology*, **133**, 23–36.

Jablonka, E. and Lamb, M.J. (1989). The inheritance of acquired epigenetic variations. *Journal of Theoretical Biology*, **139**, 69–83.

Jablonka, E. and Lamb, M.J. (1990*a*). The evolution of heteromorphic sex chromosomes. *Biological Reviews*, **65**, 249–276.

Jablonka, E. and Lamb, M.J. (1990*b*). Lamarckism and ageing. *Gerontology*, **36**, 323–332.

Jablonka E. and Lamb, M.J. (1991). Sex chromosomes and speciation. *Proceedings of the Royal Society of London Series B*, **243**, 203–208.

Jablonka, E., Goitein, R., Sperling, K., Cedar, H., and Marcus, M. (1987). 5-aza-C-induced changes in the time of replication of the X chromosomes of *Microtus agrestis* are followed by non-random reversion to a late pattern of replication. *Chromosoma*, **95**, 81–88.

Jablonka, E., Lachmann, M., and Lamb, M.J. (1992). Evidence, mechanisms and models for the inheritance of acquired characters. *Journal of Theoretical Biology*, **158**, 245–268.

Jablonka Tavory, E. (1982). Genocopies and the evolution of interdependence. *Evolutionary Theory*, **6**, 167–170.

Jaenisch, R. (1988). Transgenic animals. *Science*, **240**, 1468–1474.

Jennings, H.S. (1909). Heredity and variation in the simplest organisms. *American Naturalist*, **43**, 321–337.

Jennings, H.S. (1910). Experimental evidence of the effectiveness of selection. *American Naturalist*, **44**, 136–145.

Jennings, H.S. (1937). Formation, inheritance and variation of the teeth in *Difflugia corona*. A study of the morphogenic activities of rhizopod protoplasm. *Journal of Experimental Zoology*, **77**, 287–336.

Jerka-Dziadosz, M. and Beisson, J. (1990). Genetic approaches to ciliate pattern formation: from self-assembly to morphogenesis. *Trends in Genetics*, **6**, 41–45.

Johannsen, W. (1911). The genotype conception of heredity. *American Naturalist*, **45**, 129–159.

Johannsen, W. (1923). Some remarks about units in heredity. *Hereditas*, **4**, 133–141.

John, B. (1988). The biology of heterochromatin. In *Heterochromatin: molecular and structural aspects*, (ed. R.S. Verma), pp. 1–147. Cambridge University Press.

Johnson, F.M., Aquadro, C.F., Skow, L.C., Langley, C.H., and Lewis, S.E. (1985). Rapid mutations in mice? *Science*, **230**, 1406–1407.

Johnson, N.A., and Wu, C.-I. (1992). An empirical test of the meiotic drive models of hybrid sterility: sex-ratio data from hybrids between *Drosophila simulans* and *Drosophila sechellia*. *Genetics*, **130**, 507–511.

Jollos, V. (1921). Experimentelle Protistenstudien 1. Untersuchungen über Variabilität und Vererbung bei Infusorien. *Archiv für Protistenkunde*, **43**, 1–222.

Jollos, V. (1934). Inherited changes produced by heat-treatment in *Drosophila melanogaster*. *Genetica*, **16**, 476–494.

Jones, P.A. (1985). Altering gene expression with 5-azacytidine. *Cell*, **40**, 485–486.

Jones, R.N. and Rees, H. (1982). *B chromosomes*. Academic Press, London.

Jordanova, L.J. (1984). *Lamarck*. Oxford University Press.

Jorgensen, R. (1990). Altered gene expression in plants due to *trans* interactions between homologous genes. *Trends in Biotechnology*, **8**, 340–344.

Jorgensen, R. (1993). The germinal inheritance of epigenetic information in plants. *Philosophical Transactions of the Royal Society Series B*, **339**, 173–181.

Kafri, T., Ariel, M., Brandeis, M., Shemer, R., Urven, L., McCarrey, J., Cedar, H., and Razin, A. (1992). Developmental pattern of gene-specific DNA methylation in the mouse embryo and germ line. *Genes and Development*, **6**, 705–714.

Kahn, A.J. (1982). Alterability of development of hemoglobin concentration in mice: transmission of changes to the next generation. *Growth*, **46**, 247–258.

Kaslow, D.C. and Migeon, B.R. (1987). DNA methylation stabilizes X chromosome inactivation in eutherians but not in marsupials: evidence for multistep maintenance of mammalian X dosage compensation. *Proceedings of the National Academy of Sciences, USA*, **84**, 6210–6214.

Kauffman, S.A. (1993). *The origins of order*. Oxford University Press, New York.

Keller, E.F. (1992). Between language and science: the question of directed mutation in molecular genetics. *Perspectives in Biology and Medicine*, **35**, 292–306.

Kellum, R. and Schedl, P. (1991). A position-effect assay for boundaries of higher order chromosomal domains. *Cell*, **64**, 941–950.

Kermicle, J.L. (1970). Dependence of the *R*-mottled aleurone phenotype in maize on mode of sexual transmission. *Genetics*, **66**, 69–85.

Kermicle, J.L. (1978). Imprinting of gene action in maize endosperm. In *Maize breeding and genetics*, (ed. D.B. Walden), pp. 357–371. Wiley, New York.

Kimura, M. (1983). *The neutral theory of molecular evolution*. Cambridge University Press.

King, M. (1993). *Species evolution: the role of chromosome change*. Cambridge University Press.

Kirkwood, T.B.L. and Holliday, R. (1979). The evolution of ageing and longevity. *Proceedings of the Royal Society of London Series B*, **205**, 531–546.

Kitsberg, D., Selig, S., and Cedar, H. (1991). Chromosome structure and eukaryotic gene organization. *Current Opinion in Genetics and Development*, **1**, 534–537.

Kitsberg, D., Selig, S., Brandeis, M., Simon, I., Keshet, I., Driscoll, D.J., Nicholls, R.D., and Cedar, H. (1993). Allele-specific replication timing of imprinted gene regions. *Nature*, **364**, 459–463.

Klar, A.J.S. (1990). The developmental fate of fission yeast cells is determined by the pattern of inheritance of parental and grandparental DNA strands. *EMBO Journal*, **9**, 1407–1415.

Klekowski, E.J. (1988). *Mutation, developmental selection, and plant evolution*. Columbia University Press, New York.

Koch, A.L. (1993). Genetic response of microbes to extreme challenges. *Journal of Theoretical Biology*, **160**, 1–21.

Kohane, M.J. and Parsons, P.A. (1988). Domestication: evolutionary change under stress. *Evolutionary Biology*, **23**, 31–48.

Kornberg, R.D. and Lorch, Y. (1992). Chromatin structure and transcription. *Annual Review of Cell Biology*, **8**, 563–587.

Kricker, M.C., Drake, J.W., and Radman, M. (1992). Duplication-targeted DNA methylation and mutagenesis in the evolution of eukaryotic chromosomes. *Proceedings of the National Academy of Sciences, USA*, **89**, 1075–1079.

Kuhl, D.P.A. and Caskey, C.T. (1993). Trinucleotide repeats and genome variation. *Current Opinion in Genetics and Development*, **3**, 404–407.

Kuhn, D.T. and Packert, G. (1988). Paternal imprinting of inversion *Uab*[1] causes homeotic transformations in Drosophila. *Genetics*, **118**, 103–107.

Kunze, R., Starlinger, P., and Schwartz, D. (1988). DNA methylation of the maize transposable element *Ac* interferes with its transcription. *Molecular and General Genetics*, **214**, 325–327.

Lachmann, M. and Jablonka, E. Heredity systems in a periodically fluctuating environment. (Unpublished manuscript).

Laird, C.D. (1989). From polytene chromosomes to human embryology: connections via the human fragile-X syndrome. *American Zoologist*, **29**, 569–591.

Lamarck, J.B. (1809). *Philosophie zoologique, ou exposition des considérations relatives à l'histoire naturelle des animaux*, Dentu, Paris. English edition *Zoological Philosophy, an exposition with regard to the natural history of animals*, (trans. H. Elliot). Macmillan, London, 1914. Reprinted by the University of Chicago Press, 1984.

Lambert, D.M., Stevens, P.M., White, C.S., Gentle, M.T., Philips, N.R., Millar, C.D., Barker, J.R., and Newcomb, R.D. (1989). Phenocopies, heredity and evolution. *Evolutionary Theory*, **8**, 285–304.

Lande, R. (1979). Effective deme sizes during long-term evolution estimated from rates of chromosomal rearrangement. *Evolution*, **33**, 234–251.

Lande, R. (1985). The fixation of chromosomal rearrangements in a subdivided population with local extinction and colonization. *Heredity*, **54**, 323–332.

Landman, O.E. (1991). The inheritance of acquired characteristics. *Annual Review of Genetics*, **25**, 1–20.

Lansing, A.I. (1954). A nongenic factor in the longevity of rotifers. *Annals of the New York Academy of Sciences*, **57**, 455–464.

Latchman, D. (1990). *Gene regulation: a eukaryotic perspective*. Unwin Hyman, London.

Lederberg, J. and Lederberg, E.M. (1952). Replica plating and indirect selection of bacterial mutants. *Journal of Bacteriology*, **63**, 399–406.

Lenski, R.E. (1989). Are some mutations directed? *Trends in Ecology and Evolution*, **4**, 148–150.

Lenski, R.E. and Mittler, J.E. (1993). The directed mutation controversy and neo-Darwinism. *Science*, **259**, 188–194.

Lenski, R.E., Slatkin, M., and Ayala, F.J. (1989). Mutation and selection in bacterial populations: alternatives to the hypothesis of directed mutation. *Proceedings of the National Academy of Sciences, USA*, **86**, 2775–2778.

Levine, A., Cantoni, G.L., and Razin, A. (1991). Inhibition of promoter activity by methylation: a possible involvement of a protein mediator. *Proceedings of the National Academy of Sciences, USA*, **88**, 6515–6518.

Levine, A., Cantoni, G.L., and Razin, A. (1992). Methylation in the preinitiation domain affects gene transcription by an indirect mechanism. *Proceedings of the National Academy of Sciences, USA*, **89**, 10119–10123.

Lewin, R. (1983). How mammalian RNA returns to its genome. *Science*, **219**, 1052–1054.

Lewis, E.B. (1985). Regulation of the genes of the bithorax complex in *Drosophila*. *Cold Spring Harbor Symposia on Quantitative Biology*, **50**, 155–164.

Lewis, J. and Bird, A. (1991). DNA methylation and chromatin structure. *FEBS Letters*, **285**, 155–159.

Lewontin, R.C. (1970). The units of selection. *Annual Review of Ecology and Systematics*, **1**, 1–18.

Li, E., Beard, C., and Jaenisch, R. (1993). Role for DNA methylation in genomic imprinting. *Nature*, **366**, 362–365.

Li, W.-H. and Graur, D. (1991). *Fundamentals of molecular evolution*, Sinauer, Sunderland, Mass.

Lincoln, R.J., Boxshall, G.A., and Clark, P.F. (1982). *Dictionary of ecology, evolution and systematics.* Cambridge University Press.

Lindegren, C.C. (1949). *The yeast cell, its genetics and cytology.* Educational Publishers, Saint Louis.

Lindegren, C.C. (1966). *The cold war in biology.* Planarian Press, Ann Arbor, Michigan.

Linn, F., Heidmann, I., Saedler, H., and Meyer P. (1990). Epigenetic changes in the expression of the maize A1 gene in *Petunia hybrida*: role of numbers of integrated gene copies and state of methylation. *Molecular and General Genetics*, **222**, 329–336.

Lints, F.A. (1978). *Genetics and ageing.* Karger, Basel.

Lints, F.A. (1988). Parental age effects. In *Drosophila as a model organism for ageing studies*, (ed. F.A. Lints and M.H. Soliman), pp. 176–189. Blackie, Glasgow.

Locke, J., Kotarski, M.A., and Tartof, K.D. (1988). Dosage-dependent modifiers of position effect variegation in *Drosophila* and a mass action model that explains their effect. *Genetics*, **120**, 181–198.

Locke, M. (1988). Insect cells for the study of general problems in biology—somatic inheritance. *International Journal of Insect Morphology and Embryology*, **17**, 419–436.

Loidl, J. (1989). Effects of elevated temperature on meiotic chromosome synapsis in *Allium ursinum. Chromosoma*, **97**, 449–458.

Longo, F.J. (1987). *Fertilization.* Chapman and Hall, London.

Lowenhaupt, K., Rich, A., and Pardue, M.L. (1989). Nonrandom distribution of long mono- and dinucleotide repeats in Drosophila chromosomes: correlations with dosage compensation, heterochromatin, and recombination. *Molecular and Cellular Biology*, **9**, 1173–1182.

Lowrey, C.H, Bodine, D.M., and Nienhuis, A.W. (1992). Mechanism of DNase I hypersensitive site formation within the human globin locus control region. *Proceedings of the National Academy of Sciences, USA*, **89**, 1143–1147.

Lumsden, C.J. and Wilson, E.O. (1981). *Genes, mind, and culture. The coevolutionary process.* Harvard University Press, Cambridge, Mass.

Luria, S.E. and Delbrück, M. (1943). Mutations of bacteria from virus sensitivity to virus resistance. *Genetics*, **28**, 491–511.

Lyttle, T.W. (1991). Segregation distorters. *Annual Review of Genetics*, **25**, 511–557.

Lyttle, T.W. (1993). Cheaters sometimes prosper: distortion of mendelian segregation by meiotic drive. *Trends in Genetics*, **9**, 205–210.

Mackenzie, A. (1992). The evolutionary significance of host-mediated conditioning. *Antenna*, **16**, 141–150.

MacPhee, D. (1993). Directed evolution reconsidered. *American Scientist*, **81**, 554–561.

Mantovani, B. and Scali, V. (1992). Hybridogenesis and androgenesis in the stick-insect *Bacillus rossius-grandii benazzii* (Insecta, Phasmatodea). *Evolution*, **46**, 783–796.

Manuelidis, L. (1990). A view of interphase chromosomes. *Science*, **250**, 1533–1540.

Margulis, L. (1981). *Symbiosis in cell evolution.* Freeman, San Francisco.

Markson, Y., Weiss, D.W., and Doljanski, F. (1991*a*). Growth and proteoglycan metabolism of chick embryonic cartilaginous long bone rudiments and of isolated epiphyses. *Roux's Archives of Developmental Biology*, **200**, 202–207.

Markson, Y., Weiss, D.W., and Doljanski, F. (1991*b*). Resumption of growth of heat inactivated embryonic epiphyses by grafting: crucial morphogenetic contribution by the extracellular matrix? *Roux's Archives of Developmental Biology*, **200**, 208–212.

Martienssen, R., Barkan, A., Taylor, W.C., and Freeling, M. (1990). Somatically heritable switches in the DNA modification of *Mu* transposable elements monitored with a suppressible mutant in maize. *Genes and Development*, **4**, 331–343.

Martin, C. and Jones, J. (1992). Plant genetics flourishes. *Trends in Genetics*, **8**, 368–370.

Martin, C., Prescott, A., Lister, C., and MacKay, S. (1989). Activity of the transposon *Tam3* in *Antirrhinum* and tobacco: possible role of DNA methylation. *EMBO Journal*, **8**, 997–1004.

Martin, R. (1977). A possible genetic mechanism of aging, rejuvenation, and recombination in germinal cells. In *ICN-UCLA Symposia on molecular and cellular biology*, Vol. 7, *Molecular human cytogenetics*, (ed. R.S. Sparkes, D.E. Comings, and C.F. Fox), pp. 355–373. Academic Press, New York.

Martin, S.L. (1991). LINEs. *Current Opinion in Genetics and Development*, **1**, 505–508.

Mather, K. (1961). Nuclear materials and nuclear change in differentiation. *Nature*, **190**, 404–406.

Matsuda, Y., Hirobe, T., and Chapman, V.M. (1991). Genetic basis of X–Y chromosome dissociation and male sterility in interspecific hybrids. *Proceedings of the National Academy of Sciences, USA*, **88**, 4850–4854.

Matsuda, Y., Moens, P.B., and Chapman, V.M. (1992). Deficiency of X and Y chromosomal pairing at meiotic prophase in spermatocytes of sterile interspecific hybrids between laboratory mice (*Mus domesticus*) and *Mus spretus*. *Chromosoma*, **101**, 483–492.

Matzke, M.A. and Matzke, A.J.M. (1990). Gene interactions and epigenetic variation in transgenic plants. *Developmental Genetics*, **11**, 214–223.

Matzke, M.A., and Matzke, A.J.M. (1993). Genomic imprinting in plants: parental effects and *trans* activation phenomena. *Annual Review of Plant Physiology*, **44**, 53–76.

Matzke, M.A., Primig, M., Trnovsky, J., and Matzke, A.J.M. (1989). Reversible methylation and inactivation of marker genes in sequentially transformed tobacco plants. *EMBO Journal*, **8**, 643–649.

Maynard Smith, J. (1962). Disruptive selection, polymorphism and sympatric speciation. *Nature*, **195**, 60–62.

Maynard Smith, J. (1966). *The theory of evolution*, (2nd edn). Penguin Books, Harmondsworth, Middlesex.

Maynard Smith, J. (1988*a*). Evolutionary progress and levels of selection. In *Evolutionary progress*, (ed. M.H. Nitecki), pp. 219–230. University of Chicago Press.

Maynard Smith, J. (1988*b*). The evolution of recombination. In *The evolution of sex*, (ed. R.E. Michod and B.R. Levin), pp. 106–125. Sinauer, Sunderland, Mass.

Maynard Smith, J. (1989). Weismann and modern biology. In *Oxford surveys of evolutionary biology*, Vol. 6, (ed. P.H. Harvey and L. Partridge), pp. 1–12. Oxford University Press.

Maynard Smith, J. (1990). Models of a dual inheritance system. *Journal of Theoretical Biology*, **143**, 41–53.

Mayr, E. (1942). *Systematics and the origin of species*. Columbia University Press, New York.

Mayr, E. (1963). *Animal species and evolution*. Harvard University Press, Cambridge, Mass.

Mayr, E. (1982*a*). *The growth of biological thought*. Harvard University Press, Cambridge, Mass.

Mayr, E. (1982*b*). Processes of speciation in animals. In *Mechanisms of speciation*, (ed. C. Barigozzi), pp. 1–19. Alan R. Liss, New York.

Mayr, E. (1985). Weismann and evolution. *Journal of the History of Biology*, **18**, 295–329.

Mayr, E. (1988). On Weismann's growth as an evolutionist. In *Towards a new philosophy of biology*, pp. 491–524. Harvard University Press, Cambridge, Mass.

Mayr, E. (1992). A local flora and the biological species concept. *American Journal of Botany*, **79**, 222–238.

Mayr, E. and Provine, W.B. (ed.) (1980). *The evolutionary synthesis: perspectives on the unification of biology*. Harvard University Press, Cambridge, Mass.

McClintock, B. (1951). Chromosome organization and genic expression. *Cold Spring Harbor Symposia on Quantitative Biology*, **16**, 13–47.

McClintock, B. (1984). The significance of responses of the genome to challenge. *Science*, **226**, 792–801.

McGrath, J. and Solter, D. (1984). Completion of mouse embryogenesis requires both the maternal and paternal genomes. *Cell*, **37**, 179–183.

McKee, B.D. (1991). X-Y pairing, meiotic drive, and ribosomal DNA in *Drosophila melanogaster* males. *American Naturalist*, **137**, 332–339.

Medawar, P.B. (1957). A commentary on Lamarkism. In *The uniqueness of the individual*, pp. 79–107. Methuen, London. (Originally published in *Bulletin of the National Institute of Sciences of India*, **7**, 127, 1953).

Medawar, P. and Medawar, J. (1983). *Aristotle to zoos*. Harvard University Press, Cambridge, Mass.

Mehtali, M., LeMeur, M., and Lathe, R. (1990). The methylation-free status of a housekeeping transgene is lost at high copy number. *Gene*, **91**, 179–184.

Meins, F. (1983). Heritable variation in plant cell culture. *Annual Review of Plant Physiology*, **34**, 327–346.

Meins, F. (1985). Cell heritable changes during development. In *Plant genetics*, (ed. M. Freeling), pp. 45–59. Alan R. Liss, New York.

Meins, F. (1989*a*). A biochemical switch model for cell-heritable variation in cytokinin requirement. In *The molecular basis of plant development*, (ed. R. Goldberg), pp. 13–24. Alan R. Liss, New York.

Meins, F. (1989*b*). Habituation: heritable variation in the requirement of cultured plant cells for hormones. *Annual Review of Genetics*, **23**, 395–408.

Meins, F. and Foster, R. (1986). A cytokinin mutant derived from cultured tobacco cells. *Developmental Genetics*, **7**, 159–165.

Meistrich, M.L. and Brock, W.A. (1987). Proteins of the meiotic cell nucleus. In *Meiosis*, (ed. P.B. Moens), pp. 333–353. Academic Press, Orlando.

Mellon, I. and Hanawalt, P.C. (1989). Induction of the *Escherichia coli* lactose operon selectively increases repair of its transcribed DNA strand. *Nature*, **342**, 95–98.

Mellon, I., Spivak, G., and Hanawalt, P.C. (1987). Selective removal of transcription-blocking DNA damage from the transcribed strand of the mammalian DHFR gene. *Cell*, **51**, 241–249.

Meyer, P., Linn, F., Heidmann I., Meyer, H., Niedenhof, I., and Saedler, H. (1992). Endogenous and environmental factors influence 35S promoter methylation of a maize A1 gene construct in transgenic petunia and its colour phenotype. *Molecular and General Genetics*, **231**, 345–352.

Michie, D., (1958). The third stage in genetics. In *A century of Darwin*, (ed. S.A. Barnett), pp. 56–84. Heinemann, London.

Miklos, G.L.G. (1974). Sex chromosome pairing and male fertility. *Cytogenetics and Cell Genetics*, **13**, 558–577.

Mittler, J.E. and Lenski, R.E. (1990). New data on excisions of Mu from *E. coli* MCS2 cast doubt on directed mutation hypothesis. *Nature*, **344**, 173–175.

Mittler, J.E. and Lenski, R.E. (1992). Experimental evidence for an alternative to directed mutation in the *bgl* operon. *Nature*, **356**, 446–448.

Monk, M. (1988). Genomic imprinting. *Genes and Development*, **2**, 921–925.

Monk, M. (1990). Variation in epigenetic inheritance. *Trends in Genetics*, **6**, 110–114.

Monk, M. and Grant, M. (1990). Preferential X-chromosome inactivation, DNA methylation and imprinting. *Development*, 1990 Supplement, 55–62.

Moore, T. and Haig, D. (1991). Genomic imprinting in mammalian development: a parental tug-of-war. *Trends in Genetics*, **7**, 45–49.

Moran, N.A. (1992). The evolutionary maintenance of alternative phenotypes. *American Naturalist*, **139**, 971–989.

Mort, M.A. (1991). Bridging the gap between ecology and genetics: the case of freshwater zooplankton. *Trends in Ecology and Evolution*, **6**, 41–45.

Moss, G.I. and Mullet, J.H. (1982). Potassium release and seed vigour in germinating bean (*Phaseolus vulgaris* L.) seed as influenced by temperature over the previous five generations. *Journal of Experimental Botany*, **33**, 1147–1160.

Motara, M.A. and Rai, K.S. (1977). Chromosomal differentiation in two species of *Aedes* and their hybrids revealed by Giemsa C-banding. *Chromosoma*, **64**, 125–132.

Moxon, E.R., Rainey, P.B., Nowak, M.A., and Lenski, R.E. (1994). Adaptive evolution of highly mutable loci in pathogenic bacteria. *Current Biology*, **4**, 24–33.

Nagl, W. (1978). *Endopolyploidy and polyteny in differentiation and evolution*. North Holland, Amsterdam.

Nanney, D.L. (1958). Epigenetic control systems. *Proceedings of the National Academy of Sciences,. USA*, **44**, 712–717.

Nanney, D.L. (1960). Microbiology, developmental genetics and evolution. *American Naturalist*, **94**, 167–179.

Nanney, D.L. (1968). Cortical patterns in cellular morphogenesis. *Science*, **160**, 496–502.

Nanney, D.L. (1985). Heredity without genes: ciliate explorations of clonal heredity. *Trends in Genetics*, **1**, 295–298.

Naveira, H. and Fontdevila, A. (1986). The evolutionary history of *Drosophila buzzatii*. XII. The genetic basis of sterility in hybrids between *D. buzzatii* and its sibling *D. serido* from Argentina. *Genetics*, **114**, 841–857.

Nelsen, E.M., Frankel, J., and Jenkins, L.M. (1989). Non-genic inheritance of cellular handedness. *Development*, **105**, 447–456.

Neves, N., Barão, A., Castilho, A., Silva, M., Morais, L., Carvalho, V., Viegas, W., and Jones, R.N. (1992). Influence of DNA methylation on rye B-chromosome nondisjunction. *Genome*, **35**, 650–652.

Ng, S.F. (1990). Embryological perspective of sexual somatic development in ciliated protozoa: implications on immortality, sexual reproduction and inheritance of acquired characters. *Philosophical Transactions of the Royal Society Series B*, **329**, 287–305

Ng, S.F. and Frankel, J. (1977). 180°-rotation of ciliary rows and its morphogenetic implications in *Tetrahymena pyriformis. Proceedings of the National Academy of Sciences, USA*, **74**, 1115–1119.

Nieuwkoop, P.D. and Sutasurya, L.A. (1981). *Primordial germ cells in the invertebrates.* Cambridge University Press.

Novick, A. and Weiner, M. (1957). Enzyme induction as an all-or-none phenomenon. *Proceedings of the National Academy of Sciences, USA*, **43**, 553–566.

Nur, U. and Brett, B.L.H. (1988). Genotypes affecting the condensation and transmission of heterochromatic B chromosomes in the mealybug *Pseudococcus affinis. Chromosoma*, **96**, 205–212.

Okada, N. (1991). SINEs. *Current Opinion in Genetics and Development*, **1**, 498–504.

Oldroyd, D.R. (1983). *Darwinian impacts*, (2nd edn). Open University Press, Milton Keynes.

Opadia-Kadima, G.Z. (1987). How the slot machine led biologists astray. *Journal of Theoretical Biology*, **124**, 127–135.

Osborn, H.F. (1889). The palaeontological evidence for the transmission of acquired characters. *American Naturalist*, **23**, 559–566.

Paro, R. (1990). Imprinting a determined state into the chromatin of *Drosophila. Trends in Genetics*, **6**, 416–421.

Parsons, P.A. (1987). Evolutionary rates under environmental stress. *Evolutionary Biology*, **21**, 311–347.

Parsons, P.A. (1988). Evolutionary rates: effects of stress upon recombination. *Biological Journal of the Linnean Society*, **35**, 49–68.

Partridge, L. and Morgan, M.J. (1988). Is bacterial evolution random or selective? *Nature*, **336**, 22.

Patel, C.V. and Gopinathan, K.P. (1987). Determination of trace amounts of 5-methylcytosine in DNA by reverse-phase high-performance liquid chromatography. *Analytical Biochemistry*, **164**, 164–169.

Patterson, G.I., Thorpe, C.J., and Chandler, V.L. (1993). Paramutation, an allelic interaction, is associated with a stable and heritable reduction of transcription of the maize *b* regulatory gene. *Genetics*, **135**, 881–894.

Pfeifer, E.J. (1965). The genesis of American Neo-Lamarckism. *Isis*, **56**, 156–167.

Pfennig, D.W. (1992). Polyphenism in spadefoot toad tadpoles as a locally adjusted evolutionarily stable strategy. *Evolution*, **46**, 1408–1420.

Pillus, L. (1992). An acquired state: epigenetic mechanisms in transcription. *Current Opinion in Cell Biology*, **4**, 453–458.

Pillus, L. and Rine, J. (1989). Epigenetic inheritance of transcriptional states in *S. cerevisiae. Cell*, **59**, 637–647.

Plough, H.H. (1917). The effect of temperature on crossingover in *Drosophila. Journal of Experimental Zoology*, **24**, 147–209.

Plough, H.H. and Ives, P.T. (1935). Induction of mutations by high temperature in *Drosophila. Genetics*, **20**, 42–69.

Polikoff, D. (1981). C.H. Waddington and modern evolutionary theory. *Evolutionary Theory*, **5**, 143–168.

Pomiankowski, A. and Hurst, L.D. (1993). Genomic conflicts underlying Haldane's rule. *Genetics*, **133**, 425–432.

Popper, K. (1978). Natural selection and the emergence of mind. *Dialectica*, **32**, 339–355.

Prody, C.A., Dreyfus, P., Zamir, R., Zakut, H., and Soreq, H. (1989). *De novo* amplification within a "silent" human cholinesterase gene in a family subjected to prolonged exposure to organophosphorous insecticides. *Proceedings of the National Academy of Sciences, USA*, **86**, 690–694.

Proffitt, J.H., Davie, J.R., Swinton, D., and Hattman, S. (1984). 5-Methylcytosine is not detectable in *Saccharomyces cerevisiae* DNA. *Molecular and Cellular Biology*, **4**, 985–988.

Provine, W.B. (1971). *The origins of theoretical population genetics*. University of Chicago Press.

Puertas, M.J., Jiménez, M.M., Romera, F., Vega, J.M., and Diez, M. (1990). Maternal imprinting effect on B chromosome transmission in rye. *Heredity*, **64**, 197–204.

Raff, R.A. (1988). The selfish cell lineage. *Cell*, **54**, 445–446.

Raff, R.A. and Kaufman, T.C. (1983). *Embryos, genes, and evolution*. Macmillan, New York.

Rao, B.J. and Rao, M.R.S. (1987). DNase I site mapping and micrococcal nuclease digestion of pachytene chromatin reveal novel structural features. *Journal of Biological Chemistry*, **262**, 4472–4476.

Ratner, V.A., Zabanov, S.A., Kolesnikova, O.V., and Vasilyeva, L.A. (1992). Induction of the mobile genetic element *Dm-412* transpositions in the *Drosophila* genome by heat shock treatment. *Proceedings of the National Academy of Sciences, USA*, **89**, 5650–5654.

Razin, A. and Cedar, H. (1991). DNA methylation and gene expression. *Microbiological Reviews*, **55**, 451–458.

Read, A. and Nee, S. (1991). Is Haldane's rule significant? *Evolution*, **45**, 1707–1709.

Reeder, R.H. (1984). Enhancers and ribosomal gene spacers. *Cell*, **38**, 349–351.

Reeder, R.H. (1985). Mechanisms of nucleolar dominance in animals and plants. *Journal of Cell Biology*, **101**, 2013–2016.

Reeder, R.H. and Roan, J.G. (1984). The mechanism of nucleolar dominance in *Xenopus* hybrids. *Cell*, **38**, 39–44.

Reik, W. (1992). Genome imprinting. In *Transgenic animals*, (ed. F. Grosveld and G. Kollias), pp. 99–126. Academic Press, London.

Reik, W. and Surani, M.A. (1989). Genomic imprinting and embryonal tumours. *Nature*, **338**, 112–113.

Reik, W., Collick, A., Norris, M.L., Barton, S.C., and Surani, M.A. (1987). Genomic imprinting determines methylation of parental alleles in transgenic mice. *Nature*, **328**, 248–251.

Reik, W., Howlett, S.K., and Surani, M.A. (1990). Imprinting by DNA methylation: from transgenes to endogenous gene sequences. *Development*, 1990 Supplement, 99–106.

Renard, J.P. and Babinet, C. (1986). Identification of a paternal developmental effect on the cytoplasm of one-cell-stage mouse embryos. *Proceedings of the National Academy of Sciences, USA*, **83**, 6883–6886.

Renckens, S., De Greve, H., Van Montagu, M., and Hernalsteens, J.-P. (1992). *Petunia* plants escape from negative selection against a transgene by silencing the foreign DNA via methylation. *Molecular and General Genetics*, **233**, 53–64.

Rensch, B. (1980). Historical development of the present synthetic neo-Darwinism in Germany. In *The evolutionary synthesis: perspectives on the unification of biology*, (ed. E. Mayr and W.B. Provine), pp. 284–303. Harvard University Press, Cambridge, Mass.

Rensch, B. (1983). The abandonment of Lamarckian explanations: the case of climatic parallelism of animal characteristics. In *Dimensions of Darwinism*, (ed. M. Grene), pp. 31–42. Cambridge University Press.

Reuter, G. and Spierer, P. (1992). Position effect variegation and chromatin proteins. *BioEssays*, **14**, 605–612.

Rice, W.R. and Salt, G.W. (1990). The evolution of reproductive isolation as a correlated character under sympatric conditions: experimental evidence. *Evolution*, **44**, 1140–1152.

Richards, R.I., and Sutherland, G.R. (1992*a*). Fragile X syndrome: the molecular picture comes into focus. *Trends in Genetics*, **8**, 249–255.

Richards, R.I., and Sutherland, G.R. (1992b). Dynamic mutations: a new class of mutations causing human disease. *Cell*, **70**, 709–712.

Riedl, R. (1978). *Order in living organisms*. Wiley, Chichester.

Riggs, A.D. (1989). DNA methylation and cell memory. *Cell Biophysics*, **15**, 1–13.

Riggs, A.D. (1990). DNA methylation and late replication probably aid cell memory, and type I reeling could aid chromosome folding and enhancer function. *Philosophical Transactions of the Royal Society of London Series B*, **326**, 285–297.

Riggs, A.D. and Pfeifer, G.P. (1992). X-chromosome inactivation and cell memory. *Trends in Genetics*, **8**, 169–174.

Rivier, D.H. and Rine, J. (1992). Silencing: the establishment and inheritance of stable, repressed transcription states. *Current Opinion in Genetics and Development*, **2**, 286–292.

Robinson, G. (1979). *A prelude to genetics*. Coronado Press, Lawrence, Kansas.

Rose, S., Kamin, L.J., and Lewontin, R.C. (1984). *Not in our genes*. Penguin Books, Harmondsworth, Middlesex.

Roth, E.J., Frazier, B.L., Apuya, N.R., and Lark, K.G. (1989). Genetic variation in an inbred plant: variation in tissue cultures of soybean [*Glycine max* (L.) Merrill]. *Genetics*, **121**, 359–368.

Rubin, H. (1990). On the nature of enduring modifications induced in cells and organisms. *American Journal of Physiology*, **258**, L19–L24.

Russell, P.J., Welsch, J.A., Rachlin, E.M., and McCloskey, J.A. (1987). Different levels of DNA methylation in yeast and mycelial forms of *Candida albicans*. *Journal of Bacteriology*, **169**, 4393–4395.

Ruvinsky, A.O. (1988). Inheritance of dominant genes with variable penetrance: an evolutionary aspect. *Journal of Animal Breeding and Genetics*, **105**, 103–111.

Ruvinsky, A.O. (1990). Sex evolution and origin of multicellularity. In *Advances in Invertebrate Reproduction* 5, (ed. M. Hoshi and O. Yamashita), pp. 55–60. Elsevier, Amsterdam.

Ruvinsky, A.O. and Agulnik, A.I. (1990). Gametic imprinting and the manifestation of the Fused gene in the house mouse. *Developmental Genetics*, **11**, 263–269.

Ruvinsky, A.O., Lobkov, Y.I., and Belyaev, D.K. (1983*a*). Spontaneous and induced activation of genes affecting the phenotypic expression of glucose 6-phosphate dehydrogenase in *Daphnia pulex*. I. Intraclonal variations in the electrophoretic mobility of G6PD. *Molecular and General Genetics*, **189**, 485–489.

Ruvinsky, A.O., Lobkov, Y.I., and Belyaev, D.K. (1983*b*). Spontaneous and induced activation of genes affecting the phenotypic expression of glucose 6-phosphate dehydrogenase in *Daphnia pulex*. II. Glucose-induced changes in the electrophoretic mobility of G6PD. *Molecular and General Genetics*, **189**, 490–494.

Ruvinsky, A.O., Lobkov, Y.I., and Belyaev, D.K. (1986). Spontaneous and induced activation of genes affecting the phenotypic expression of glucose 6-phosphate dehydrogenase in *Daphnia pulex*. 3. Occurrence frequencies of the alternative electrophoretic variants of G6PD in a natural population. *Theoretical and Applied Genetics*, **72**, 811–815.

Ryan, F.J. (1955). Spontaneous mutation in non-dividing bacteria. *Genetics*, **40**, 726–738.

Ryan, F.J. (1959). Bacterial mutation in a stationary phase and the question of cell turnover. *Journal of General Microbiology*, **21**, 530–549.

Ryan, F.J., Nakada, D., and Schneider, M.J. (1961). Is DNA replication a necessary condition for spontaneous mutation? *Zeitschrift für Vererbungslehre*, **92**, 38–41.

Sager, R. and Kitchin, R. (1975). Selective silencing of eukaryotic DNA. *Science*, **189**, 426–433.

Sano, H., Kamada, I., Youssefian S., and Wabiko H. (1989). DNA demethylation by 5-azacytidine induces dwarlism in rice. *Journal of Cellular Biochemistry*, Supplement 13D, p. 214.

Sano, H., Kamada, I., Youssefian, S., Katsumi, M., and Wabiko, H. (1990). A single treatment of rice seedlings with 5-azacytidine induces heritable dwarfism and undermethylation of genomic DNA. *Molecular and General Genetics*, **220**, 441–447.

Sapienza, C. (1989). Genome imprinting and dominance modification. *Annals of the New York Academy of Sciences*, **564**, 24–38.

Sapienza, C. (1990). Sex-linked dosage-sensitive modifiers as imprinting genes. *Development*, 1990 Supplement, 107–113.

Sapienza, C., Peterson, A.C., Rossant, J., and Balling, R. (1987). Degree of methylation of transgenes is dependent on gamete of origin. *Nature*, **328**, 251–254.

Sapienza, C., Paquette, J., Tran, T.H., and Peterson, A. (1989). Epigenetic and genetic factors affect transgene methylation imprinting. *Development*, **107**, 165–168.

Sapp, J. (1987). *Beyond the gene*. Oxford University Press, New York.

Sarkar, S. (1990). On the possibility of directed mutations in bacteria: statistical analyses and reductionist strategies. In *PSA-1990*, Vol. 1, (ed. A. Fine, M. Forbes, and L. Wessels), pp. 111–124. Philosophy of Science Association, East Lansing, Mich.

Sarkar, S. (1991). Lamarck *contre* Darwin, reduction versus statistics: conceptual issues in the controversy over directed mutagenesis in bacteria. In *Organism and the origins of self*, (ed. A.I. Tauber), pp. 235–271. Kluwer, Dordrecht.

Sasaki, H., Hamada, T., Ueda, T., Seki, R., Higashinakagawa, T., and Sakaki, Y.

(1991). Inherited type of allelic methylation variations in a mouse chromosome region where an integrated transgene shows methylation imprinting. *Development*, **111**, 573–581.

Savić, D.J. and Kanazir, D.T. (1972). The effect of a histidine operator constitutive mutation on UV-induced mutability within the histidine operon of *Salmonella typhimurium*. *Molecular and General Genetics*, **118**, 45–50.

Scarbrough, K., Hattman, S., and Nur, U. (1984). Relationship of DNA methylation level to the presence of heterochromatin in mealybugs. *Molecular and Cellular Biology*, **4**, 599–603.

Scharloo, W. (1991). Canalization: genetic and developmental aspects. *Annual Review of Ecology and Systematics*, **22**, 65–93.

Scheid, O.M., Paszkowski, J., and Potrykus, I. (1991). Reversible inactivation of a transgene in *Arabidopsis thaliana*. *Molecular and General Genetics*, **228**, 104–112.

Schmalhausen, I.I. (1949). *Factors of evolution: the theory of stabilizing selection*, (trans. I. Dordick). Blackiston, Philadelphia.

Schmid, B. (1992). Phenotypic variation in plants. *Evolutionary Trends in Plants*, **6**, 45–60.

Schmidt, B.R. (1993). Are hybridogenetic frogs cyclical parthenogens? *Trends in Ecology and Evolution*, **8**, 271–273.

Schmidtke, J., Kuhl, P., and Engel, W. (1976). Transitory hemizygosity of paternally derived alleles in hybrid trout embryos. *Nature*, **260**, 319–320.

Schneeberger, R.G. and Cullis, C.A. (1991). Specific DNA alterations associated with the environmental induction of heritable changes in flax. *Genetics*, **128**, 619–630.

Schultz, R.J. (1977). Evolution and ecology of unisexual fishes. *Evolutionary Biology*, **10**, 277–331.

Searle, A.G. and Beechey, C.V. (1990). Genome imprinting phenomena on mouse chromosome 7. *Genetical Research*, **56**, 237–244.

Selig, S., Okumura, K., Ward, D.C., and Cedar, H. (1992). Delineation of DNA replication time zones by fluorescence *in situ* hybridization. *EMBO Journal*, **11**, 1217–1225.

Selker, E.U. (1990*a*). Premeiotic instability of repeated sequences in *Neurospora crassa*. *Annual Review of Genetics*, **24**, 579–613.

Selker, E.U. (1990*b*). DNA methylation and chromatin structure: a view from below. *Trends in Biochemical Sciences*, **15**, 103–107.

Serfling, E. (1989). Autoregulation—a common property of eukaryotic transcription factors? *Trends in Genetics*, **5**, 131–133.

Serrano, M., Salas, M., and Hermoso, J.M. (1993). Multimeric complexes formed by DNA-binding proteins of low sequence specificity. *Trends in Biochemical Sciences*, **18**, 202–206.

Shapiro, J.A. (1984). Observations on the formation of clones containing *araB lacZ* cistron fusions. *Molecular and General Genetics*, **194**, 79–90.

Shapiro, J.A. (1988). Bacteria as multicellular organisms. *Scientific American*, **258**, (6), 62–69.

Shapiro, J.A. (1991). Genomes as smart systems. *Genetica*, **84**, 3–4.

Shapiro, J.A. (1992). Natural genetic engineering in evolution. *Genetica*, **86**, 99–111.

Shaposhnikov, G.K. (1965). Morphological divergence and convergence in an experiment with aphids (Homoptera, Aphidinea). *Entomological Review*, **44**, 1–12.

Shaposhnikov, G.K. (1966). Origin and breakdown of reproductive isolation and the criterion of the species. *Entomological Review*, **45**, 1–18.

Shaposhnikov, G.K. (1985). The main features of the evolution of aphids. *Proceedings of the International Aphidological Symposium at Jabłonna 1981*, pp. 19–99. Polska Akademia Nauk, Wrocław.

Sharman, G.B. (1971). Late DNA replication in the paternally derived *X* chromosome of female kangaroos. *Nature*, **230**, 231–232.

Signoret, J. and David, J.C. (1986). DNA-ligase activity in axolotl early development: evidence for a multilevel regulation of gene expression. *Journal of Embryology and Experimental Morphology*, **97** (Supplement), 85–95.

Signoret, J., David, J.C., Lefresne, J., and Houillon, C. (1983). Control of DNA ligase molecular forms in nucleocytoplasmic combinations of axolotl and *Pleurodeles. Proceedings of the National Academy of Sciences, USA*, **80**, 3368–3371.

Silander, J.A. (1985). Microevolution in clonal plants. In *Population biology and evolution of clonal organisms*, (ed. J.B.C. Jackson, L.W. Buss, and R.E. Cook), pp. 107–152. Yale University Press, New Haven.

Silva, A.J. and White, R. (1988). Inheritance of allelic blueprints for methylation patterns. *Cell*, **54**, 145–152.

Simpson, G.G. (1944). *Tempo and mode in evolution*. Columbia University Press, New York.

Simpson, G.G. (1953). The Baldwin effect. *Evolution*, **7**, 110–117.

Simpson, V.J., Johnson, T.E., and Hammen, R.F. (1986). *Caenorhabditis elegans* DNA does not contain 5-methylcytosine at any time during development or aging. *Nucleic Acids Research*, **14**, 6711–6719.

Sinervo, B. and McEdward, L.R. (1988). Developmental consequences of an evolutionary change in egg size: an experimental test. *Evolution*, **42**, 885–899.

Sladden, D.E. and Hewer, H.R. (1938). Transference of induced food-habit from parent to offspring. III. *Proceedings of the Royal Society of London Series B*, **126**, 30–44.

Slatkin, M. (1984). Somatic mutations as an evolutionary force. In *Evolution: essays in honour of John Maynard Smith*, (ed. P.J. Greenwood, P.H. Harvey, and M. Slatkin), pp. 19–30. Cambridge University Press.

Sober, E. (1984). *The nature of selection*. MIT Press, Cambridge, Mass.

Soll, D.R., Morrow, B., and Srikantha, T. (1993). High-frequency phenotypic switching in *Candida albicans. Trends in Genetics*, **9**, 61–65.

Solomon, F. (1979). Detailed neurite morphologies of sister neuroblastoma cells are related. *Cell*, **16**, 165–169.

Solomon, F. (1981). Specification of cell morphology by endogenous determinants. *Journal of Cell Biology*, **90**, 547–553.

Solter, D. (1988). Differential imprinting and expression of maternal and paternal genomes. *Annual Review of Genetics*, **22**, 127–146.

Sonneborn, T.M. (1930). Genetic studies on *Stenostomum incaudatum*. II. The effects of lead acetate on the hereditary constitution. *Journal of Experimental Zoology*, **57**, 409–439.

Sonneborn, T.M. (1964). The differentiation of cells. *Proceedings of the National Academy of Sciences, USA*, **51**, 915–929.

Spiegelman, S., Lindegren, C.C., and Lindegren, G. (1945). Maintenance and increase of a genetic character by a substrate-cytoplamic interaction in the absence of the specific gene. *Proceedings of the National Academy of Sciences, USA*, **31**, 95–102.

Spofford, J.B. (1976). Position-effect variegation in *Drosophila*. In *The genetics and biology of Drosophila*, Vol. 1c, (ed. M. Ashburner and E. Novitski), pp. 955–1018. Academic Press, London.

Spradling, A.C., Karpen, G., Glaser, R., and Zhang, P. (1993). Evolutionary conservation of developmental mechanisms: DNA elimination in *Drosophila*. In *Evolutionary conservation of developmental mechanisms*, (ed. A.C. Spradling), pp. 39–53. Wiley-Liss, New York.

Spurway, H. (1955). The causes of domestication: an attempt to integrate some ideas of Konrad Lorenz with evolution theory. *Journal of Genetics*, **53**, 325–362.

Stack, S.M. (1984). Heterochromatin, the synaptonemal complex and crossing over. *Journal of Cell Science*, **71**, 159–176.

Stahl, F.W. (1988). A unicorn in the garden. *Nature*, **335**, 112–113.

Stahl, F.W. (1992). Unicorns revisited. *Genetics*, **132**, 865–867.

Stearns, S.C. (1989). The evolutionary significance of phenotypic plasticity. *BioScience*, **39**, 436–445.

Steele, D.F. and Jinks-Robertson, S. (1992). An examination of adaptive reversion in *Saccharomyces cerevisiae. Genetics*, **132**, 9–21.

Steele, E.J. (1979). *Somatic selection and adaptive evolution: on the inheritance of acquired characters.* Williams-Wallace, Toronto.

Steele, E.J., Gorczynski, R.M., and Pollard, J.W. (1984). The somatic selection of acquired characters. In *Evolutionary theory: paths into the future*, (ed. J.W. Pollard), pp. 217–237. John Wiley, Chichester.

Stent, G.S. and Calendar, R. (1978). *Molecular genetics: an introductory narrative*, (2nd edn). Freeman, San Francisco.

Stöger, R., Kubička, P., Liu, C.-G., Kafri, T., Razin, A., Cedar, H., and Barlow, D.P. (1993). Maternal-specific methylation of the imprinted mouse *Igf2r* locus identifies the expressed locus as carrying the imprinted signal. *Cell*, **73**, 61–71.

Stuart, J.J. and Hatchett J.H. (1988). Cytogenetics of the Hessian fly: II. Inheritance and behavior of somatic and germ-line-limited chromosomes. *Journal of Heredity*, **79**, 190–199.

Sueoka, N. (1992). Directional mutation pressure, selective constraints, and genetic equilibria. *Journal of Molecular Evolution*, **34**, 95–114.

Sumner, A.T. (1990). *Chromosome banding*. Unwin Hyman, London.

Surani, M.A.H., Barton, S.C., and Norris, M.L. (1984). Development of reconstituted mouse eggs suggests imprinting of the genome during gametogenesis. *Nature*, **308**, 548–550.

Surani, M.A., Kothary, R., Allen, N.D., Singh, P.B., Fundele, R., Ferguson-Smith, A.C., and Barton, S.C. (1990). Genome imprinting and development in the mouse. *Development*, 1990 Supplement, 89–98.

Swain, J.L., Stewart, T.A., and Leder, P. (1987). Parental legacy determines methylation and expression of an autosomal transgene: a molecular mechanism for parental imprinting. *Cell*, **50**, 719–727.

Symonds, N. (1989). Anticipatory mutagenesis? *Nature*, **337**, 119–120.

Symonds, N. (1991). A fitter theory of evolution? *New Scientist*, **131**, (1787), 30–34.

Szathmáry, E. (1994). Toy models for simple forms of multicellularity, soma and germ. *Journal of Theoretical Biology*, **169**, 125–132.

Takagi, N. and Sasaki, M. (1975). Preferential inactivation of the paternally derived X chromosome in the extraembryonic membranes of the mouse. *Nature*, **256**, 640–642.

Tamame, M. and Santos, E. (1989). Demonstration of 5-methyl cytosine occurrence in the genome of *Aspergillus nidulans. Journal of Cell Biochemistry*, Supplement 13D, p. 214.

Tamame, M., Antequera, F., and Santos, E. (1988). Developmental characterization and chromosomal mapping of the 5-azacytidine-sensitive *fluF* locus of *Aspergillus nidulans. Molecular and Cellular Biology*, **8**, 3043–3050.

Tartof, K.D. and Bremer, M. (1990). Mechanisms for the construction and developmental control of heterochromatin formation and imprinted chromosome domains. *Development*, 1990 Supplement, 35–45.

Tartof, K.D., Bishop, C., Jones, M., Hobbs, C.A., and Locke, J. (1989). Towards an understanding of position effect variegation. *Developmental Genetics*, **10**, 162–176.

Tate, W.P. and Marshall, C.J. (1991). Post-dormancy transcription and translation in the brine shrimp. In *Artemia biology*, (ed. R.A. Browne, P. Sorgeloos, and C.N.A. Trotman), pp. 21–36. CRC Press, Boca Raton, Florida.

Tauber, C.A. and Tauber, M.J. (1989). Sympatric speciation in insects: perception and perspective. In *Speciation and its consequences*, (ed. D. Otte and J.A. Endler), pp. 307–344. Sinauer, Sunderland, Mass.

Templeton, A.R. (1989). The meaning of species and speciation: a genetic perspective. In *Speciation and its consequences*, (ed. D. Otte and J.A. Endler), pp. 3–27. Sinauer, Sunderland, Mass.

Thomas, B.J. and Rothstein, R. (1991). Sex, maps, and imprinting. *Cell*, **64**, 1–3.

Thomson, K.S. (1988). *Morphogenesis and evolution*. Oxford University Press, New York.

Thomson, K.S. (1992). Macroevolution: the morphological problem. *American Zoologist*, **32**, 106–112.

Thuriaux, P. (1977). Is recombination confined to structural genes on the eukaryotic genome? *Nature*, **268**, 460–462.

Tourte, Y., Kuligowski-Andres, J., and Barbier-Ramond, C. (1980). Comportement différentiel des chromatines paternelles et maternelles au cours de l'embryogenèse d'une fougère: Le Marsilea. *European Journal of Cell Biology*, **21**, 28–36.

Trut, L.N. (1987). Problems associated with the arisal of new forms and the integrity of the organism in the context of destabilizing selection. *Soviet Genetics*, **23**, 671–681.

Turner, J.R.G. (1988). The evolution of mimicry: a solution to the problem of punctuated equilibrium. *American Naturalist*, **131** (Supplement), S42–S66.

Urieli-Shoval, S., Gruenbaum, Y., Sedat, J., and Razin, A. (1982). The absence of detectable methylated bases in *Drosophila melanogaster* DNA. *FEBS Letters*, **146**, 148–152.

Van der Woude, M.W., Braaten, B.A., and Low, D.A. (1992). Evidence for global regulatory control of pilus expression in *Escherichia coli* by Lrp and DNA methylation: model building based on analysis of *pap. Molecular Microbiology*, **6**, 2429–2435.

Van Holde, K.E. (1989). *Chromatin*. Springer-Verlag, New York.

Van Valen, L.M. (1987). Non-Weismannian evolution. *Evolutionary Theory*, **8**, 101–107.

Van Valen, L.M. (1988). Is somatic selection an evolutionary force? *Evolutionary Theory*, **8**, 163–167.

Vinogradov, A.E., Borkin, L.J., Günther, R., and Rosanov, J.M. (1990). Genome elimination in diploid and triploid *Rana esculenta* males: cytological evidence from DNA flow cytometry. *Genome*, **33**, 619–627.

Vogt, P. (1990). Potential genetic functions of tandem repeated DNA sequence blocks in the human genome are based on a highly conserved "chromatin folding code". *Human Genetics*, **84**, 301–336.

Vuillaume, M. and Berkaloff, A. (1974). LSD treatment of *Pieris brassicae* and consequences on the progeny. *Nature*, **251**, 314–315.

Waddington, C.H. (1942). Canalization of development and the inheritance of acquired characters. *Nature*, **150**, 563–565.

Waddington, C.H. (1953). Genetic assimilation of an acquired character. *Evolution*, **7**, 118–126.

Waddington, C.H. (1957). *The strategy of the genes.* Allen and Unwin, London.

Waddington, C.H. (1961). Genetic assimilation. *Advances in Genetics*, **10**, 257–293.

Waddington, C.H. (1968). The basic ideas of biology. In *Towards a theoretical biology. 1. Prolegomena*, (ed. C.H. Waddington), pp. 1–32. Edinburgh University Press.

Wade, M.J. (1978). A critical review of the models of group selection. *Quarterly Review of Biology*, **53**, 101–114.

Wake, D.B., Roth, G., and Wake, M.H. (1983). On the problem of stasis in organismal evolution. *Journal of Theoretical Biology*, **101**, 211–224.

Wallace, B. (1990). Norms of reaction: do they include molecular events? *Perspectives in Biology and Medicine*, **33**, 323–334.

Wallace, H. and Langridge, W.H.R. (1971). Differential amphiplasty and the control of ribosomal RNA synthesis. *Heredity*, **27**, 1–13.

Warner, A.H. and Bagshaw, J.C. (1984). Absence of detectable 5-methylcytosine in DNA of embryos of the brine shrimp, *Artemia. Developmental Biology*, **102**, 264–267.

Watson, J.D., Hopkins, N.H., Roberts, J.W., Steitz, J.A., and Weiner, A.M. (1987). *Molecular biology of the gene*, (4th edn), Benjamin/Cummings, Menlo Park, California.

Weinberg, J.R., Starczak, V.R., and Jörg, D. (1992). Evidence for rapid speciation following a founder event in the laboratory. *Evolution*, **46**, 1214–1220.

Weintraub, H. (1979). Assembly of an active chromatin structure during replication. *Nucleic Acids Research*, **7**, 781–792.

Weintraub, H. (1985). Assembly and propagation of repressed and derepressed chromosomal states. *Cell*, **42**, 705–711.

Weismann, A. (1885). The continuity of the germ-plasm as the foundation of a theory of heredity. In *Essays upon heredity and kindred biological problems*, (ed. and trans. E.B. Poulton, S. Schönland, and A.E. Shipley 1889), pp. 163–248. Clarendon Press, Oxford.

Weismann, A. (1893). *The germ-plasm: a theory of heredity*, (translated from the 1892 German edn by W. Newton Parker and H. Rönnfeldt). Walter Scott, London.

Weismann, A. (1902). *On germinal selection as a source of definite variation*, (2nd edn), (trans. T.J. McCormack). Open Court, Chicago.

Weismann, A. (1904). *The evolution theory*, Vol. 1 and 2, (trans. J.A. Thomson and M.R. Thomson). Edward Arnold, London.

Werren, J.H. (1991). The paternal-sex-ratio chromosome of *Nasonia. American Naturalist*, **137**, 392–402.

West-Eberhard, M.J. (1986). Alternative adaptations, speciation, and phylogeny (A review). *Proceedings of the National Academy of Sciences, USA*, **83**, 1388–1392.

West-Eberhard, M.J. (1989). Phenotypic plasticity and the origins of diversity. *Annual Review of Ecology and Systematics*, **20**, 249–278.

White, M.J.D. (1968). Models of speciation. *Science*, **159**, 1065–1070.

White, M.J.D. (1978). *Modes of speciation*. Freeman, San Francisco.

Whitham, T.G. and Slobodchikoff, C.N. (1981). Evolution by individuals, plant-herbivore interactions, and mosaics of genetic variability: the adaptive significance of somatic mutations in plants. *Oecologia*, **49**, 287–292.

Whitt, G.S. (1981). Developmental genetics of fishes: isozymic analyses of differential gene expression. *American Zoologist*, **21**, 549–572.

Whitt, G.S., Cho, P.L., and Childers, W.F. (1972). Preferential inhibition of allelic isozyme synthesis in an interspecific sunfish hybrid. *Journal of Experimental Zoology*, **179**, 271–282.

Williams, G.C. (1966). *Adaptation and natural selection*. Princeton University Press, New Jersey.

Williams, G.C. (1992). *Natural selection: domains, levels, and challenges*. Oxford University Press, New York.

Wills, C. (1984). The possibility of stress-triggered evolution. In *Evolutionary dynamics of genetic diversity*, (ed. G.S. Mani), pp. 299–312. Springer-Verlag, Berlin.

Wills, C. (1991). *The wisdom of the genes*. Oxford University Press.

Wilson, A.C., Bush, G.L., Case, S.M., and King, M.-C. (1975). Social structuring of mammalian populations and rate of chromosomal evolution. *Proceedings of the National Academy of Sciences, USA*, **72**, 5061–5065.

Wilson, A.C., White, T.J., Carlson, S.S., and Cherry, L.M. (1977). Molecular evolution and cytogenetic evolution. In *Molecular human cytogenetics*, (ed. R.S. Sparkes, D.E. Comings, and C.F. Fox), pp. 375–393. Academic Press, New York.

Wilson, C., Bellen, H.J., and Gehring, W.J. (1990). Position effects on eukaryotic gene expression. *Annual Review of Cell Biology*, **6**, 679–714.

Wilson, D.S. and Sober, E. (1989). Reviving the superorganism. *Journal of Theoretical Biology*, **136**, 337–356.

Wolfe, K.H., Sharp, P.M., and Li, W.-H. (1989). Mutation rates differ among regions of the mammalian genome. *Nature*, **337**, 283–285.

Wolffe, A.P. (1991). Developmental regulation of chromatin structure and function. *Trends in Cell Biology*, **1**, 61–66.

Wolffe, A.P. (1991). Activating chromatin. *Current Biology*, **1**, 366–368.

Wolpert, L. (1990). The evolution of development. *Biological Journal of the Linnean Society*, **39**, 109–124.

Wright, S. (1945). Genes as physiological agents. General considerations. *American Naturalist*, **74**, 289–303.

Wright, W.E. and Shay, J.W. (1992). Telomere positional effects and the regulation of cellular senescence. *Trends in Genetics*, **8**, 193–197.

Wu, C.-I. and Davis, A.W. (1993). Evolution of postmating reproductive isolation: the composite nature of Haldane's rule and its genetic bases. *American Naturalist*, **142**, 187–212.

Wu, C.-I. and Hammer, M.F. (1991). Molecular evolution of ultraselfish genes of meiotic drive systems. In *Evolution at the molecular level*, (ed. R.K. Selander, A.G. Clark, and T.S. Whittam), pp. 177–203. Sinauer, Sunderland, Mass.

Wynne-Edwards, V.C., (1962). *Animal dispersion in relation to social behaviour*. Oliver and Boyd, Edinburgh.

Yeom, Y.I., Abe, K., and Artzt, K. (1992). Evolution of the mouse H-2K region: a hot spot of mutation associated with genes transcribed in embryos and/or germ cells. *Genetics*, **130**, 629–638.

Zakian, S.M., Nesterova, T.B., Cheryaukene, O.V., and Bochkarev, M.N. (1991). Heterochromatin as a factor affecting X-inactivation in interspecific female vole hybrids (*Microtidae, Rodentia*). *Genetical Research*, **58**, 105–110.

Zentall, T.R. and Galef, B.G. (ed.) (1988). *Social learning: psychological and biological perspectives*, Lawrence Erlbaum, Hillsdale, New Jersey.

Zhimulev, I.F., Belyaeva, E.S., Bolshakov, V.N., and Mal'ceva, N.I. (1989). Position-effect variegation and intercalary heterochromatin: a comparative study. *Chromosoma*, **98**, 378–387.

Zirkle, C. (1946). The early history of the idea of the inheritance of acquired characters and of pangenesis. *Transactions of the American Philosophical Society*, **35**, 91–151.

Zlatanova, J. (1990). Histone H1 and the regulation of transcription of eukaryotic genes. *Trends in Biochemical Sciences*, **15**, 273–276.

Zolan, M.E. and Pukkila P.J. (1986). Inheritance of DNA methylation in *Coprinus cinereus*. *Molecular and Cellular Biology*, **6**, 195–200.

Zuckerkandl, E. (1986). Polite DNA: functional density and functional compatibility in genomes. *Journal of Molecular Evolution*, **24**, 12–27.

Zuckerkandl, E. (1992). Revisiting junk DNA. *Journal of Molecular Evolution*, **34**, 259–271.

Zuckerkandl, E. and Pauling, L. (1962). Molecular disease, evolution, and genic heterogeneity. In *Horizons in biochemistry*, (ed. M. Kasha and B. Pullman), pp. 189–225. Academic Press, New York.

Index

Achwal, C.W. 131 (n.17)
Adams, R.L.P. 109 (n.20), 210–11
adaptability
  and directed mutation 24–5
  and genetic assimilation 31–7, 51, 285, 286
  in Lamarck's theory 6, 27 (n.6)
adaptation
  in bacteria 23–5, 58–66, 82–3, 108 (n.3), 154
  by Darwinian and Lamarckian mechanisms 9
  through directed genomic responses 72, 74, 171–2
  through epigenetic changes 106, 153–4, 159 (n.18), 219–23, 237–8, 277–8
  to fluctuating environments 192–5
  through genetic assimilation 31–7
  and imprinting 216–19
  through polyphenism 238–41
  through use and disuse 3–6, 9, 23, 230
adaptive mutation, *see* directed mutation
Aedes spp. (mosquitoes) 248
ageing 146–7, 189 (n.18), 215
Agulnik, A.I. 126, 130 (n.9), 131 (n.11)
Alberts, B. 108 (n.10), 109 (n.25)
Albrecht-Buchler, G. 89
algae 113, 200–1, 202, 226 (n.10)
allelic exclusion 218–19
Alleman, M. 158 (n.10)
Allen, G.E. 28 (n.21)
Allen, N.D. 128, 129, 132 (n.29), 136
Allison, C. 226 (n.13)
alternative splicing 108 (n.6)
Alu sequences 185
Ambystoma mexicanum (axolotl) 113, 246
amphibian oocytes 69, 93, 120
amplification, DNA 65, 67–8, 69–71, 103, 125–6, 144–5, 172, 187
androgenetic embryos 114, 217, 252
Angelman syndrome 116, 118
anisogamy 157, 216, 219
annelids 202, 236
antigen switching 68, 75, 195
Antirrhinum (snapdragon) 159 (n.17)
ants 7, 209
aphids
  insecticide resistance 136, 144–5
  life cycle 212, 239, 241
  methylation in 144–5, 210–12
  speciation in 237–8, 242–3
Aphis fabae (aphid) life cycle 239
Arabidopsis thaliana (crucifer) 135
Aristotle 191, 225 (n.1)
Artemia (brine shrimp) 210, 226 (n.18)
arthropods 202
Ascaris (nematode) 52 (n.13), 68, 209, 226 (n.16)
asexual reproduction
  in ciliates 67–8
  and species richness 242–3
  transmitted variations during 144–5, 219–20, 237– 8, 243
  see also parthenogenesis; sex, functions of
Ashby, R. 281
Ashley, T. 165
Aspergillus nidulans (fungus) 134, 148
Asplanchna sieboldi (rotifer) 136
Atchley, W.R. 145, 159 (n.13)
Aufderheide, K.J. 87, 89
Ausio, J. 95, 109 (n.15)
autoregulation
  of gene activity 84, 108 (n.6)
  in Schmalhausen's theory 36–7
axolotl 113, 246
5-azacytidine 97, 105, 148, 261

Babinet, C. 252
Bacillus grandii (stick-insect) 248
Bacillus subtilis (bacterium) 204, 226 (n.13)
bacteria
  adaptation of 23–5, 58–66, 82–3, 108 (n.3), 195
  colony organization of 204, 226 (n.13)
  directed mutation in 22, 58–66, 72, 76 (n.9, 10)
  DNA methylation in 192, 195
Badino, G. 136
Bagshaw, J.C. 210
Baker, T.G. 270 (n.23)
Baldwin, J.M. 12, 31
Baldwin effect 31
Baltimore, D. 110 (n.28)
Bander, S.A.A. 130 (n.9)
Banks, J.A. 159 (n.16)
barley 211, 248
Barlow, D.P. 131 (n.11), 132 (n.27), 186
barnacles 240
Barr body 91
Bartolomei, M.S. 131 (n.11)
Barton, N.H. 269 (n.8)
basal bodies 86–7, 108 (n.9)

Bateman, K.G. 52 (n.5)
Bateson, G. 281
B chromosomes 113, 261–2
Beale, G.H. 158 (n.5)
beans 135, 139
Beardmore, J.A. 136
Beechey, C.V. 116
beetles 243, 269 (n.15)
behaviourally-mediated evolution 6, 284
Beisson, J. 87, 108 (n.8)
Bell, G. 171, 200–1, 214, 215, 226 (n.8,10,11), 227 (n.20,21), 269 (n.14)
Bell, L.R. 108 (n.6)
Belyaev, D.K. 136, 141–2, 158 (n.8), 162, 231–2, 268, 269 (n.2)
Bengtsson, B.O. 271 (n.34)
Ben-Jacob, E. 226 (n.13)
Berkaloff, A. 136
Bernardi, G. 184–5
Bernstein, H. 189 (n.7), 227 (n.19), 253
Berrill, N.J. 41–2
Bestor, T.H. 192
Beukeboom, L.W. 262
Bickmore, W.A. 99, 109 (n.21)
Biémont, C. 113
biophores 38
Bird, A. 109 (n.20), 174, 175
birds 120, 211, 232, 235, 255; *see also* chicken
Blacher, L.I. 12, 27 (n.2,6,8), 31, 51 (n.2), 53 (n.24)
Blackman, R.L. 158 (n.12), 239, 269 (n.11)
Blau, H.M. 108 (n.6), 109 (n.28)
bluebell 211
Bodnar, J.W. 189 (n.10)
Boe, L. 77 (n.20)
Bohr, V.A. 162, 163, 188 (n.3)
Bonner, J.T. 200, 202, 208
Borodin, P.M. 162, 269 (n.2)
Borst, P. 77 (n.23), 225 (n.2)
Bottema, S. 269 (n.4)
Boulikas, T. 164, 188 (n.2)
boundary elements 93–4
Bowler, P.J. 5, 27 (n.2), 28 (n.15), 76 (n.3,5)
Bownes, M. 162, 164
Boyd, R. 289 (n.9)
Boyes, J. 174, 175
bracken 211
Brandeis, M. 131 (n.15,16)
Breese, E.L. 220
Bremer, M. 189 (n.13)
Brenner, S. 24–5
Brent, L. 53 (n.22)
Brett, B.L.H. 261
Brettell, R.I.S. 135, 159 (n.17)
brine shrimp 210, 226 (n.18)
Brink, R.A. 135, 137, 143, 158 (n.9)
Brock, W.A. 120
Brockdorff, N. 190 (n.19)
Brown, C.J. 190 (n.19)
Brown, D.P. 109 (n.25)
Brown, S.W. 113, 130 (n.1)
Brun, J. 136, 145
Bucci, S. 247, 270 (n.20)
Buffon 7
Burdon, R.H. 210–11
Burgoyne, P.S. 270 (n.23)
Burt, A. 189 (n.7)
Bush, G.L. 263, 269 (n.6,9), 271 (n.34)
Buss, L.W. 12, 42–3, 45, 46–7, 53 (n.17,18), 154, 206–7, 208, 213, 214, 226 (n.15), 241–3, 283, 289 (n.6)
Bussey, H. 108 (n.5)
butterflies 54, 136, 240, 255

Caenorhabditis elegans (nematode) 130 (n.2), 136, 145–6, 242
Cairns, J. 12, 62–4, 66, 77 (n.12,21), 162
Cairns-Smith, A.G. 192
Caledia captiva (grasshopper) 264
Calendar, R. 60, 61, 108 (n.3)
Campbell, J.H. 72–4, 78 (n.26), 137, 142–3
canalization 36–7, 207, 240, 253, 283
Candida albicans (yeast) 195
Cannon, H.G. 31, 51 (n.1)
Capdevila, M.P 106
Carausius morosus (stick-insect) 136, 237, 269 (n.10)
Caskey, C.T. 190 (n.21)
Castro-Sierra, E. 246
caterpillar epidermis 89
Cattanach, B.M. 92, 130 (n.7)
Caulobacter (bacterium) colonies 204
Cavalli-Sforza, L.L. 289 (n.9)
C bands 98–9, 165, 177–81, 248; *see also* heterochromatin, constitutive
Cedar, H. 109 (n.20)
cell lineages
  competition and co-operation between 46–8, 205–16
  as units of evolution 204–5
cell memory 79, 80, 102, 223
  evolution of 191–5
  length of 101, 161, 173–8, 186–7, 205, 276–7
  mechanisms underlying, *see* epigenetic inheritance systems
  see also Dauermodifikations
cell organelles 21–2, 216, 226 (n.12)
cells in culture
  age changes in 146
  DNA repair in 163
  epimutations in 103–5

gene amplification in 69–70
structural inheritance in 88–9
tobacco 106, 139–40
cell specialization 202–7
cell turnover 205, 210–13, 225
cell types, numbers of 184, 202
cellular inheritance 16–17, 25, 191–5; *see also* cell memory; epigenetic inheritance systems
central dogma 17, 26, 29 (n.25), 49
centrioles 88, 108 (n.9)
Chandler, V.L. 159 (n.17)
Chandley, A.C. 165
Chandra, H.S. 130 (n.1), 219
Charlesworth, B. 270 (n.30)
Charlesworth, D. 77 (n.15)
chiasma 165, 189 (n.7)
chicken 133–4, 211, 246
α-globin gene methylation 97
lysozyme gene 96
Chironomus thummi (insect) 254
Chlamydomonas (alga) 113, 201, 226 (n.10)
CHO cells 163
Chomet, P.S. 159 (n.17)
chromatin diminution 38–41, 52 (n.13), 67, 68–9, 70, 209, 226 (n.16,17)
chromatin looped domains 93–4, 98, 109 (n.17), 118, 163, 183
chromatin-marking systems 90–102, 134–7, 158 (n.2), 192, 212–13
and imprinting 119–22
chromatin marks 121–2, 152, 153, 206
defined 90
divergence of 234–5, 267, 268
erasure of 119, 159 (n.22), 215–16, 221
and genetic assimilation 167–8
inheritance of 100–2, 147–51, 159 (n.19), 168–71
in populations 168–71, 195–9, 280
stubborn 156–7, 186, 219, 221–2, 243
transgenerational transmission of 106–7, 121–2, 128–9, 133–7, 147–52, 154–7, 221–3, 237–8, 261
see also imprints
chromatin structure 90, 94, 98, 175
active and inactive 91–8, 161–71
changes during ageing 146–7
inheritance of 100–2, 176–7
restructuring during embryogenesis 120–1
restructuring during gametogenesis 119–22, 133, 154–6, 216, 253–7, 260–2, 267, 268, 270 (n.26), 278
see also chromosome bands; DH sites; DNase-I sensitivity; methylation patterns
chromocentres 177, 271 (n.35)
chromosome bands 98–9, 109 (n.21), 177, 183–6, 277; *see also* heterochromatin, constitutive
chromosome domains, inactivation of 181–3, 277
chromosome elimination 217–18
of *B* chromosomes 261–2
in coccids 112, 113, 119
in hybridogenesis 247, 248, 250–2
in *Sciara* 68–9, 111–12, 113, 119, 183
chromosome inactivation 91–3, 112; *see also* heterochromatin; PEV; position effects; X chromosome inactivation
chromosome puffs 93
chromosome rearrangements 67–8
and imprinting 114–16
and meiotic drive 258–60
and position effects 92–3, 144–5, 178–80, 228 (n.31), 254
and speciation 235–6, 254, 255, 262–6, 271 (n.33,34)
Churchill, F.B. 28 (n.21), 51 (n.2)
cilia 86–7, 202, 226 (n.11)
ciliates 67–8, 86–8, 106, 201, 226 (n.9); *see also Paramecium*
Clermont, L. 70
clines 280
clonal memory 173–87; *see also* Dauermodifikations
clonal variation 86–8, 144–5, 219–20
Cnidaria 202, 221
coccids 112, 113, 119, 183, 210, 211, 212; *see also* mealy bugs
colchicine 136, 139
complexity, levels of 282–4
concerted evolution 264–5, 271 (n.35)
Conklin, K.F. 109 (n.22), 158 (n.1)
Cook, P.R. 81
Cooper, D.W. 130 (n.5)
Cope, E.D. 10, 28 (n.12)
copepods 209
copia-like elements 164
Coprinus cinereus (fungus) 134, 147–8
Corbet, S.A. 238
corn 211
cortical inheritance 86–8, 105, 108 (n.9), 133
Coturnix coturnix japonica (quail) 246
Coyne, J.A. 255, 270 (n.27,30)
CpG dinucleotides 100–1, 109 (n.23), 276
and mutation 131 (n.24), 166, 187
density of 174–5, 189 (n.9), 195, 206
Creationists 27 (n.1)
Crepis spp. (hawk's beard) 248
Crick, F.H.C. 29 (n.25)
Crosby, J.L. 159 (n.21)
Crouse, H.V. 69, 111, 113, 183
Crow, J.F. 270 (n.28)
Cullis, C.A. 78 (n.24), 162

cultural inheritance 25, 281, 284–5
Cyanobacteria 202
Cyclops strenuus (copepod) 209
cystic fibrosis gene 109 (n.27)
cytokinin 106, 108 (n.5), 135, 139–40
cytoplasm, role in heredity 14, 81–2, 89, 158 (n.2), 273
cytoplasmic organelles and neo-Darwinism 21–2
cytosine deamination 131 (n.24), 166, 212
cytosine methylation, *see* methylation
cytoskeleton 89
cytotaxis 87–90; *see also* structural inheritance

Daphnia pulex (crustacean) 136, 144, 158 (n.11)
daphnids 136, 144, 158 (n.11), 241
Darlington, C.D. 42
Darwin 2, 27 (n.7), 37, 53 (n.23), 229
  on domestication 230–1, 233, 267–8, 269 (n.16)
  on hybrids 244, 269 (n.16)
  and pangenesis 7, 12, 230–1
Darwin's theory of evolution 8, 9, 22, 279
Dauermodifikations 134, 137–9, 158 (n.3,4), 198, 226 (n.14)
David, J.C. 113, 246
Davies, D.R. 248
Davies, K. 190 (n.21)
Davis, A.W. 270 (n.25)
Davis, B.D. 64, 77 (n.18)
Dawkins, R. 18–19, 20, 27 (n.6), 29 (n.27), 52 (n.15), 56, 288 (n.1,5)
Dean, A.C.R. 108 (n.3)
De Beer, G.R. 285, 289 (n.13)
DeChiara, T.M. 122, 131 (n.11)
Delage, Y. 10–11
Delbrück, M. 13, 58–60, 62, 82
Dennis, E.S. 135, 159 (n.17)
determinants 38, 40–1, 54–5
Detlefsen, J.A. 28 (n.10)
Deumling, B. 70
development 244
  evolution of 46–7, 205–16, 222–4
  genomic changes during 38–41, 66–9
  importance of, in evolution theory 16–17, 28 (n.22), 29 (n.24), 285–8
  maternal control of 43, 46, 208–13, 218, 225, 237–8
  from a single cell 207–8
developmental clocks 173
De Vries, H. 28 (n.15), 53 (n.16)
DHFR (dihydrofolate reductase) gene 69–70, 121, 163
DH sites (DNase-I hypersensitive sites) 95, 96, 106, 133–4, 164
Dictyostelium mucoroides (slime mould) 208
Diehl, S.R. 269 (n.9)
Difflugia corona (protozoan) 134
Dileptus (ciliate) 88
diplogenesis 10
directed assembly 87, 89, 179, 191–2; *see also* structural inheritance
directed changes in eukaryote genomes 66–72
directed epigenetic changes 12–13, 80, 104–6, 107, 143, 144, 193, 279–82
directed mutation 12, 187, 266, 273–4, 279–80
  as an adaptive strategy 72–4, 172
  in bacteria 22, 58–66, 72, 76 (n.9,10)
  meaning of term 22–5, 56–8, 76 (n.8)
  mechanisms of 63–6
directed variation 8, 54–5, 75 (n.1), 273–5, 281–2
Dittrich, B. 131 (n.16)
division of labour 202–5, 209
DNA amplification 65, 67–8, 69–71, 103, 125–6, 144–5, 172, 187
DNA-binding proteins 72, 152, 182, 234–5, 245, 258
  and cell memory 174–7, 277
  and imprints 126–8, 218, 219
  and methylated DNA 98, 109 (n.26), 121–2
  and position effects 178–81
  in steady-state systems 84–6
DNA, foreign 124, 284; *see also* transgenes
DNA methylation, *see* methylation; methylation patterns
DNA-protein complexes, inheritance of 101–2, 143, 148, 176–7, 179
DNA rearrangements 66, 69, 71–2, 124, 195; *see also* chromosome rearrangements
DNA repair 64–5, 77 (n.18), 162–4, 171, 284
  and meiosis 214–15, 227 (n.19), 253–4
DNA replication, time of 93, 95, 99, 100, 109 (n.16), 118, 184–5, 218, 250
DNA sequence motifs 73, 284
DNase-I sensitivity 95–6, 99, 120, 163; *see also* DH sites
Dobzhansky, T. 229, 254
Doerfler, W. 131 (n.21)
Doljanski, F. 108 (n.11)
domestication 141, 142, 230–3, 236, 267–8, 269 (n.16)
Dooner, H.K. 131 (n.23), 135, 158 (n.10)
dormant genes 140–2, 144, 158 (n.7), 231–2, 235
Dorn, R. 136, 143, 180
dosage compensation 217, 225; *see also* X chromosome inactivation

Dougherty, E.C. 227 (n.19)
Dover, G.A. 264–5
Drake, J.W. 77 (n.14,22)
Drosophila 21, 142
  autoregulation 84
  chorion genes 69, 103
  chromatin elimination 52 (n.13), 226 (n.17)
  epigenetic inheritance 143, 180
  hybrids 255, 257, 260, 270 (n.26)
  meiotic drive 178, 258–60
  and methylation 98, 131 (n.17), 183, 210–12
  mutation rate 55
  polytene chromosomes 69, 93, 181, 226 (n.17), 257
  position effect variegation 92–3, 113, 136, 143, 145, 178–80, 228 (n.31)
  puffing 93
  recombination 165
  transposition 164
  X chromosomes 183, 257, 260
Drosophila buzzatii 257
Drosophila melanogaster
  Antennapedia complex 181–2
  Bithorax complex 181–2
  bithorax phenocopy 106, 136, 146, 167–8
  crossveinless phenocopy 32–5
  dumpy phenocopy 189 (n.8)
  imprinting 113, 143, 180
  induced mutation 55, 138
  Lansing effects 136
  nucleolar organizer region 245–9
  Polycomb 182
  segregation distortion 178, 258–60
  sex-lethal 108 (n.6)
  white 178–9
Drosophila serido 257
Drosophila simulans 245–9
ducks 232
Durham, W.H. 289 (n.9)
Durica, D.S. 249
Durrant, A. 162
Dworkin, N.B. 161
Dysaphis spp. (aphids) 237–8
Dyson, F.J. 191

ecdysone 93
echinoderms 202, 209–10
ectopic pairing 177, 181, 186
EIS, *see* epigenetic inheritance systems
Eissenberg, J.C. 109 (n.18)
Elgin, S.C.R. 109 (n.18)
enduring modifications 134, 137–9, 158 (n.3,4), 198, 226 (n.14)
Engler, P. 128, 136
enhancers 84, 94, 174, 259
environmentally-induced variation, *see* directed variation
Ephrussi, B. 79
epialleles 102, 137, 147–8, 192, 220–3, 243, 267
  coordinated changes in 219, 224, 225
  defined 91, 108 (n.12)
  in populations 195–9, 244–5, 276, 280–2
  stability of 103–5, 156, 276–8
epidermal cells, caterpillar 89
epigenetic engineering 275
epigenetic inheritance systems 25, 79–110, 108 (n.1), 134–52, 157, 158 (n.2), 274–9
  compared with genetic system 80–1, 102–7
  defined 80
  evolutionary origin of 191–2, 276
  and fluctuating environments 192–5, 276
  interactions with genetic system 160–88, 206–7, 234–5, 266–7, 276–7
  and meiosis 81, 106–7, 111
  and modes of development 210–13
  in speciation 233–8, 243, 244–5, 266–7, 278
  and the transition to multicellularity 198–205
epigenetic marks, *see* chromatin marks
epigenetic memory 156, 173–87, 191–5, 267, 276–7; *see also* cell memory
epigenetics, definitions of 79–80
epimutation 102, 145, 153, 192, 215, 280
  rate of 104, 105, 139–40, 144, 147, 192–5
Epstein, H. 183
Escherichia coli
  colony organization 204, 226 (n.13)
  directed mutations 24, 59–63
  lac operon 61–2, 82–3, 161
  maintenance of induced state 82–3
  mutation rate 161
  pilus type 195
Eshel, I. 270 (n.29)
Ettinger, L. 108 (n.11), 215, 254
Euchlanis triquetra (rotifer) 136
euchromatin 91, 98–9, 161, 177, 256
Euglena (protozoan) 21
eukaryotes, origin of 199
Euplotes (protozoan) 134
Evenson, D.P. 162
evolutionary epistemology 56, 76 (n.7)
evolutionary transitions 198–205, 282–4
evolutionary trends 8–9, 54–5, 75, 76 (n.3), 279–80
evolution of evolutionary systems 72–4, 171–2, 284–5
extracellular matrix 89–90, 108 (n.11), 204–5
extraembryonic tissues 109 (n.16), 114–15, 130 (n.4)

Falk, R. 28 (n.15,22)
Fedor, M.J. 109 (n.15)
Fedoroff, N.V. 135, 149–50, 159 (n.16), 162
Feldman, M.W. 289 (n.9)
Field, L.M. 136, 145, 211
Fieldes, M.A. 108 (n.5)
Filipski, J. 184
Finch, C.E. 211
Fincham, J.R.S. 78 (n.25)
fish 136, 211, 247–8, 251–2, 270 (n.17)
Fisher, R.A. 55
Fisher, R.C. 269 (n.15)
Fitch, W.M. 21–2, 145, 159 (n.13)
flatworms 136
Flavell, R.B. 135, 147, 245
flax 71, 75, 108 (n.5), 172
flexible genome 72–5
fluctuating environments 192–5, 224, 276
fluctuation test 58–9, 60, 77 (n.10,16), 102
flukes 241
Fontdevila, A. 257
Forejt, J. 127, 131 (n.11), 246
Foster, P.L. 64, 65, 77 (n.14,21,22)
Foster, R. 158 (n.6)
Fothergill, P.G. 11
founder effect 234–6, 269 (n.8)
foxes 142–3, 231–2
fragile-X syndrome 116, 125–6, 187
fragmentation 207–8
Francis, A. 135, 139
Frank, S.A. 260, 270 (n.27)
Frankel, J. 88, 108 (n.8,9)
frogs 247, 250–51
fungi 21, 29 (n.28), 134, 147–8, 200, 202, 221, 258
  ripping in 71–2, 124–5
Futuyma, D.J. 271 (n.32)

Galef, B.G. 289 (n.9)
gall midges 239
Gallus gallus domesticus (chicken) 246
gall wasps 239
gametogenesis 119–21, 125, 133–4, 165–6, 216
  in hybrids 250–62, 268
  quality control in 253–4, 268
Garcia-Bellido, A. 106
Garrard, W.T. 95, 96, 109 (n.15)
Garstang, W. 285
Gaunt, S.J. 189 (n.15)
Gause, G.F. 31
G bands 98–9, 165, 183–6, 277
gemmules 7, 12, 39
gene activity
  and allele frequencies 168–71
  in bacterial operons 72, 82–3
  and chromatin structure 90–9, 102, 224
  and chromosome structure 109 (n.27), 177–86
  and DNA repair 65, 77 (n.18), 161–4
  in gametocytes 125, 165–6
  and methylation 96–8, 105, 123, 128–9, 189 (n.9)
  and mutation 64, 161–4, 167–71, 184, 187
  and recombination 165–6, 189 (n.7)
  regulatory networks of 82, 84–5
  and replication time 100, 109 (n.16), 118, 218
  self-regulated 82–4
  in species hybrids 245–50
  and transposition 164–5
gene conversion 68, 186, 264
gene phenotypes 121, 133, 161, 196–7, 241
  of chicken α-globin gene 97
  of chicken lysozyme gene 96
  defined 90–1
  see also methylation patterns
genetic assimilation 32–7, 167–8, 189 (n.8), 207, 221–2, 277–8, 283–4
genetic load 187, 220
genetic maps 166
genetic revolution 234–5, 266
genocopy 167
genome
  reorganization in eukaryotes 66–72, 232–3, 266
  as a response system 22, 24–5, 26, 72–4, 171–2
  size 166, 212
genomic imprinting, *see* imprinting
genotrophs 71
genotype-phenotype distinction 15–17, 22, 28 (n.18,21)
  in molecular terms 17–20
Gerbi, S.A. 131 (n.17), 211
germinal selection, Weismann's theory of 47–8, 54–5, 75 (n.1)
germ line
  determination, modes of 42–6, 53 (n.17), 213, 241–3, 267, 278
  quality control 253–5, 268
  repair 164, 188 (n.5)
germ line–soma segregation 37–48, 51, 151–2, 221, 225
  advantages of 213–4
  and speciation 241–2, 267
germ-line-specific genes 154, 221–3, 255
germ plasm 72
  theory 37–42, 52 (n.9,10)
Gilbert, S.F. 226 (n.16), 286–7, 289 (n.15)
Glavinic, R. 51, 53 (n.24)
globin genes 53 (n.21), 97, 106, 121
Gluecksohn-Schoenheimer, S. 286, 289 (n.15)

glutenin proteins 147
Glycine max (soybean) 71
Gold, J.R. 247
golden shiner 247
Goldman, M.A. 95, 109 (n.15)
Goldschmidt, R. 222–4, 228 (n.29,31), 240, 285, 286
Goldsmith, M. 10–11
Gopinathan, K.P. 131 (n.17), 211
Gottlieb, G. 289 (n.14)
Gottschling, D.E. 182
Gould, C.G. 227 (n.21)
Gould, J.L. 227 (n.21)
Gould, S.J. 27 (n.9), 226 (n.7), 228 (n.30)
Grafen, A. 72, 270 (n.29)
grafting 50–1, 53 (n.23)
Graham, C. 227 (n.24,25)
Grant, M. 123
grasshopper 264
Grauer, D. 288 (n.4)
Greaves, D.R. 77 (n.23), 225 (n.2)
Green, M.C. 159 (n.13)
Gregorová, S. 127, 131 (n.11), 246
Grell, R.F. 162, 165
Griesemer, J.R. 52 (n.15)
Grimes, G.W. 87, 89
Grimes, S.R. 119
Groeters, F.R. 264
Gross, D.S. 95, 96, 109 (n.15)
Groudine, M. 106, 109 (n.22), 158 (n.1)
group selection 204–5, 277, 282, 285
Groves, C.P. 265
Grüneberg, H. 145, 159 (n.13)
gynogenetic embryos 114, 217, 252

Haack, H. 130 (n.2)
habitat selection 236–7
Hadchouel, M. 128, 131 (n.18), 136
Hägele, K. 254
Haig, D. 218, 227 (n.24,25), 270 (n.29)
Haigh, L.S. 97
Haldane, J.B.S. 39, 55, 269 (n.15)
Haldane's rule 255–7, 260–1, 268, 270 (n.24,26)
Hall, B.G. 12, 57, 62, 64–5, 74, 77 (n.12,19), 162
Hall, B.K. 52 (n.4), 108 (n.2), 228 (n.30)
Hall, J.G. 116, 117, 130 (n.3,8), 131 (n.13), 142
Hamburger, V. 29 (n.24)
Hammer, M.F. 270 (n.28)
Hämmerling, J. 158 (n.4)
Hanawalt, P.C. 162, 163–4, 188 (n.3)
Hansen, J.C. 95, 109 (n.15)
hard inheritance 30, 231, 273, 288 (n.1)
Harris, M. 110 (n.29)
Harwood, J. 28 (n.22), 285
Hatchett, J.H. 113
hawk's beard 248
Hayes, J.J. 109 (n.19)
Hayes, W. 76 (n.9)
Hayman, D.L. 189 (n.7)
Hayward, M.D. 220
Heitz, E. 91
Hemmer, H. 233, 269 (n.3)
Henikoff, S. 189 (n.12)
heredity
  concept of 1–2, 15–17, 28 (n.23), 272–5, 286–8
  early theories of 7–11, 37–9, 48–9
  see also multiple inheritance systems
Hergersberg, M. 109 (n.20), 189 (n.9), 211, 212, 213
heritability 275
Herman, R.K. 161
herring 211
Heslop-Harrison, J.S. 135, 148, 250, 270 (n.19)
hessian fly 113
heterochromatin, constitutive 91–2, 98–9, 226 (n.17), 234, 271 (n.35)
  and gene activity 177–81, 195, 206, 223–4, 232, 249–50, 277
  and recombination 165, 166, 261
heterochromatin, facultative 91–2, 111–12, 182–3, 184, 256–7; *see also* X chromosome inactivation
heterochromatinization 123, 182–3, 189 (n.18), 232, 245, 250, 270 (n.18); *see also* heterochromatin, facultative; inactive X chromosome
heterochrony 222–4
heterogametic sex 120, 165, 189 (n.7), 255–7
Hewer, H.R. 136, 269 (n.10)
Highkin, H.R. 135, 138–9
Hinshelwood, C.N. 13, 24, 82, 108 (n.3)
Hippocrates 7, 12
histones 94–6, 119–20, 178, 216, 254, 262
history of genetics, stages in 272–4
Hitzeroth, H. 247
Hjelm, K.K. 89
Ho, L. 162
Ho, M.W. 52 (n.5), 136, 146, 167
Hodgkin, J. 130 (n.2)
Hoffmann, A.A. 188 (n.1)
Holliday, R. 77 (n.19), 91, 102, 109 (n.22), 110 (n.29), 145, 173, 188 (n.5), 215, 218, 254
Holmquist, G.P. 99, 109 (n.21,27), 184
homogametic sex 189 (n.7), 255, 270 (n.26)
Homo sapiens, *see* humans
hopeful monsters 222–4, 228 (n.29), 282
Hordeum spp. (barley) 248

hormone-induced changes 96, 137, 142–3, 221–3, 231–2, 235, 243
host races 236–8
housekeeping genes 90, 97, 133–4, 154, 166, 184, 189 (n.9)
Hovasse, R. 31
Hsieh, C.L. 162
Hughes, C. 226 (n.13)
Hull, D.L. 28 (n.14,23)
Hulten, M.A. 131 (n.13)
humans 211
  fragile-X syndrome 116, 125–6, 187
  gene amplification 70
  genetic maps 166
  imprinting 116–17, 142
  methylation in 137, 148
Hurst, L.D. 227 (n.23), 260, 270 (n.27)
Huxley, J. 30
hybrid inviability 229, 234, 244–52, 268, 270 (n.26), 278
hybridogenesis 247, 248, 250–2
hybrids 229, 234, 244–60, 268, 278
hybrid sterility 229, 234, 244–5, 252–7, 268, 269 (n.16), 270 (n.30), 278
Hydra 242
hydrocortisone 141–2
hypermutable state 64–5

ids 38, 41
ILC environment 192–5
immune system 49, 69, 75, 103
imprinting
  in adaptive evolution 216–19, 225, 227 (n.25, 27)
  defined 111
  during gametogenesis 119–22, 129
  in mammals 112–18, 120–2
  occurrence of 112–18, 141–2, 143, 150, 156, 261
  of transgenes 122–6, 128–9
imprints
  and chromatin structure 119–22, 180, 186, 212, 260, 261, 274
  complementary 114, 252
  erasure of 119–22, 128, 130
  establishment of 119–22, 129
  and modifiers 126–9
  and pedigree analysis 116–17, 142
  species differences in 245, 246, 250, 252, 268
inactive X chromosome, *see* X chromosome inactivation
inbred lines
  and repair 215
  selection in 32, 52 (n.5), 167
  variation in 71, 145–6, 147, 159 (n.18), 231

induced epiallelic variants, *see* directed epigenetic changes; directed variation
induced mutation 55, 58–66, 70–1; *see also* directed mutation
inheritance of acquired characters
  Dawkins on 18–19
  examples of 134–51
  Johanssen on 16
  Lamarck on 3–6, 7
  meaning of term 6, 10–14, 26
  and molecular genetics 17–20, 49
  and the origin of organelles 21–2
  reasons for lack of evidence 20–5, 152–4
  through genetic assimilation 31–7
  through somatic selection 49–51
  Weismann on 7–8, 28 (n.13), 29 (n.27), 54
  where likely to be found 21, 144, 153–4, 274
  see also epigenetic inheritance systems; Lamarckian inheritance
inheritance systems, *see* cultural inheritance; epigenetic inheritance systems; multiple inheritance systems
insect host-plant conditioning 236–8
insecticide resistance 70, 144–5, 154
intercalary heterochromatin 181
isochores 184–6, 280, 288 (n.3)
isolating mechanisms
  post-zygotic 229, 234, 236, 237, 244–57, 262–4, 268, 278; *see also* Haldane's rule
  pre-zygotic 229, 233–4, 237–8, 255, 257, 278
Ives, P.T. 55, 162

Jablonka, E. 12, 57, 77 (n.18), 90, 109 (n.24), 131 (n.13,14), 136, 146, 155, 189 (n.7), 193, 216, 218, 225 (n.6), 244, 254, 256, 270 (n.18,23,26), 289 (n.10)
Jablonka Tavory, E. 52 (n.7)
Jaenisch, R. 162, 164
Japanese quail 246
Jennings, H.S. 15, 28 (n.20), 134
Jerka-Dziadosz, M. 87
Jinks-Robertson, S. 66
Johannsen, W. 15–17, 28 (n.18,19,21,23), 272
John, B. 109 (n.13), 189 (n.6)
Johnson, F.M. 159 (n.13)
Johnson, N.A. 270 (n.30)
Jollos, V. 55, 76 (n.6), 134, 137–8, 158 (n.4), 198, 226 (n.14)
Jones, J. 131 (n.25)
Jones, P.A. 105
Jones, R.N. 135, 139, 270 (n.31)
Jordanova, L.J. 27 (n.2)

Jorgensen, R. 108 (n.12), 131 (n.23), 156
jumping genes 21, 67, 72–3, 125; *see also* transposable elements

Kafri, T. 120–1, 158 (n.1)
Kahn, A.J. 137
Kanazir, D.T. 163
karyotype differences 235–6, 244, 249–50, 262–6, 268, 271 (n.33), 280–1; *see also* chromosome rearrangements
Kaslow, D.C. 122
Kauffman, S.A. 289 (n.16)
Kaufman, T.C. 226 (n.12)
Keller, E.F. 23
Kellum, R. 109 (n.18)
Kermicle, J.L. 91, 113, 158 (n.10)
Kimura, M. 288 (n.4)
King, M. 255, 263, 266, 269 (n.8), 271 (n.34)
Kirkwood, T.B.L. 188 (n.5)
Kitchin, R. 113
Kitsberg, D. 109 (n.17), 118
Klar, A.J.S. 113
Klekowski, E.J. 12, 46–7, 154
Koch, A.L. 171
Kohane, M.J. 269 (n.5)
Kornberg, R.D. 109 (n.15)
Koufopanou, V. 200–1
Kricker, M.C. 78 (n.25), 124, 131 (n.24)
Krider, H.M. 249
Kuhl, D.P.A. 190 (n.21)
Kuhn, D.T. 113
Kunze, R. 159 (n.17)

Lachmann-Tarkhanov, M. 193
Laird, C.D. 132 (n.28)
Lamarckian inheritance 7, 11–14, 28 (n.14,23), 29 (n.26,27), 30–1, 58, 240
  and directed mutation 22–5
  see also inheritance of acquired characters
Lamarck's theory of evolution 2–6, 9, 27 (n.3,4)
Lamb, M.J. 12, 77 (n.18), 90, 109 (n.24), 131 (n.13,14), 136, 146, 189 (n.7), 216, 218, 244, 254, 256, 270 (n.23,26)
Lambert, D.M. 106
lampbrush chromosomes 93, 120
Lande, R. 271 (n.33)
Landman, O.E. 29 (n.29)
Langridge, W.H.R. 248
Lansing, A.I. 136, 146, 159 (n.15)
Lansing effects 146–7, 159 (n.14,15)
Latchman, D. 101
Lederberg, E.M. 59–61
Lederberg, J. 59–61
Lenski, R.E. 23, 77 (n.13,14,15,17,22)
Lepidoptera 54, 136, 240, 255
Lepomis spp. (fish) 247
Levine, A. 98
Lewin, R. 49
Lewis, E.B. 189 (n.14)
Lewis, J. 109 (n.20)
Lewontin, R.C. 288 (n.5)
Li, E. 122
Li, W.-H. 288 (n.4)
Lieber, M.R. 162
Lincoln, R.J. 80
Lindegren, C.C. 21, 29 (n.28), 53 (n.24), 77 (n.10)
LINEs 99, 185–6
lingering modifications 134, 137–8, 158 (n.3,4), 198, 226 (n.14)
Linn, F. 132 (n.26)
Lints, F.A. 136, 137, 159 (n.14,15)
Linum spp. (flax) 71, 75, 108 (n.5), 172
Liu, C.K. 41–2
Locke, J. 179, 189 (n.13)
Locke, M. 89
locust 210, 211
Loidl, J. 162
Lolium perenne (perennial rye-grass)
  clonal variation 220
  colchicine-induced variation 135, 139
Longo, F.J. 120
Lorch, Y. 109 (n.15)
Lowenhaupt, K. 183
Lowrey, C.H. 109 (n.18)
Lumsden, C.J. 289 (n.9)
Luria, S.E. 58–60, 62
Lysenko, T.D. 51, 231
Lyttle, T.W. 189 (n.11), 270 (n.28)

McClintock, B. 269 (n.5)
  on gene activity 94
  on genomic stress 74, 160, 171, 172, 233
  and transposable elements 145, 149
McEdward, L.R. 222
McGrath, J. 130 (n.6)
McKee, B.D. 260
Mackenzie, A. 158 (n.12), 269 (n.10)
McLaughlin, P. 27 (n.3)
MacPhee, D. 77 (n.13,19)
macroevolution 222–4, 228 (n.29), 286
macromutations 286
macronucleus 67–8, 226 (n.9)
Mahadevaiah, S.K. 270 (n.23)
maize 21
  B locus 135, 143
  imprinting 113, 150
  paramutation 135, 143
  R locus 113, 135, 143, 156

maize (*continued*)
Spm element 135, 149–51, 153, 156, 164
mammals 112–18, 124, 136–7, 146, 152, 211, 270 (n.26), 263–4; *see also* humans, mouse, voles, X chromosome inactivation
Mantovani, B. 248
Manuelidis, L. 109 (n.21)
Margulis, L. 202, 226 (n.11)
Markovian process 280
marks, *see* chromatin marks
Markson, Y. 108 (n.11)
Marshall, C.J. 211
Marsilea vestita (water fern) 113
marsupials 114–15, 122, 189 (n.7)
Martienssen, R. 135, 159 (n.17)
Martin, C. 131 (n.25), 159 (n.17)
Martin, R. 253
Martin, S.L. 190 (n.20)
Mather, K. 158 (n.2)
mating behaviour 229, 233, 253
mating-type switching 68, 182
Matsuda, Y. 257
Matzke, A.J.M. 131 (n.23), 135, 159 (n.18)
Matzke, M.A. 131 (n.23), 135, 159 (n.18)
Mayer, G.C. 271 (n.32)
Mayetiola destructor (hessian fly) 113
Maynard Smith, J. 25, 52 (n.14), 108 (n.1), 159 (n.22), 207, 208, 227 (n.19), 269 (n.9), 282–3
Mayr, E. 27 (n.2,8), 269 (n.7), 289 (n.11)
and founder effects 234–6
on hard and soft inheritance 13–14, 273, 288 (n.1)
on sympatric speciation 365
on species 229, 269 (n.1,8)
on Weismann 52 (n.12), 76 (n.4)
mealy bugs
chromosome elimination 112, 113, 261
methylation 122, 131 (n.17)
Medawar, J. 79
Medawar, P.B. 9, 13–14, 23–4, 37, 79
Mehtali, M. 132 (n.26)
Meins, F. 80–1, 106, 108 (n.5), 135, 139–40, 158 (n.6)
meiosis
and the evolution of multicellularity 214–16
functions of 214–16, 253–4
in hybrids 253, 262
and the transmission of epigenetic variants 81, 106–7, 154–6, 220, 221; *see also* imprinting
meiotic drive 257–62, 263, 264, 268, 270 (n.27–31)
meiotic pairing 119, 124–5, 131 (n.13), 215, 251–2, 253–5, 256–7, 261, 268
Meistrich, M.L. 120
Mellon, I. 163, 188 (n.4)
Mendelian genetics and evolutionary ideas 7, 14–17, 28 (n.17), 55, 231, 272
methotrexate resistance 69–70
methylation 96–7, 143, 261
and cell turnover 210–13, 225, 226 (n.18)
changes during development 120–2, 133–4, 146
changes following selection 128–9
evolutionary origin of 192
and gene activity 95, 96–7, 105, 123, 128–9, 144–5, 195, 237
and mutation 131 (n.24), 162, 166, 187, 212, 225; *see also* epigenetic inheritance systems, interaction with genetic system
and RIGS 124–5, 131 (n.24)
of transgenes 124–5, 128–9, 136, 150–1
methylation patterns 281–2
changes during development 120–2, 150–1, 159 (n.17)
and clonal memory 173, 174–5
and gene activity 96–7, 189 (n.9)
inheritance of 100–1, 134–7, 147–51, 157, 159 (n.19)
methyltransferase 101, 122
Meyer, P. 135, 151
Mexican fish 247
Michie, D. 53 (n.24), 272
micrococcal nuclease sensitivity 95
Microtus spp. (voles) 246, 249–50, 270 (n.18)
Migeon, B.R. 122
Miklos, G.L.G. 270 (n.23)
Miller, S. 64
mites 242
Mittler, J.E. 77 (n.13,14,17,22)
model
Delbrück's, steady-state 82
of genetic assimilation 34–5
of induced changes in germ cells 221–3
of mark transmission through the germ line 154–6
of marks in populations 196–9
of mutational assimilation 168–71
of optimal mutation rates 193–4
of self-maintaining regulatory networks 84–5, 108 (n.5)
Modern Synthesis 14–15, 17, 30, 56, 231, 273, 285–6
modifiers 21, 131 (n.10), 141, 152
of imprinting 123–4, 126–9
of meiotic drive 260–1
of PEV 178–9
molecular clock 280, 288 (n.3)
molecular drive 186, 264–5, 271 (n.35)
molecular genetics
and acquired characters 17–20, 49
and neo-Darwinism 17–18, 288

molluscs 202, 208, 210
Monk, M. 123, 130 (n.3), 219
Moore, T. 227 (n.24)
Moran, N.A. 241
Morgan, C.L. 31
Morgan, M.J. 77 (n.15)
Morgan, T.H. 16
Mort, M.A. 158 (n.11)
mosaic development 208–13
mosquitoes 211, 248
Moss, G.I. 135, 139
Motara, M.A. 248
mouse 21, 211
  epialleles 148
  Fused gene 118, 126–8, 136, 141–2, 156, 232, 136
  H19 locus 117–18
  hybrids 127, 245, 246, 257
  Igf2 gene 117–18, 122, 227 (n.25)
  Igf2r gene 117–18, 186, 227 (n.25)
  imprinting 114–18, 252
  insulin II gene 118
  methylation deficient 122
  MHC complex 164
  species 127, 245, 246, 257
  t-complex 118, 259, 260
  Tme locus 117–18, 127, 246
  transgenic 116–17, 128–9, 136, 156
  tyrosine hydroxylase gene 118
  variation in inbred lines 145
Moxon, E.R. 171
MTI (*metallothionein I*) gene 163
MTOC 202, 226 (n.11)
Mucor (fungus) 202
Muller, H.J. 229
Mullet, J.H. 135, 139
multicellularity 102, 202
  evolution of 198–205, 214–15, 226 (n.12), 277, 283
  and sex 214–16, 227 (n.22)
multiple inheritance systems 19, 25, 79, 274, 282–4
Mus, *see* mouse
mutational assimilation 167–9, 189 (n.8), 279
mutational bias 64, 167, 184, 276, 280, 288 (n.3)
mutation rates 55, 64–5, 103–5, 139, 145, 161–4, 168– 71, 193–4, 279–80
mutilations, inheritance of 6, 20
myxobacteria 202, 204
Myzus persicae (aphid) 136

Nagl, W. 67
Nanjundiah, V. 219
Nanney, D.L. 13, 80, 86, 88, 103, 107, 108 (n.7), 134, 158 (n.2), 192
Nasonia vitripennis (wasp) 261
Naveira, H. 257
Nee, S. 270 (n.24)
Nelsen, E.M. 134
Nelson-Rees, W.A. 113
nematodes 40–1, 52 (n.13), 68, 130 (n.2), 136, 146, 208, 209, 210–12, 226 (n.16), 242, 269 (n.14)
neo-Darwinism 8–10, 37, 56–7, 272–3, 279, 282, 286
neo-Lamarckism 6–14
Nereis acuminata (annelid) 236
Neurospora (fungus) 21, 29 (n.2), 124
Neves, N. 261
Ng, S.F. 108 (n.8), 134
Nicotiana tabacum, *see* tobacco
Nieuwkoop, P.D. 42–3, 52 (n.13), 53 (n.17,18)
non-disjunction 260, 261
non-Mendelian segregation 257–62
NORs (nucleolar organizer regions) 245–9
Notemigonus crysoleucas (fish) 247–8
Novick, A. 82
nuclear matrix 93
nuclear transplantation 114
nucleolar dominance 245–9
nucleosomes 94–6, 163
Nur, U. 261

O'Dell, M. 135, 147, 245
Ohno, S. 246
Okada, N. 190 (n.20)
Oldroyd, D.R. 27 (n.2)
oogenesis 120, 125, 165–6, 221–3
Opadia-Kadima, G.Z. 77 (n.12)
operons, bacterial 61–2, 72, 82–3, 161
Orr, H.A. 255, 270 (n.27,30)
orthogenesis 54–5, 76 (n.3)
Oryza sativa (rice) 135, 148
Osborn, H.F. 9, 31
Oschmann, B. 254
Overbaugh, J. 64
Oxytricha (protozoan) 67

Packert, G. 113
pairing failure 254, 255–7, 260–1
pangenesis 7, 12, 39, 49, 230
parallel induction 10, 11, 12, 28 (n.13), 48, 55, 189 (n.8), 240
Paramecium
  cortical inheritance 86–8, 134
  Dauermodifikations 134, 137–8, 226 (n.14)
  selection in pure lines 15

Paramecium (*continued*)
  senescence 215
paramutation 143, 158 (n.9)
parasites 195, 216, 225 (n.5)
Paraurostyla (protozoan) 134
parental-age effects 146–7, 159 (n.15)
parental conflict 218, 227 (n.25)
parental dominance 250–2
Paro, R. 181, 189 (n.15)
Parsons, P.A. 162, 188 (n.1), 269 (n.5)
parthenogenesis 144, 212, 215, 226 (n.18), 237–8, 239, 241–3, 269 (n.14)
Partridge, L. 77 (n.15)
Pascal 272
Patel, C.V. 131 (n.17), 211
Patterson, G.I. 135, 143
Pauling, L. 140–1, 158 (n.7)
peas 135, 138–9
pedigrees 116–17
P elements 164
penetrance 112, 116, 126–9, 131 (n.10), 141–3, 152, 153
peripatric speciation 234–6
Perkins, P. 137, 142–3
Petunia hybrida 70, 132 (n.26), 135, 151
PEV 92–3, 113, 145, 178–80, 228 (n.31)
Pfeifer, E.J. 28 (n.11)
Pfeifer, G.P. 100
Pfennig, D.W. 240
Phaseolus vulgaris (bean) 135, 139
phenocopies 32–7, 106, 167
phenotype 16; *see also* gene phenotypes; genotype–phenotype distinction
Philodina citrina (rotifer) 136
Pieris brassicae (butterfly) 136
pigeon 211
Pillus, L. 107, 134, 189 (n.17)
pine 211
Pisum sativum (pea) 135, 138–9
Planococcus citri (mealy bug) 113
plants 135, 202, 210, 211, 248, 250
  carry-over effects in 135, 138–40
  clonal variation in 145, 220–1
  cuttings of 220–1
  grafting of 50–1, 53 (n.23)
  methylation in 189 (n.9), 211
  paramutation in 143, 158 (n.9)
  somatic selection in 46, 151–2, 159 (n.18)
  transgenic 124, 132 (n.26), 150–1
  transposable elements in 149–50, 159 (n.17)
  and Weismann's doctrine 39–41
  see also individual genera
Pleurodeles waltlii (salamander) 246
Plough, H.H. 55, 162
Poecilia reticulata (fish) 136
Poeciliopsis monacha-lucida (fish) 248, 251–2
Polikoff, D. 52 (n.4)
polychaetes 236
polymorphism 164, 173, 254, 263
polyphenism 238–41, 243
polyploidy 67, 226 (n.9), 236
polytene chromosomes
  in ciliates 67–8
  in *Drosophila* 69, 93, 181, 226 (n.17), 257
Pomiankowski, A. 260, 270 (n.27)
Popper, K. 76 (n.7)
Porifera 221
position effects
  in mammals 92–3, 123, 182, 189 (n.18)
  and speciation 254, 260–1, 262, 264
  in yeast 182
  see also PEV
Prader-Willi syndrome 116, 118
Prescott, D.M. 162
Prody, C.A. 70
Proffitt, J.H. 211
protamines 120, 262
Proteus mirabilis (bacterium) 204, 226 (n.13)
Protozoa 21, 67–8, 134, 226 (n.9), 227 (n.20); *see also* ciliates, *Paramecium*, *Tetrahymena*
Provine, W.B. 28 (n.19), 289 (n.11)
Pseudococcus affinis (mealy bug) 261
pseudogenes 49, 53 (n.21), 73
Pseudomonas 226 (n.13)
PSR in *Nasonia vitripennis* 262
Puertas, M.J. 113, 261
Pukkila, P.J. 134, 147
pure lines, selection in 15–16, 28 (n.19,21), 144

rabbit 137, 211
Raff, R.A. 53 (n.20), 226 (n.12)
Rai, K.S. 248
Rana spp. (frogs) 247, 250–1
random mutation 56–61, 76 (n.8,9)
Rao, B.J. 120
Rao, M.R.S. 120
rates of evolution 263–4, 279
Ratner, V.A. 162, 164
rats 120, 137
Razin, A. 109 (n.20)
Read, A. 270 (n.24)
recombination
  as a defence system 270 (n.29)
  and gene activity 165–6, 189 (n.7)
  and heterochromatin 165, 260, 261
  and meiotic drive 258–60
  and repeat clustering 180–1
  and ripping 71, 124
  as a source of variation 30, 74, 165

redear sunfish 247
Reeder, R.H. 245, 247
Rees, H. 270 (n.31)
regeneration 38, 139, 209–13
regional control 102, 109 (n.27), 118, 177, 181–6
regulative development 208–13
Reik, W. 116, 130 (n.3), 131 (n.18,20), 132 (n.29)
Renard, J.P. 252
Renckens, S. 132 (n.26)
Rensch, B. 10, 12, 28 (n.11), 55
repeated DNA sequences
and chromosome bands 98–9
and clonal memory 173, 175–7, 180–1, 186–7, 189 (n.10), 195, 206, 223, 276–7
and heterochromatin 180–1, 183, 262
in human diseases 125–6, 187
loss and amplification of 69–71, 125–6, 187, 264–5
and methylation 124–5, 212
and molecular drive 264–5
and nucleolar dominance 245
and ripping 71–2, 124–5, 131 (n.24)
and segregation distortion 178, 258
in transgenes 123, 124
replica plating 59–61
replication domains 93–4, 99, 109 (n.17,27), 118; *see also* DNA replication
reproductive isolation 229, 233–8, 244–57, 262–4, 278
reproductive rate and size 201–2
restriction-modification system 192
retrotransposons 185
retrovirus 49–50, 164
Reuter, G. 189 (n.12)
reverse transcription 29 (n.25), 49–50, 63
Rhabditis nigrovenosa (nematode) 40–1
ribosomal DNA 69, 71, 172; *see also* NORs
rice 135, 148
Rice W.R. 237
Richards, R.I. 132 (n.28)
Richerson, P.J. 289 (n.9)
Riedl, R. 289 (n.16)
Riggs, A.D. 100, 109 (n.20,22)
RIGS 124–5, 131 (n.24), 166, 254
Rine, J. 107, 109 (n.27), 134
ripping 71–2, 78 (n.25), 124–5, 131 (n.24), 172
Rivier, D.H. 109 (n.27)
Roan, J.G. 247
Robinson, G. 53 (n.16,23)
Robotti, C. 136
Romanes, G.J. 50, 53 (n.23)
Rose, S. 288 (n.2)
Rosenberger, R.F. 77 (n.19)
Roth, E.J. 71
Rothstein, R. 162, 166, 189 (n.7)
rotifers 136, 146, 241
Rubin, H. 108 (n.3), 158 (n.3)
rudd 247
Russell, P.J. 195
Ruvinsky, A.O. 126, 130 (n.9), 131 (n.11), 136, 137, 141, 144, 158 (n.8,11), 214
Ryan, F.J. 60–1, 77 (n.11)
rye 113, 261
rye-grass 135, 139, 220

Saccharomyces (yeast) 66, 68, 134
Sager, R. 113
salamander 246
salmon 211
Salmonella typhimurium (bacterium) 163
Salmo spp. (trout) 247
Salt, G.W. 237
Salvelinus fontinalis (trout) 247
Sano, H. 135, 148
Santos, E. 134
Sapienza, C. 123, 128, 131 (n.18), 142
Sapp, J. 17, 20, 28 (n.16)
Sarkar, S. 76 (n.8), 77 (n.16)
Sasaki, H. 131 (n.20), 136, 148
Sasaki, M. 130 (n.4)
satellite DNA 73, 99, 177–8, 258; *see also* repeated DNA sequences
Savić, D.J. 163
Scali, V. 248
Scarbrough, K. 131 (n.17), 211
Scardinus erythrophthalmus (fish) 247
Scharloo, W. 52 (n.4)
Schedl, P. 109 (n.18)
Scheid, O.M. 135
Schizosaccharomyces pombe (yeast) 113
Schmalhausen, I.I. 12, 31, 36–7, 285
Schmid, B. 220
Schmidt, B.R. 246, 270 (n.20)
Schmidtke, J. 247
Schneeberger, R.G. 78 (n.24)
Schultz, R.J. 248, 270 (n.21)
Sciara coprophila (midge)
chromosome elimination in 68–9, 111–12, 113, 119, 183
imprinting in 111–12, 113, 119, 182–3
methylation of 131 (n.17), 183, 211
Scilla siberica (squill) 70
sea urchins 211
Searle, A.G. 116
Secale cereale (rye) 113, 261
segregation distortion 178, 257–61
selection
in bacterial cultures 64–5
between cells, *see* somatic selection
between groups 204–5, 277, 282, 285
in inbred lines 32, 52 (n.5), 145–6, 167

selection (*continued*)
intracellular 23–4, 63
levels of 23–4, 47–8, 51, 282
in pure lines 15–16, 144
units of 23–4, 204–5, 226 (n.14), 282–3
selfish DNA 76 (n.2), 215, 254, 257–62
self-perpetuating metabolic patterns, *see* steady-state systems
Selig, S. 109 (n.27)
Selker, E.U. 71–2, 78 (n.25), 109 (n.26), 131 (n.22), 162
senescence in protozoa 215
Serfling, E. 108 (n.4)
Serrano, M. 189 (n.16)
sex, functions of 171, 214–16, 227 (n.19,21), 270 (n.29); *see also* meiosis
sex chromosomes
changes during gametogenesis 120, 256–7, 268
and meiotic drive 259–60
and speciation 249–50, 255–7, 260–1, 268
see also X chromosome inactivation; Y chromosomes
sex differences in recombination 165–6, 189 (n.7)
sex ratio 259–60, 262
Shami, S.A. 136
Shapiro, J.A. 61–2, 73–4, 77 (n.12), 78 (n.27), 162, 226 (n.13)
Shaposhnikov, G.K. 237, 269 (n.11)
Sharman, G.B. 130 (n.5)
Shaw, D.D. 264
Shay, J.W. 189 (n.18)
Signer, E.R. 131 (n.25)
Signoret, J. 113, 246
Silander, J.A. 145, 220, 227 (n.27)
silkworm moth 211
Silva, A.J. 137, 148
Simpson, G.G. 52 (n.3), 285, 289 (n.12)
Simpson, V.J. 210
Sinervo, B. 222
SINEs 99, 185–6
Singh, P.B. 189 (n.15)
single-celled origin of development 207–8
Sinoto 51, 53 (n.24)
size
of eggs 221–3
of embryos and imprinting 218, 227 (n.25)
of gametes 157, 216, 219
and multicellularity 200–2, 207–8, 226 (n.10)
skin-thickening 6, 10, 27 (n.7), 31–2
Sladden, D.E. 136, 269 (n.10)
Slatkin, M. 53 (n.19)
slime moulds 200, 202, 203, 208
slippage 186, 264
Slobodchikoff, C.N. 46, 220
Sober, E. 288 (n.5)
social insects 7, 209, 239–40
soft inheritance 13–14, 30, 273, 288 (n.1)
Soll, D.R. 225 (n.4)
Solomon, F. 89
Solter, D. 130 (n.3,6)
somatic induction 10, 11
somatic mutations 46–7, 49–50, 53 (n.19)
somatic selection 69
in plants 46, 53 (n.19), 159 (n.18), 220
and the segregated germ line 43–8, 205–7, 214
in Steele's hypothesis 12, 48–51
Sonneborn, T.M. 13, 87, 88, 108 (n.8), 134, 136
Soviet Union 31, 36, 51, 141, 231
soybean 71
speciation
allopatric 233–6
domestication as a model for 230–3
and karyotype differences 235–6, 262–6
and post-mating isolation 234, 244–57, 260–1, 268, 278
role of epigenetic inheritance in 234–5, 237–8, 240–1, 266–7, 268, 278
sympatric 236–8, 263, 268, 269 (n.9)
through polyphenism 238–41, 243
species definition 229, 269 (n.1)
species-richness 241–4
spermatogenesis 119–20
in hybrids 254, 257
in mealy bugs 69, 112, 261
and recombination 165–6
and RIGS 125
in *Sciara* 111–12
Spiegelman, S. 134
Spierer, P. 189 (n.12)
Spofford, J.B. 113
sponges 200, 202
Spradling, A.C. 52 (n.13), 226 (n.17)
Spurway, H. 253
Stack, S.M. 189 (n.6)
stage-specific genes 90, 174, 184, 277
Stahl, F.W. 77 (n.14,20)
stasipatric speciation 263
stasis 283–4
steady-state systems 13, 81–6, 106, 108 (n.6), 109 (n.28), 134–5, 157
and DNA replication 100, 102, 118
in evolution 204, 216
origin of 191
Stearns, S.C. 269 (n.13)
Steele, D.F. 66
Steele, E.J. 12, 48–50, 51, 53 (n.22)
Stenostomum incaudatum (flatworm) 136
Stent, G.S. 60, 61, 108 (n.3)
stick-insects 136, 237, 248
Stöger, R. 131 (n.15,16), 186

stress
  and domestication 141, 232–3
  and hereditary changes 71, 72–3, 145, 162, 171–2
  and induced mutation in bacteria 63, 64–5, 171
  and speciation 235, 266, 278
structural inheritance 13, 86–90, 108 (n.9,11), 134, 157
  in evolution 204, 216
  origin of 191–2
Stuart, J.J. 113
stubborn marks 156, 186, 219, 243
Stylonychia (protozoan) 67, 134
Sueoka, N. 184
Sumner, A.T. 99, 109 (n.21)
Surani, M.A. 116, 130 (n.6), 132 (n.29)
Sutasurya, L.A. 42–3, 52 (n.13), 53 (n.17,18)
Sutherland, G.R. 138 (n.28)
Swain, J.L. 117, 123, 131 (n.18)
Symonds, N. 77 (n.11,13,19,21)
systemic mutations 222
Szathmáry, E. 214

Takagi, N. 130 (n.4)
Tamame, M. 134, 148
Tartof, K.D. 189 (n.13)
Tate, W.P. 211
Tauber, C.A. 237
Tauber, M.J. 237
Tautz, D. 265
t-complex 118, 259, 260
T-DNA 135, 150–1
telomeres 68, 181, 182, 189 (n.18)
Templeton, A.R. 269 (n.1)
Tetrahymena (protozoan) 86, 87, 88, 134
Teutophrys trisulca (protozoan) 88
theories of inheritance 7–11, 37–8
Thomas, B.J. 162, 166, 189 (n.7)
Thomson, K.S. 222
Thuriaux, P. 166
tissue-specific genes 90, 97, 174, 184, 189 (n.9), 277
toads 240, 245, 247
tobacco
  cytokinin autotrophy 106, 135, 139–40
  transgenic 135, 150–1
totipotency 208, 213–14
Tourte, Y. 113
transgenes
  imprinting of 116–17, 122–5, 128–9, 133
  methylation of 124–5, 128–9, 136, 150–1, 156
transposable elements
  and gene activity 164–5
  in LINEs and SINEs 185
  in maize 149–51, 159 (n.17), 164–5
  and stress responses 72–3, 105, 172, 232–3, 266
trematodes 222 (n.5), 239, 241, 242
trinucleotide repeats 125–6, 187
Triticale 135, 148
Triticum (wheat) 135, 147, 211, 245
trout 247
Trut, L.N. 269 (n.2)
Trypanosoma brucei (protozoan) 68
trypanosome phase variation 68, 75
Turner, J.R.G. 289 (n.8)

Ulva (green alga) 202
unequal crossing-over 186, 264–5
units of selection 23–4, 204–5, 226 (n.14), 282–4
Urieli-Shoval, S. 131 (n.17)
use and disuse 3–6, 9, 13, 23, 230

Van de Woude, M.W. 225 (n.3)
Van Holde, K.E. 95, 109 (n.15)
Van Valen, L.M. 52 (n.8), 53 (n.20)
variable expressivity 112, 131 (n.10), 153
  and imprinting 116–17, 126–9
vertebrates 109 (n.21), 183–6, 189 (n.9), 202, 210, 277
Vinogradov, A.E. 270 (n.20), 247
Vogt, P. 175, 189 (n.10)
voles 246, 249–50, 270 (n.18)
Volvocales 200–1, 226 (n.10)
Volvox (green alga) 200, 202, 226 (n.10)
Vuillaume, M. 136

Wachtiella persicariae (gall midge) 209
Waddington, C.H. 12, 30, 31–7, 52 (n.4), 167–8, 189 (n.8), 207, 240, 270 (n.22), 285, 286, 289 (n.15)
Wade, M.J. 288 (n.5)
Wake, D.B. 289 (n.8)
Walbot, V. 159 (n.17)
Wallace, A.R. 272
Wallace, B. 24–5
Wallace, H. 248
Ward, D.C. 189 (n.10)
warmouth 247
Warner, A.H. 210
Wassermann, K. 162, 163, 188 (n.3)
water fern 113
Watson, J.D. 77 (n.23)
Weinberg, J.R. 236
Weiner, M. 82

Weintraub, H. 106, 109 (n.15,25)
Weismann, A.
  diagrams 39–41, 52 (n.11,15)
  germ plasm theory 8, 28 (n.18), 37–42, 52 (n.9,10,11), 68, 226 (n.16)
  and the inheritance of acquired characters 7–11, 12, 20, 28 (n.13), 29 (n.27), 48, 54, 240
  theory of germinal selection 47–8, 54–5, 75 (n.1,2), 240, 269 (n.12)
Weismann's doctrine 37–48, 51, 53 (n.8,16)
Werren, J.H. 262
West-Eberhard, M.J. 238, 240–1
Westoby, M. 227 (n.24)
wheat 135, 147, 211, 245
White, M.J.D. 236, 262–3, 266, 269 (n.6), 271 (n.32)
White, R. 137, 148
Whitham, T.G. 46, 220
Whitt, G.S. 247, 270 n.17
Williams, G.C. 35–6, 52 (n.6,15), 288 (n.5)
Wills, C. 74, 78 (n.27,28) 171, 269 (n.5)
Wilson, A.C. 263, 271 (n.34), 280
Wilson, C. 109 (n.14), 131 (n.19), 189 (n.12)
Wilson, D.S. 289 (n.5)
Wilson, E.O. 289 (n.9)
Wimsatt, W.C. 52 (n.15)
Wolfe, K.H. 184–5
Wolffe, A.P. 95, 109 (n.15,19)
Wolpert, L. 53 (n.20), 203
Wright, S. 55, 81–2
Wright, W.E. 189 (n.18)
Wu, C.-I. 270 (n.28), 270 (n.25,30)
Wynne-Edwards, V.C. 282

X chromosome inactivation
  and DNA replication 93, 109 (n.16), 118, 130 (n.4)
  and gametogenesis 119, 120, 218, 255–7, 260
  in mammals, non-random 109 (n.16), 112–15, 118, 122, 130 (n.4), 189 (n.7), 246, 249–50, 270 (n.18)
  in mammals, random 91–2, 93, 103, 109 (n.28), 146, 182, 189 (n.9)
  in *Sciara* 111–12, 182–3
X chromosomes and sterility 119, 255–7, 268, 270 (n.26), 259–60
Xenopus spp. 211
  species hybrids 245, 247

Y chromosomes
  and gametogenesis 120, 256
  and meiotic drive 259
  and PEV 143, 179–80
yeast
  directed mutation in 66
  imprinting 113
  mating-type switching in 68, 182
  position effects 182
  transmission of functional states in 107, 133, 134, 195
Yeom, Y.I. 164

Zakian, S.M. 246, 249, 270 (n.18)
Zea mays, *see* maize
Zentall, T.R. 289 (n.9)
Zhimulev, I.F. 181
Zirkle, C. 7, 12, 27 (n.5,8)
Zlatanova, J. 109 (n.15)
Zolan, M.E. 134, 147
Zuckerkandl, E. 140–1, 158 (n.7), 178, 185